# PC Systems, Installation and Maintenance

# PC Systems, Installation and Maintenance

**SECOND EDITION**

**Robert Beales**

AMSTERDAM • BOSTON • HEIDELBERG • LONDON • NEW YORK • OXFORD
PARIS • SAN DIEGO • SAN FRANCISCO • SINGAPORE • SYDNEY • TOKYO
Newnes is an imprint of Elsevier

ELSEVIER

Newnes

Newnes
An imprint of Elsevier
Linacre House, Jordan Hill, Oxford OX2 8DP
200 Wheeler Road, Burlington, MA 01803

First published 1999

Second edition 2004

**British Library Cataloguing in Publication Data**
A catalogue record for this book is available from the British Library

**Library of Congress Cataloguing in Publication Data**
A catalogue record for this book is available from the Library of
Congress

ISBN 0 7506 6074 0

For information on all Newnes publications visit our
website at www.newnespress.com

Typeset by Charon Tec Pvt. Ltd, Chennai, India.
Printed and bound in Great Britain.

# Contents

# Preface

The prime objective of this second edition of *PC Systems, Installation and Maintenance* is to help the reader to develop a good knowledge base and practical skills base to assemble, upgrade and troubleshoot modern PC Systems. To achieve this aim I have tried to keep the text free of advanced technical jargon while still providing a sound fundamental understanding of the principles and practices involved in PC systems support. As a firm believer in the adage that a 'picture is worth a thousand words', each section contains numerous photographs and illustrations to complement and augment the text.

The knowledge and skills encompassed by the book also provide the second objective that of providing a good source of information for training lead body examinations in PC systems support. This book and on-line resources, will prove useful to students studying for BTEC, City & Guilds e-Quals, CompTIA A+ core, and similar examinations.

The more informative sections of the book contain a full set of self-assessment questions. In response to many requests I have not included the answers in the book however answer sheets are available on request.

I have purposely focused the book on the subject rather than on training lead body objectives as I feel the latter approach usually results in a staccato read that is only suitable for class or homework reference. The former approach was adopted as it allows the reader to develop a more fluent knowledge and skills base.

I wish to thank my family, friends and colleagues who gave me 'space' during the preparation of this book. Many thanks also to Rachel and Doris at Elsevier for providing the impetus and encouragement and to the busy people at Charon Tec in India, for their patience during the proofing stage to complete this second edition.

Robert Beales
2003

# Part 1 PC hardware technology

# Unit 1 The PC defined

*An introduction to the digital computer*

# Introduction to digital computers

The world's first digital computers began to appear at the end of the Second World War. By today's standards they were enormous – taking up an area approximately equivalent to that of a small village hall. Their computing power was similar to that of a modern electronic calculator. But despite the shortcomings this type of computer represented a radical change in thinking – instead of using analogue techniques based on electro-mechanical systems, the architecture was based on digital electronics. These early digital designs were composed of thousands of pairs of electronic valves[1] each one similar in size and appearance to a small light bulb.

Over the decades the size of a digital computer has reduced dramatically while the processing power has increased by a reciprocal amount. This huge increase in the power-to-size ratio was due to the invention of the bipolar transistor and later the field-effect transistor. A transistor in discreet form is about 1/20th the size of a valve and consumes far less energy. Not surprisingly, the transistor quickly replaced the valve in virtually all electronic equipment including computers.

Transistors make use of the properties of a special group of elements called semiconductors. The two main semiconductors used in electronics are germanium and silicon. The first transistors were manufactured singly onto small pieces of semiconductor. But as the technology developed, full circuits containing more than one transistor could be fabricated onto a thin chip of semiconductor. These miniscule but nonetheless powerful circuits are called 'integrated circuits' (ICs). Today, at the upper end of what we can physically squeeze onto a single chip, complex single chip circuits composed of tens of millions of transistors are in production (Figure 1.1).

Most ICs or 'chips' as they are popularly called are etched onto slices of pure silicon. The transistors making up each circuit are so small that they are invisible to the naked eye and yet each one performs a similar task to the light-bulb sized valves used in the first digital computers. Imagine the effort required to build one of these early computers – with thousands of valves, miles of wiring and the huge framework to house it. Now compare this to a modern computer circuit composed of tens of millions of transistors fabricated onto a tiny slice of silicon no bigger than your small finger nail! This incredible advance in technology has made it possible to incorporate huge memories and enormous processing power onto tiny mass produced ICs, enabling today's powerful computers to be built both reliably and at low cost.

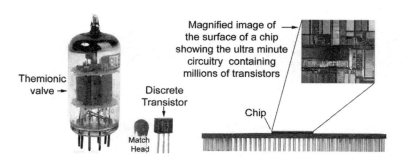

[1] Known as 'tubes' in the USA.

**Figure 1.1**  *Size comparison of a valve, transistor and IC*

# What is a computer?

In essence, a digital computer is a device to help us process, store and present data. However, before a computer can perform a task, it has to be told what to do and how to do it. Also the data we input must be in a form the computer can handle.

The heart of a digital computer is made up from complex arrangements of basic transistor circuits called logic gates and flip-flops.[2] These devices use input and output signals based on just two voltage levels or states. They behave in a similar way to an on–off switch. When the device is 'on' it goes to one voltage state (e.g. +3 V) and when it is 'off' it goes to the other state (e.g. 0 V). These two states conveniently represent the binary numbers 1 and 0, respectively, or 0 and 1, it does not matter which. What matters is that the same convention is adhered to throughout the computer system. So when binary 1 or 0 is presented to an input it is actually just a high- or low-voltage level. The binary digits '0' and '1' are referred to as *bits*.

To the uninitiated it is often difficult at first to understand how information is represented in binary numbers. The way logical values are defined by a computer, such as yes/no, on/off, hot/cold, up/down, are easy to grasp because they are easily represented in binary by two voltage levels. But how do we represent more awkward values like 52 per cent say 'yes', partly on, tepid, half way up, etc.?

In the real world qualities and quantities tend to span an infinite range of possible values (i.e. they are analogue in nature). For example, take a look at a typical daylight scene and try to count how many different colours and shades you see. A computer cannot accurately reproduce a photographic image by simply telling it that parts of the image are bright and other parts are dull, because there are literally thousands of different hues and brightness levels we perceive in a typical scene.

The secret to representing any quality or quantity in binary is to use groups of binary digits. For example, if we use just four bits, in other words four switching circuits, we have 16 different combinations, from all 'off' (0000), to all 'on' (1111), as shown below:

0000, 0001, 0010, 0011, 0100, 0101, 0110, 0111,
1000, 1001, 1010, 1011, 1100, 1101, 1110, 1111.

Each one of the combinations can be used to represent a different quantity or quality. The 16 combinations above would not be enough to represent the different hues in a photographic image. But if we use more switches, we gain a greater number of combinations.

For example, if we use 8 bits we get 256 different on–off combinations and these can be used to represent a reasonable range of values (Table 1.1).

This is more like it but whether it is enough or not depends on the application: 8 bits will represent the colours and brightness levels of a scene far better than 4. But what if we use 16 or even 32 bits? Now we have a far greater range of possible values to represent real-world qualities and quantities: 16 bits will give us a possible 65,536 different combinations and 32 bits will give us 4,294,967,295!

When data from the outside world is input to a computer it must be converted to binary with enough bits to accurately represent the

[2] Information on basic logic functions can be obtained from the Elsevier website at http://books.elsevier.com/companions/0750660740

**Table 1.1**   *With 8 bits there are 256 possible combinations*

| | | | | | | | |
|---|---|---|---|---|---|---|---|
| 00000000 | 00100000 | 01000000 | 01100000 | 10000000 | 10100000 | 11000000 | 11100000 |
| 00000001 | 00100001 | 01000001 | 01100001 | 10000001 | 10100001 | 11000001 | 11100001 |
| 00000010 | 00100010 | 01000010 | 01100010 | 10000010 | 10100010 | 11000010 | 11100010 |
| 00000011 | 00100011 | 01000011 | 01100011 | 10000011 | 10100011 | 11000011 | 11100011 |
| 00000100 | 00100100 | 01000100 | 01100100 | 10000100 | 10100100 | 11000100 | 11100100 |
| 00000101 | 00100101 | 01000101 | 01100101 | 10000101 | 10100101 | 11000101 | 11100101 |
| 00000110 | 00100110 | 01000110 | 01100110 | 10000110 | 10100110 | 11000110 | 11100110 |
| 00000111 | 00100111 | 01000111 | 01100111 | 10000111 | 10100111 | 11000111 | 11100111 |
| 00001000 | 00101000 | 01001000 | 01101000 | 10001000 | 10101000 | 11001000 | 11101000 |
| 00001001 | 00101001 | 01001001 | 01101001 | 10001001 | 10101001 | 11001001 | 11101001 |
| 00001010 | 00101010 | 01001010 | 01101010 | 10001010 | 10101010 | 11001010 | 11101010 |
| 00001011 | 00101011 | 01001011 | 01101011 | 10001011 | 10101011 | 11001011 | 11101011 |
| 00001100 | 00101100 | 01001100 | 01101100 | 10001100 | 10101100 | 11001100 | 11101100 |
| 00001101 | 00101101 | 01001101 | 01101101 | 10001101 | 10101101 | 11001101 | 11101101 |
| 00001110 | 00101110 | 01001110 | 01101110 | 10001110 | 10101110 | 11001110 | 11101110 |
| 00001111 | 00101111 | 01001111 | 01101111 | 10001111 | 10101111 | 11001111 | 11101111 |
| 00010000 | 00110000 | 01010000 | 01110000 | 10010000 | 10110000 | 11010000 | 11110000 |
| 00010001 | 00110001 | 01010001 | 01110001 | 10010001 | 10110001 | 11010001 | 11110001 |
| 00010010 | 00110010 | 01010010 | 01110010 | 10010010 | 10110010 | 11010010 | 11110010 |
| 00010011 | 00110011 | 01010011 | 01110011 | 10010011 | 10110011 | 11010011 | 11110011 |
| 00010100 | 00110100 | 01010100 | 01110100 | 10010100 | 10110100 | 11010100 | 11110100 |
| 00010101 | 00110101 | 01010101 | 01110101 | 10010101 | 10110101 | 11010101 | 11110101 |
| 00010110 | 00110110 | 01010110 | 01110110 | 10010110 | 10110110 | 11010110 | 11110110 |
| 00010111 | 00110111 | 01010111 | 01110111 | 10010111 | 10110111 | 11010111 | 11110111 |
| 00011000 | 00111000 | 01011000 | 01111000 | 10011000 | 10111000 | 11011000 | 11111000 |
| 00011001 | 00111001 | 01011001 | 01111001 | 10011001 | 10111001 | 11011001 | 11111001 |
| 00011010 | 00111010 | 01011010 | 01111010 | 10011010 | 10111010 | 11011010 | 11111010 |
| 00011011 | 00111011 | 01011011 | 01111011 | 10011011 | 10111011 | 11011011 | 11111011 |
| 00011100 | 00111100 | 01011100 | 01111100 | 10011100 | 10111100 | 11011100 | 11111100 |
| 00011101 | 00111101 | 01011101 | 01111101 | 10011101 | 10111101 | 11011101 | 11111101 |
| 00011110 | 00111110 | 01011110 | 01111110 | 10011110 | 10111110 | 11011110 | 11111110 |
| 00011111 | 00111111 | 01011111 | 01111111 | 10011111 | 10111111 | 11011111 | 11111111 |

**ASCII**

There are standard sets of alphanumeric characters and commands used by keyboards. The most common being the American Standard Code for Information Interchange (ASCII), pronounced 'askey'. Using the 8-bit ASCII code a total of 256 different characters and control functions are available.[3]

[3] Information on ASCII character set can be obtained from the Elsevier website at http://books.elsevier.com/companions/0750660740.

original form. This is achieved by devices capable of converting human activities or artefacts into a form the computer can handle. These are called *input devices* or *input peripherals*.

## Input devices

The simplest input peripherals allow the user to physically interact with the computer in some way. For example, a computer mouse is an input device that converts the movement of the user's hand and wrist into binary data representing the position of the mouse at any instant. If 8 bits are used to represent the position along the perpendicular axes, 256 steps can be resolved in each direction. A mouse of greater accuracy would of course require more bits.

A keyboard is another important input device: when a key or group of keys are depressed, a unique set of binary numbers are generated and sent to the computer. The computer then uses this data to represent a particular alphanumeric character or keyboard command.

Generally programs loaded from disk are referred to as *software* while the programs built into the system ROM are called the *firmware*. The electronic parts of the computer system, the disk drives, printed circuit boards, ICs, monitors, printers, modems, scanners, digital cameras, etc. are known as *hardware*.

The Operating System (OS) is a key piece of software that performs several important roles:

- To provide an interface with the user.
- To provide an interface for add-on devices.
- To provide a range of input output services to link the user and application programs to the hardware.
- To provide system integrity, security and management.

The OS therefore has a major influence on how the computer behaves and the type of software it can run.

# Program

To enable a computer to accept and act upon data from the outside world, it must be told what to do and how to do it. The user does this by running a chosen program or suite of programs. In a nutshell a program is a sequence of binary instructions placed in the computer's memory to carry out a particular task or set of tasks.

Even if the user does nothing except press the computer's 'on' button, a set of built-in programs are invoked. These programs – known as the firmware – are permanently stored on a special chip called the *system read only memory* (*ROM*). The first program in the system ROM to execute is the *power on self-test* (*POST*) which checks the integrity of the electronic hardware in the system and then automatically runs a second much smaller program called the *bootstrap loader*. Its job is to search each disk drive in turn for a bootable disk. The bootable disk has a tiny program called the *bootstrap* stored on a special sector of the disk. The bootstrap loader copies the bootstrap program into the computer's memory and then runs it. This in turn loads and runs the operating system (OS), from the disk. Once the OS is up and running the computer is 'alive and kicking' and ready to respond to the users wishes … well almost.

# Output devices

To be of use, any action carried out by the computer must result in some useful outcome. The result of running a program must be made available to the outside world in a form we can understand and use. *Output peripherals* perform this task by converting binary data back into something tangible – effectively the exact opposite action of an input device. For example, a printer converts binary data from the computer onto a printed page as text or as a graphical image. The monitor a.k.a.[4] visual display unit (VDU), is the main output peripheral. It provides the user with immediate visual feedback to his/her input actions.

Naturally the question, 'how many bits?', applies to output devices too. For example, for graphical output, the greater the number of bits used to represent the output data, the higher the resolution. To output a full colour photographic image on a monitor or printer requires a lot of bits. This is necessary to truly represent the full range of hues and brightness levels in the scene.

# Making use of computers

We have seen that a digital computer is an electronic device with an input output interface that can be programmed to carry out useful tasks. The operating system and firmware on the whole make the computer what it is, … a Windows PC, Apple Mac, etc.

The actual processes taking place inside the computer while it is executing a program can be broken down into a sequence of basic switching actions.

The major feature of a computer is that it can carry out these actions quickly, reliably and consistently. However, it can only serve us well when it is fed the right information. Tasks from the outside world must be broken down into a suitable sequence of actions that the computer can handle.

[4] a.k.a.: alternatively known as.

While executing a program, the computer has to be ready to respond to input requests and supply feedback in a form that people can use. In most computing tasks the data being processed at any instant in time comes from a variety of sources within the computer system (Figure 1.2).

A good example of computerized data processing is demonstrated in a weather forecasting system. The computer 'predicts' the weather patterns a few days in advance, from data gathered from current weather behaviour, past weather patterns and from a mathematical model of 'wave front' behaviour. See Figure 1.3.

One of the most powerful digital computers ever built, the CRAY supercomputer, is infinitely less powerful than the human brain; yet it can solve problems that would take a skilled person more than a lifetime to complete manually. We even use computers to do simple tasks, rather than use manual methods. The reason we use computers is that they perform tasks *rapidly*, *consistently* and *reliably*.

As long as a task can be broken down into a series of binary instructions, the computer offers the following advantages over manual methods:

- speed of operation,
- permanent storage,
- consistent performance,
- accuracy.

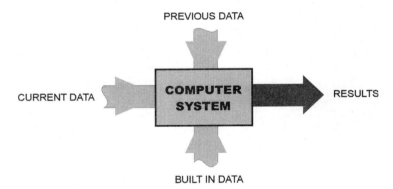

**Figure 1.2**   *A digital computer processes binary data and supplies results*

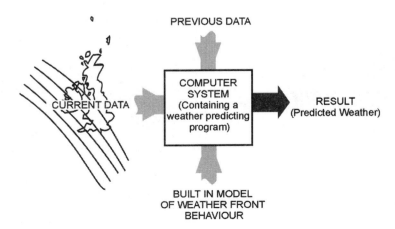

**Figure 1.3**   *A computer-aided short-term weather forecasting system*

*Quoting Spock from the U.S.S. Enterprise*:

> A computer is not influenced by human frailties such as tiredness, boredom, absent mindedness and forgetfulness.

# Five stage computer model

All computers however complicated can be represented by a block diagram composed of the following five elements (Figure 1.4):

1. input unit,
2. central processor unit (CPU),
3. primary storage (ROM and read *and* write memory (RAM)),
4. secondary or auxiliary storage (disks/tape),
5. output unit.

Each element will now be discussed briefly.

### 1. Input

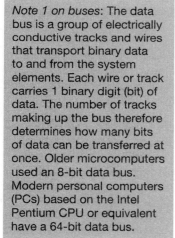

*Note 1 on buses*: The data bus is a group of electrically conductive tracks and wires that transport binary data to and from the system elements. Each wire or track carries 1 binary digit (bit) of data. The number of tracks making up the bus therefore determines how many bits of data can be transferred at once. Older microcomputers used an 8-bit data bus. Modern personal computers (PCs) based on the Intel Pentium CPU or equivalent have a 64-bit data bus.

The primary function of the input unit is to convert data from the outside world into computer readable form (i.e. binary data). Input data can come from a variety of sources and in many different forms. The raw data can be in the form of written characters, numbers, bar codes, punched tape, spoken words and 'keyed in' characters, to name just a few. Only one input device can send data to the computer at any instant in time, even though many input devices are physically connected to the system. In fact all I/O devices share the same data highway. This highway is known as the *data bus*.

Whenever an input device is connected to the system, the computer must be informed so that it can connect the device to one of its data input channels. This is usually carried out automatically by the OS.

The process of enabling an active I/O device onto a data channel is known as placing the device *on-line*. Once on-line, the channel can only connect to the computer's data bus when the CPU is ready to handle it. In other words the CPU dictates when and what device to connect to the bus at any instant in time.

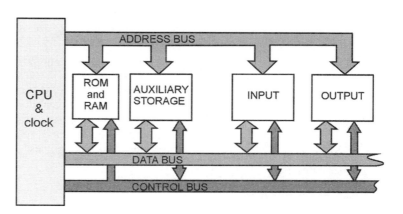

**Figure 1.4** *Five stage model of a computer system*

**Figure 1.5**   *The top and bottom pin view of a modern CPU*

*\*Note 2 on buses*: The CPU has three main buses used for transporting data around the system, the *data bus*, *address bus* and *control bus*.

## 2.  Central processor

The CPU is an electronic chip containing a full sub-system of digital circuits made up from millions of transistors. The CPU and memory form the 'intelligence' or 'brains' of the system. The CPU carries out all the data processing functions for the whole system. In Figure 1.5, the rectangular area in the middle of the CPU is the actual silicon chip. The reverse side of the CPU carries the device's pinouts. These form the various external buses, etc. of the CPU.

The CPU can perform several basic operations on binary data as follows:

- *Arithmetic operations*, for example add, subtract, multiply, divide, increment and decrement.
- *Logical operations*, for example AND, OR, EXOR and NOT functions.[5]
- *Sorting*, for example greater than, less than and equal to.
- *Data transfer*: copying or moving data around the system, for example between the CPU and memory or between the CPU and an I/O device.

Modern CPUs process data at very high speeds, typically thousands of millions of operations per second. This takes a lot of electrical power which is dissipated as heat in the CPU. A large metal heat sink and cooling fan is therefore clamped to the CPU to keep it cool.

The powerful CPU used on most office or home PCs is based on a single electronic chip (IC). It has hundreds of connecting pins which form the connections to the various buses\* used in the system.[6]

- The *data bus* transports instructions and information between the CPU and all the system devices including memory, I/O and ancillary hardware. The more wires used for this bus the greater the amount of data that can be transported at a time. Modern Pentium style CPUs have a 64-bit wide data bus. Some of the early CPUs only had an 8-bit data bus.
- The *address bus* carries the unique address of the memory location or I/O device currently being accessed. The more wires used for this bus the greater the number of memory locations the CPU can access. Modern CPUs usually have 32 address wires which allow the CPU to address up to 4 GB of memory.
- The *control bus* is a group of signal connections that synchronize the transfer of data around the system. Along with the address bus they control when, how and what device to connect to the CPU's data bus.
- Physically the buses are groups of conductors in the form of wires and printed circuit tracks that connect to the CPU and pervade the whole system. If we could freeze the signal state of the individual conductors making up each bus, each one would carry a binary digit in the form of a high or low voltage. Examined over a short-time period (e.g. a few tens of nanoseconds) they appear as a series of voltage pulses.

## 3.  Primary storage

Primary storage is the fast access memory used by the system for storing the firmware and current programs and data. It is fast because

[5] Information on basic logic functions can be obtained from the Elsevier website at http://books.elsevier.com/companions/0750660740
[6] The basic architecture of a CPU is discussed in a later unit.

**Figure 1.6**   *The short term memory of a computer is composed of Random Access Memory (RAM).*

It is imperative that users regularly save their current work to a more permanent storage system. This is what happens when you click on File then Save after typing a letter in a word processor or after partly completing a game. It copies your work from RAM to the currently selected disk-storage system – usually the hard disk inside the system-unit.

The powerful CPUs used in today's PCs have a small amount of super fast but very expensive, electronic memory known as 'cache memory'*. One form of cache memory is incorporated into the CPU chip itself. The CPU copies the current and most likely to be used data from the slightly slower RAM into the cache, thus speeding up data transfer even further.

The cache itself is of course also volatile (*pronounced '*cash*').

it uses electronic memory rather than the slower magnetic or optical storage technology used by disks. Being electronic, this form of memory is packaged in IC form. Primary storage is composed of two distinct types of memory: *short-term memory* and *permanent memory*.

### Short-term memory (RAM)

All programs and data currently being used by system must be stored in a form of memory that allows instructions and data to be fetched quickly by the CPU when required. This eliminates the need to continually input new data for processing – which would result in an extremely inefficient system. In addition the CPU needs a scratch pad – an area of fast memory for storing partly processed data.

This type of memory is the computer's *short-term* memory. It is fast but only holds data temporarily. In other words it is volatile. As soon as the computer is switched off the data is erased irretrievably. This effectively wipes the slate clean for the next session.

Short-term memory is composed of *Random Access Memory chips* (Figure 1.6).

Nowadays RAM is supplied in memory modules which slot into the motherboard inside the system-unit. There are several types available – Figure 1.6 shows a  dual in-line memory module (DIMM).

As well as the user's current work, RAM also holds a temporary copy of the OS and applications.[7] The OS and essential applications are automatically loaded into RAM from disk when the computer is powered up. Other applications can be loaded into RAM from disk at the whim of the operator while the computer is running.

### Permanent memory

Computers also use permanent memory chips called Read Only Memory (ROM). As their title suggests, this type of memory cannot be erased by switching off the computer or by removing batteries. It holds essential programs including the POST, bootstrap loader and a special suite of programs called the *basic input output system* (BIOS).[8] These programs are collectively termed the *firmware* and reside in a chip called the *system ROM*.

### 4. Secondary storage

*Secondary storage* – also termed auxiliary storage – is much slower than RAM and ROM but importantly it uses a non-volatile storage technique based on magnetic or optical disk technology or sometimes both. *Non-volatile* memory uses permanent storage technology to keep the stored data intact even when the computer system is switched

[7] Programs installed by the user are called *applications* (apps), for example games, word processors, etc.
[8] The POST and BIOS are covered fully in later units.

When you purchase a new application it is usually supplied on *compact disc read only memory* (CD-ROM). During program installation the software from the CD is stored onto the system's hard disk. This is subsequently loaded into RAM when the user executes the program. In Windows this is accomplished by simply double clicking an icon or file name.

off. These devices provide long-term data storage. Typical secondary storage devices widely used today are floppy disks, hard disks and recordable compact discs (CDs) – CD-R, CD-RW and DVD-R.[9]

## 5. Output

The function of the output sub-system is the exact opposite of the input unit. It converts the binary data processed by the computer to a form that can be permanently stored or directly used by people.

Output data can be delivered in many forms including images and text on a monitor screen or paper (hard copy). Today's PCs can also process and output audio and video files.

All output devices share the system's *data bus* in a similar way to input devices, so whenever an output device is connected to the system, the computer must be informed so that it can connect the device to a suitable output channel. The process of enabling an active output device onto a data channel is known as placing the device *on-line*.

The data bus can only be connected to an I/O channel at times dictated by the CPU, for example when it is not reading or writing to memory. The CPU automatically determines when and what device to connect to the bus at any instant in time.

If you look at the control panel of a printer – one of the most commonly used output devices – you may see the words 'On Line' next to an indicator light. When the light is illuminated, the device is on-line and ready to receive data from the computer.

## CPU

We will now examine the CPU in a little more detail (Figure 1.7). A basic CPU contains three essential elements, namely a set of internal registers, an arithmetic logic unit (ALU) and a control unit.

### Internal registers

Built into the CPU chip are a set of special temporary storage devices used by the CPU to hold data and instructions while they are being processed. These are not part of the main primary and secondary storage areas or the cache memory but are purely part of the CPU architecture. Each register can store a group of binary digits (bits), usually 8, 16, 32 or 64 bits – depending on the power of the CPU. The registers are usually identified using alphabetical characters. In the basic CPU (Figure 1.7), the data registers are labelled 'A'–'L'.

[9]Notice the variant spelling of disk and disc – the latter being reserved for optical drives.

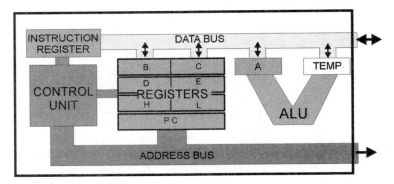

**Figure 1.7**  *A basic block diagram of a CPU*

## Control unit

The control unit is capable of electronically recognizing certain binary patterns to reconfigure the flow of data inside the CPU to perform a specific task. The binary patterns are called instructions and a list of specific instructions is known as a program.

Instructions are simply groups of binary numbers (patterns of '0's and '1's) that the CPU can understand and act upon. For example, Table 1.2 shows a few instruction codes for the ZILOG Z80 8-bit CPU. The Z80 was used in wide variety of early microcomputers like the Sinclair ZX Spectrum circa 1982. The Z80 CPU was chosen as an example here because its architecture fits in with simplified block diagram above. Humans cannot understand CPU instructions in binary form so programmers use a mnemonic form called *assembly language*. The programmer uses a special program called an *assembler* to convert his list of mnemonics into machine code (binary instructions). Notice the actual binary instruction in the right-hand side column in Table 1.2.

When an instruction code is fetched into the instruction register, for example, the binary pattern 10000000 (mnemonic ADD A,B) which means 'Add whatever binary value is held in register B to whatever binary value is held in register A'. The CPU's control unit recognizes the instruction and configures the CPU to carry it out. In this case the binary number stored in register B is first copied into the temporary register and then the ALU *adds* the contents of the temporary register to register A.

Even in a relatively simple CPU like the Z80, there are over 500 instructions (commands) available, some are two or more bytes long. The control unit recognizes them all and is able to configure the CPU to carry them out at lightning speed.

Unless you are a computer programmer, you do not need to understand or use instruction mnemonics (machine code). However it is important to be aware that a computer program in its raw form (machine code) is just a list of binary numbers stored in the computers primary memory area that represent specific instructions to the CPU. The CPU fetches each instruction from memory in turn and executes it before fetching the next instruction and so on, working its way through the list in sequence. Some instructions, called *conditional* and *unconditional branch instructions*, can force the CPU to jump higher or lower down the list but generally the

**Table 1.2**

| A list of arbitrary instructions | Mnemonic | The actual binary instruction recognized by the Z80 CPU |
|---|---|---|
| COPY register A into register B | LD B,A | 01000111 |
| ADD register B to register A | ADD A,B | 10000000 |
| SUBTRACT register B from register A | SUB B | 10010000 |
| INCREASE register A by 1 | INC A | 00111100 |
| DECREASE register A by 1 | DEC A | 00111101 |

instructions follow on from one another in the order they are stored in memory.

### Arithmetic Logic Unit (ALU)

As its name suggests, the ALU can be configured to perform a variety of Arithmetic or logic operations on data held in the registers or in memory. Closely associated with the ALU is the 'A' register, known as the 'accumulator', because it accumulates the result of an arithmetic or logic command. You may be wondering, what happens to the result of an arithmetic operation? It cannot be left in the accumulator for ever! This is true, to utilize the result temporarily stored in the accumulator it can be saved to memory or placed in another register temporarily using a suitable instruction. There are many instructions available to copy the contents of the registers to the main memory or to output devices, etc. Using the vast range of instructions available, programmers are free to do as they please with the data.

## Computer programs

You have probably gathered by now that a program is simply a list of instructions in the form of binary numbers or patterns, stored in successive memory locations.

Each location has a unique address number. The instructions (data) are just sets of binary digits (bytes) stored in the computer's memory as groups of binary digits that the CPU can recognize and act upon. These are fetched one by one into the CPU to be interpreted and executed. Fetching and executing each instruction from memory is an automatic process and is carried out by a special register called the *program counter* a.k.a. *instruction pointer* (*IP*). Before the program is executed the IP is loaded with the start address of the program held in memory. The CPU then fetches the instruction from the address held in the IP and executes the first instruction. It then automatically increments the address held in the IP so that it can fetch the next instruction and so on. This continues until the program has been completed. A special command at the end of the program either signals the CPU to halt or return to what it was doing before the program was executed.

Instructions are executed using a strict timing regime, determined by a quartz crystal controlled clock. The faster the clock, the quicker the execution. The clock generates a series of regular voltage pulses which control the timing of everything in the computer system. Data can only be transferred from one device to another or processed at specific times corresponding to each pulse. In a modern CPU the clock speed can be well over a thousand million cycles per second (1 GHz).

### Pipelining

The type of CPU we are interested in on this course – based on Intel's range of CPUs and alternatives – uses a special architecture called a pipeline. Rather than using the basic 'fetch from memory and execute' sequence, a pipelined CPU can fetch instructions from memory and store them in the 'pipe' while an existing instruction is executing, thereby saving time. To do this the CPU architecture is split into two separate units. Intel calls them the 'execution unit' (EU) and 'bus interface unit' (BIU). The EU contains the ALU and data registers and the BIU contains the IP and pipe. Advanced super-pipelining techniques use several EUs in parallel so that more than one instruction can be executed at a time. However the instructions are essentially still ordered in the same sequence as they occur in memory.

## Summary

The CPU sounds complicated, it is! But you should now be able to see just how simple – compared to a human brain – a computer is.

Everything going on in the system takes place in sympathy with a clock pulse. The CPU *fetches* each instruction from memory into the control unit. This configures the CPU to do a certain task which then *executes*. Instructions are carried out one after another as they occur in memory, in a fetch and execute sequence. This idea of processing commands one by one in sequence was first proposed by the Hungarian mathematician John Von Neumann 1903–1957 (pronounced 'Noyman').

Most modern digital computers use the sequential, Von Neumann, method of program execution despite the clever techniques used by today's CPUs.

# Computer types

Digital computers have become an important part of everyday life. Their ability to process and store data quickly and consistently has made them totally indispensable in our technological age.

Nowadays there is at least one computer system in use in the average UK household. There is probably one in your washing machine. Its purpose is to process input signals, from the push buttons, temperature sensors and level sensors, and provide appropriate output responses to control water heaters, inlet valves, pumps and motors. A selection of 'washing programs' are held in the computer's permanent memory and any one of these can be selected at the touch of a button. Such a computer is known as a *dedicated computer* as it only performs one useful set of task.

The type of computer we are interested in on this course is called a *multipurpose computer* as it can be programmed to perform many different tasks. Multipurpose computers at one time were categorized into four main groups as supercomputers, mainframes, minicomputers and PCs. However because computer technology advances so rapidly, the distinction between these groups is becoming quite blurred. For example, a top mainframe of the late 1980s is not as powerful, in raw processing terms, as a top PC of today. We will therefore make only a general comparison between the different types (Table 1.3). Generally mainframes and supercomputers (Figure 1.8) cost hundreds of times that of a PC and are therefore only used by large organizations.

## Mainframes

Mainframes are used for high-volume data processing in banks, government departments and large multinational companies where ultra-reliable performance and multi-user access are essential. For example, imagine the massive database required to issue driving licences and reminders to all the drivers in the UK, or the stock handling and shipping requirement of a large mail order catalogue company. Mainframes are also increasingly being used as scalable, secure servers for large Internet shopping companies (e-businesses).

**Table 1.3**   *Comparison of the major types of computer*

| Type | Size (approximately) | Processing power |
|------|---------------------|------------------|
| Supercomputer | Size varies from desktop size to that of a large freezer | Exceedingly high: Can support many users simultaneously but often the enormous processing power is put to use solving a single problem at a time |
| Mainframe | Size varies from desktop size to that of a large freezer | Very high: Is able to support many users and programs simultaneously |
| Minicomputer | Size varies from desktop size to that of a filing cabinet | High: Is able to support several users and programs simultaneously; can be regarded as a cut down version of a mainframe |
| PC (desktop, laptop and palmtop) | Small enough to sit on or under a desk; some portable types can sit on your lap or in your hand | Medium to high: Supports one user at a time |

**Figure 1.8**   *The NEC SX5/15A supercomputer. Courtesy NEC corporation*

## Minicomputers

A minicomputer can be though of a cut down, less expensive alternative to a mainframe, used for medium- to high-volume data processing. This type of computer is rapidly disappearing and being superseded by networked PCs or mainframes.

## PCs

PCs are designed to be used by one person at a time. They are used everywhere and for everything, from running small businesses to general purpose home and office use as a word processor, spreadsheet,

**Figure 1.9** *A typical PC system*

database and games system. Increasingly they are becoming an everyday household utility.

The PC – acronym of personal computer and a colloquial name for the old 1980s international business machines (IBM)[10] XT/AT PC (AT is the acronym for advanced technology) is the world's most popular computer type. The number of PCs in the world is now measured in hundreds of millions! Some top-end PCs now offer a performance on par with minicomputers.

The category of use for each computer type is not written in concrete. For example, one company might insist on using a minicomputer for a certain task while another may be quite content using a top-end PC for the same purpose. Another factor blurring the distinction between the various computer types is the increasing use of computer networks. Several interconnected PCs linked to one or more central administration computers (servers) can often provide the same functionality as a mainframe.

The PC is the type we are interested in on this course. A basic PC – at the very least – contains a system-unit, monitor, keyboard, mouse and printer. The system-unit also known as the base-unit is the heart of the PC system and contains the CPU, memory and disk storage systems as well as a series of I/O connectors that interface to the rest of the system.

A minimum home PC also contains extra I/O devices for entertainment such as a sound and video players. These PCs are known generically as multimedia PCs.

A typical PC system is illustrated in Figure 1.9.

All multipurpose computer systems are open systems as they depend on human input to generate useful output. We as users are part of the system. To interact with the computer we use the mouse, keyboard and visual cues on screen. And just like interacting with friends and colleagues, we prefer to communicate using our native language.

Unlike humans, digital computers communicate most efficiently when data is presented in binary.

[10] IBM is the largest and most influential computer manufacturing company in the world.

**Table 1.4**  *A beginner's first assembly language program*

| Machine Code in binary | Hexadecimal | | | | Mnemonic form (assembly language) |
|---|---|---|---|---|---|
| 11101011, 00001101, 01001000, 01100101 | EB | 0D | 48 | 65 | message DB '"Hello World!"' |
| 01101100, 01101100, 01101111, 00100000 | 6C | 6C | 6F | 20 | MOV DX, message    ; Set the address |
| 01010111, 01101111, 01110010, 01101100 | 57 | 6F | 72 | 6C | MOV AX, 0900h    ; Set function WRITE |
| 01100100, 00100001, 00100100, 10111010 | 64 | 21 | 24 | BA | INT 21h    ; CALL DOS Print Routine |
| 00000010, 00000001, 10111000, 00000000 | 02 | 01 | B8 | 00 | RET    ; End Program |
| 00001001, 11001101, 00100001, 11000011 | 09 | CD | 21 | C3 | |
| 10001000, 00001110, 10000000, 00000000 | 88 | 0E | 80 | 00 | |
| 00110100, 00000000, 01100001, 00010000 | 34 | 00 | 61 | 10 | |
| 11011110, 10111110, 00011010, 11010100 | DE | BE | 1A | D4 | |
| 10111010, 11111111, 11111111, 10111000 | BA | FF | FF | B8 | |
| 00000000, 10101110, 11001101, 00101111 | 00 | AE | CD | 2F | |
| 00111100, 00000000, 11000011, 10100000 | 3C | 00 | C3 | A0 | |

For example, the program listed in Table 1.4 prints the message 'Hello World!', on the screen.[11]

A program in binary, like the example in Table 1.4, is known as *machine code* because it is in a form the computer 'understands'. The programmer could enter the list of binary numbers manually into memory but this would be extremely tedious and error prone. Instead a suite of programming tools called an *editor*, *assembler* and *linker* are used. The editor allows the programmer to type out the program using *assembler* mnemonics. These are easier to use than pure machine code. The assembler then converts the completed program from mnemonic form into machine code. The linker them automatically configures the machine code to fit into the computer's available memory space.

Even so, it is still not easy to write large programs in this manner. Each assembler command is based on rudimentary tasks carried out by the CPU, not commands that the programmer uses in day-to-day conversation. Unfortunately, at one time machine code and assembly language were the only programming languages available – highlighting the problem faced by the early digital computer designers, how to improve the operator/machine interface.

The ultimate programming language would allow the programmer to communicate in his/her native language but this has yet to be achieved. Programming languages close to our own level of communication are termed *high-level* languages whereas machine code is classed as *low-level* language.

# THE IBM-PC (A mini-history of its development)

[11] 'Hello World' – the words traditionally used for a beginner's first program.

Mass produced computers first started to appear in the mid-1970s after the development of the microprocessor chip – the forerunner of today's powerful CPUs. One of the most successful of this era was the Apple II microcomputer introduced in 1977 by two young computer enthusiasts: Steve Wozniak, the designer and Steve Jobs, the marketing manager. The computer was so successful that their company went public in 1980 and just 3 years later had a stock value approaching 1 billon dollars. In the late 1970s 'BM' (otherwise known as 'Big Blue') – the world's largest company in terms of the value of its stock, worth more than the combined value of all the stock of all companies listed on the American Stock Exchange at the time – decided to make a stake in the growing PC market and design their own machine. As their expertise up to this point was in the mainframe computer market, they built their PC from standard components, rather than using their own technology. The result, introduced in August, the 12th 1981, was a PC with an *open architecture*. The open design was a major contributing factor to the huge success of the IBM-PC. It meant that the system could easily be expanded, reproduced and upgraded by other manufacturers.

The IBM-PC's operating system was designed by what was then a tiny American software company called 'Microsoft'. It was a *disk*-based *operating system* (DOS). The design brief for the new DOS was that it had to be similar in feel and use to the then highly popular CP/M OS, so that existing computer users could quickly gain familiarity. The IBM version of Microsoft's DOS was marketed as 'PC-DOS'. For other vendors, Microsoft named their product 'MS-DOS'. The new machine and its OS quickly gained popularity and after just 4 months from launch, more than 100,000 machines had been sold.

The early machines were based on Intel's 8088 CPU. This was an innovative chip, which featured a 16-bit internal data bus and an 8-bit external data bus. By choosing a CPU with an 8-bit external bus, all the additional peripheral chips associated with the CPU could also be 8-bit devices. This made the PC relatively inexpensive to build, as 8-bit chips were readily available and less expensive than 16-bit versions. The entry level PC had a tiny (by today's standards), 16 kB of memory, and a magnetic cassette tape storage system. A few months later, 64 kB and 256 kB memory versions were announced.

A few months after the first PC's launch, IBM announced the XT-PC (XT is IBM's term for a PC with eXTended features). This has two floppy-disk drives, more memory, upgradeable to 640 kB and a hard-disk drive option – incidentally in this era, a 5 MB hard disk cost well over a thousand pounds! which is why it was just an option. However, in spite of the costs of these early machines, the IBM-PC soon became the world's most popular PC.

The PC was designed from the outset to be *modular*, and *upgradeable*. It was based on a plug-in expansion bus architecture. This took the form of a bank of double sided 62-pin edge connectors (slots). The slots connect via an interface to the data, address and control bus of the CPU and its associated circuitry.

The expansion bus made the system easy to upgrade. For example, new devices could be added by plugging an appropriate adapter card into one of the slots. Importantly to accommodate this feature, MS-DOS was adaptable enough to accept the additional software (drivers) required to operate the add-on. The drivers were simply installed from floppy disk.

The open design of the PC made it easy for other manufacturers to sell add-on cards and even produce their own PC compatible machines. For example, some manufacturers used the Intel 8086 in place of the 8088. This was a true 16-bit CPU and therefore processed data more quickly. They also incorporated better display technology. These PC compatibles were faster and produced better displays than the original IBM-PC and often sold at prices well below the standard IBM offering.

## THE IBM-PC-AT

The growing sales of XT compatible machines, falling 16-bit chip prices, and the appearance of compatible 16-bit PCs on the competitive PC market, prompted IBM to upgrade the XT architecture. In August 1984 they launched their AT machine. This was based on Intel's 80286 CPU. These early AT machines could address up to 16 MB of memory (the 8088 and 8086 could only address a maximum of 1 MB), operate at a higher-clock speed, and most importantly, provide full downward compatibility to the earlier PCs.

The PC-AT also sported a higher-capacity 5.25 in. 1.2 MB floppy drive that could also read and write to the 360k floppy disks used on the PC-XT. The graphics display adapter was also upgraded to provide 16 screen colours at a resolution of 640 by 350 pixels, compared to the four screen colours and 320 × 200 pixels, of the XT-CGA card. This was also downwards compatible to earlier monitors and software driver.

The PC-AT was an immediate winner, mainly due to its greatly increased performance and downward compatibility. Other manufacturers were soon at it again, producing cheaper- and higher-performance add-ons and compatibles.

As the years passed, Intel developed increasingly more powerful CPUs, all downward compatible with their older 8086/8088 series CPU and PC manufacturers quick to incorporate the new technology into their latest AT compatible models.

In 1987 IBM introduced the Personal System 2 (PS/2) computer. It offered better graphics, a neater 3.5 in. high-capacity floppy-disk drive and a faster expansion bus system. However, this proved to be a major mistake. The expansion bus, known as the micro-channel architecture (MCA) – a technically brilliant design – offered no downward compatibility. This meant that readily available and cheaper AT cards could not be used to upgrade the machine. Also data stored on PC-AT 5.25 in. floppy disks could not be readily accessed on the PS/2. The PS/2 was therefore never fully accepted by the now extensive PC fraternity.

Despite the meagre enthusiasm for the PS/2, IBM has maintained an important role in the development of PCs, in addition to dominating the corporate mainframe world. Offshoots from their impressive research facilities have made a huge impact on the advancement of computer technology.

### Today's PC compatibles

PC-AT compatible PCs still abound in huge numbers but they are gradually being replaced by more advanced systems. New standards, updates and recommendations are continually being announced by the various lead bodies responsible for the development of the PC. The last major update to the basic layout and design of the PC resulted in the extended AT recommendation ATX. The ATX design however still maintains a measure of downwards compatibility to the AT standard. This is a fitting testament to the success of the open architecture of the original IBM designs.

The term 'compatible' in the computer sense means that two machines of different manufacture are able to exchange hardware and software without modification to the machines or the software. The success of the IBM compatible computer is largely due to the ability of a later machine to accept data and hardware from an older machine – this is known as 'downward compatibility'.

The early IBM-PC-XT and AT architectures have been superseded by modern standards but the basic compatibility has still been retained. A PC-AT machine, for example, has a 16-bit expansion bus arranged as an extension to the 8-bit XT slot, so old 8-bit adapter cards can still be used. This 16-bit bus is known as the industry standard architecture (ISA) bus. Likewise the later AT and ATX motherboards still provide slots for the older 8- and 16-bit ISA cards.

The need to offer full downward compatibility to legacy standards has slowly diminished over the years as consumer demand dictates new and more advanced features. To keep pace with these market forces, manufactures introduce increasingly more powerful hardware requiring increasingly more powerful software in a seemingly never-ending spiral of advancing technology. The open architecture of the modern PC however will ensure that it continues to go from strength to strength.

### Alternative PC platforms

Intel and Microsoft have dominated the PC world from the time when IBM deviated from the XT/AT standard and introduced the PS/2, right up to today. However, in parallel with the development of the IBM-PC compatible, Apple has managed to sustain a large following particularly among the desktop publishing fraternity for their non-IBM-PC compatible PCs.

The popular Apple-Macintosh range of computers have a style and functionality still revered by many PC users. The first serious competitor to the PC from Apple was the 'Lisa' launched in 1985 and a year later the 'Macintosh'. These machines owed their success to Apple's graphical user interface (GUI) known as 'Finder'. Rather than typing commands on a keyboard to do basic tasks like opening a file, the MAC user could choose files from pull-down menus using an on-screen mouse-driven pointer. This type of OS interface is referred to as a Windows–Icons–Mouse–Pull-down-menu system or WIMP. Microsoft had also started work on a WIMP system known as 'Windows' but although it was announced in 1983 it only started to appear on shelves in 1985 as Windows 1.0. This early version of Windows was not as functional as the Apple WIMP system and sales were sluggish. It was not until the launch of Windows 3.1 that the PC fraternity had a comparable product.

Today, most PCs run a WIMP-based OS. The popular Microsoft GUIs, Windows 95/98/ME/2000/XP, in chronological order have improved enormously from version to version in usability, reliability and performance since the launch of the first Windows offering in 1985.

### The PC defined

Throughout this book the following PC definitions are used:

- The acronym *PC* describes an open-architecture PC with a CPU that is downwards compatible with the Intel 8086/8088 CPU and capable of running MS-DOS applications as well as modern Windows applications. It may however, *not* have provision to accommodate 8- or 16-bit ISA cards and is therefore not truly IBM-PC compatible. Most modern PC systems fall into this category.
- An *IBM-PC compatible* is a PC capable of running MS-DOS/PC-DOS and corresponding application programs and is able to accommodate 8- and 16-bit ISA cards.

- *Alternative PC systems* using CPUs that are not downwards compatible with the 8086/8088 or capable of running MS-DOS applications will be referred to by their brand name, for example Apple Mac.

# The PC system

We are now in a position to examine the components of a modern PC system in more detail. The PC system will be analysed under the following major subdivisions:

- the system-unit,
- input peripherals,
- output peripherals,
- networks.

Although single stand-alone PC only encompasses the first three subdivisions above almost everybody with a PC today connects to some form of computer network either via a modem to the Internet or via a local area network (LAN) or wide area network (WAN).

# Self-assessment questions

**SAQ1.1**   What device revolutionized the development of all electronic equipment including digital computers?

**SAQ1.2**   How many states can be represented by:

(a) 1 binary digit?
(b) 4 binary digits?
(c) 8 binary digits?
(d) 16 binary digits?

**SAQ1.3**   What generic name refers to programs and data stored in ROM?

**SAQ1.4**   What two parts of a computer can be regarded as the 'brains' of the system?

**SAQ1.5**   There are several buses used in a computer. Name three of them.

**SAQ1.6**   Name four basic operations a CPU can perform on data.

**SAQ1.7**   What are the main features of *primary storage* and *secondary storage*?

**SAQ1.8**   Apart from the CPU what else can perform arithmetic and logic operations?

**SAQ1.9**   (a) What is the function of the *pipeline* in a CPU and (b) what is a *super-pipeline*?

**SAQ1.10**   What is a PC?

**SAQ1.11**   What does the term *downward compatible* mean when applied to a PC?

**SAQ1.12**   What made the IBM-PC and the latter implementations, the PC, so popular?

# Unit 2 The system-unit

*The heart of the PC system*

# The system-unit

The main parts of the system-unit are illustrated in Figure 2.1.

As you can see from figure, the personal computer (PC) system-unit is modular and therefore highly upgradeable. Each part is designed to be physically fitted or replaced with the minimum of tools, for example just a screwdriver will often suffice.

The motherboard is the main circuit board in the system-unit. It fits along one of the side panels of the case and integrates the various parts into a working system. It contains key components including the *central processor unit* (CPU), the *system ROM, RAM*, the *expansion bus* and a basic *input/output* (*I/O*) panel.

The main feature that makes the PC so popular is its expansion and upgrade capability. Extra parts and upgrades can be added to increase the system functionality and to keep the system up to date.

To facilitate this feature, all the major motherboard components connect via sockets and plugs or printed circuit board (PCB) edge connectors (slots). For example, the expansion bus will accept a wide range of adapter cards including sound, graphics, modem and network cards. Each adapter card has a series of gold-plated edge contacts. These connect with the expansion bus slot contacts thereby interfacing the card's electronics to the motherboard system. The basic connection scheme within the system-unit is shown in Figure 2.2.

The main parts that make up the system-unit will now be discussed in detail as follows:

- System-unit case
- Power supply unit (PSU)
- Motherboard
- Disk drives
- Interface cards

## System-unit case

There is a bewildering array of case sizes, styles and colours to choose from when building a new PC but thankfully they all conform loosely to a standard set of sizes and layouts. The main sizes are categorized as *full tower*, *midi-tower*, *mini-tower* and *desktop*. The smaller midi and mini-tower cases are usually placed at the side of the monitor on top of the desk. The full tower is designed to stand on the floor at the side of the user's desk. Desktop cases are designed to sit underneath the monitor (Figure 2.3).

Apart from holding everything together, the case reduces radio frequency interference (RFI) and electromagnetic interference (EMI) emanating from the electronic boards and components inside the system-unit. In fact system manufacturers have to ensure that levels of RFI and MFI fall below a minimum level specified by CE regulations.[1]

### Form factors

The system-unit case design has evolved slowly compared to other PC hardware advances due mainly to the industry insistence on maintaining downward compatibility. Excluding laptop and low-profile designs, only three main case/motherboard/PSU fitting standards – *form factors* – have been announced since the launch of the first PC.

[1] There is more on this subject in Unit 7.

**Figure 2.1** *The system-unit components: (1) motherboard, (2) CPU, (3) RAM, (4) adapter cards, (5) floppy-disk drive, (6) hard-disk drive, (7) CD/DVD/CDR/CDRW drive, (8) PSU, (9) ribbon cables and (10) case*

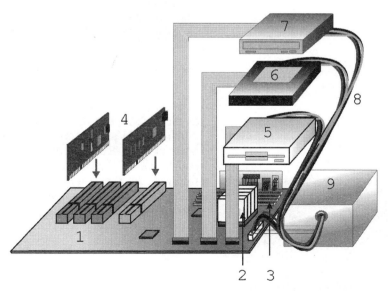

**Figure 2.2** *The basic connection system inside a PC system-unit: (1) motherboard (2) CPU, (3) RAM, (4) adapter cards and expansion bus slots, (5) floppy-disk drive, (6) hard-disk drive, (7) compact disk drive, (8) low-voltage power cables and (9) PSU*

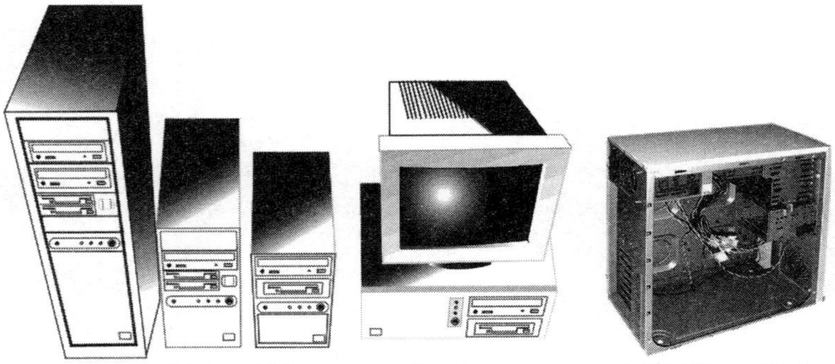

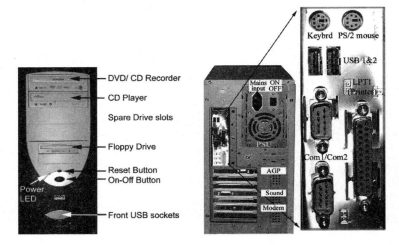

Full tower  Midi tower  Mini tower  Desktop  ATX Midi tower interior

60x18x55cm  38x17x41cm  32x17x40cm  40x15x40cm

**Figure 2.3**  *System-unit case.* Note: *The case sizes shown here are only typical and vary from one manufacturer to another*

DVD/ CD Recorder

CD Player

Spare Drive slots

Floppy Drive

Reset Button
On-Off Button

Power
LED

Front USB sockets

Keybrd  PS/2 mouse

USB 1&2

Mains ON
input OFF

LPT1
(Printer)

PSU

AGP

Sound

Modem

Com1/Com2

**Figure 2.4**  *ATX midi tower case showing the front and rear panels*

In July 1995 Intel introduced the ATX form factor (revision 1.0) as an evolution of the advanced technology (AT) design. It was a much needed specification that went some way in tidying-up the basic internal design of the PC to meet the requirements of current and future PC hardware technology. Several revisions have since followed. ATX revision 2.1 was announced in 2002. Most full size PC systems sold today conform to some degree to the ATX recommendations (Figures 2.4 and 2.5).

## PSU

The power supply is easily recognizable as the rectangular tin box usually located at the back of the system-unit case. It is fixed to the rear panel with four screws and facilitates easy removal for servicing or replacement. Today's CPUs' motherboards and graphics cards cram a lot of processing power into energy hungry chips so a good quality adequately rated power supply is essential (Figure 2.6).

In the quest to capture a wider share of the home computer market several PC manufacturers are now offering very compact in vogue case designs based on unconventional shapes and sizes. Unfortunately this move away from accepted standards will inevitably reduce the upgradeability and serviceability of the product.

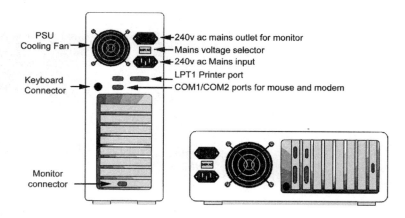

**Figure 2.5**   *Rear view of an AT midi tower and desktop case*

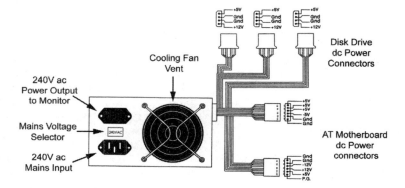

**Figure 2.6**   *The layout of a typical PC-AT PSU*

The power supply has two main functions:

- it electrically isolates the system from the dangerous mains supply,
- it provides safe well-regulated, low-voltage direct current (DC) supplies for the motherboard and disk drives.

The power output capability of the PSU is surprisingly high for such a small unit, typically providing between 200 and 400 W. This is achieved using a *switched mode power supply unit* (SMPSU) design instead of the more conventional *linear* power supply design used on mini- and mainframe computers.

In a linear PSU, the mains alternating current (AC) is stepped down to provide a set of low-voltage AC feeds using a mains transformer. These are then rectified to AC, smoothed and regulated to provide a range of stable DC outputs. The method is simple to implement but the drawback is the size and weight of the components needed to transform and smoothen the low-frequency mains supply. The higher the AC supply frequency the smaller and lighter these components could be but the mains frequency is fixed at a low 50 Hz.

In an SMPSU a different approach is used: first the 240 V AC mains is rectified to its peak value (approximately +340 V DC). It is then converted back to AC electronically at a much higher frequency

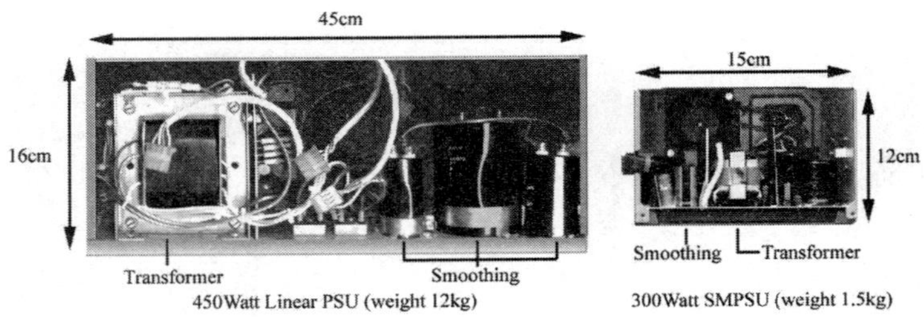

**Figure 2.7**  *A bulky linear power supply compared with its neater SMPS sibling*

than the mains supply – usually above 30,000 Hz. This high-frequency AC is then transformed down to a series of low-voltage feeds, rectified, smoothed and regulated in a similar way to the linear PSU. As the AC frequency is now much greater than 50 Hz, the transformer and smoothing components can be a lot smaller in size and weight, compared to the linear design.

The only shortcomings of an SMPSU compared to its bulky cousin are the increased circuit complexity and possible RFI.

RFI is generated by high-frequency currents in the transformer and surrounding circuitry. To reduce RFI to acceptable limits, electronic filter components are used on the mains input connections inside the PSU and the whole unit is encased in a steel enclosure. These measures stop RFI leaking onto the mains supply and also reduce radiated RFI.

Figure 2.7 compares a bulky linear 450 W PSU removed from an old minicomputer, with a 300 W SMPSU from a PC. The cover has been removed from the SMPSU revealing the relatively tiny transformer and smoothing components.

As well as the main system-unit PSU, SMPSUs are used to power all manner of computer peripherals, including monitors, printers, scanners, modems and plotters. Their small size and high efficiency give designers more freedom to incorporate sleek styles, as well as functionality, into their products.

## ATX power supply

The original ATX specification announced in 1995 suggested an alternative PSU fan arrangement. Instead of placing a separate fan on top of the CPU, a bigger PSU fan was proposed. The idea being that the power supply would be positioned close to the CPU so that cool air from the PSU fan will be blown across the CPU and heat–sink. Unfortunately this idea failed to take into account the staggering evolution in CPU speeds and the consequent heat generation. So a separate CPU fan is still necessary.

Figure 2.8 shows a typical PSU conforming to the ATX 2.1 specification. The main motherboard power connection from the PSU is a single 20-way female connector (Molex 39-01-2200). The corresponding motherboard mounted connection is a 20-way male connector (Molex 39-29-9202). Each contact on the PSU connector has a plastic shroud of square cross section protruding from the main shell. Several of these are rounded on one side so they only locate correctly with the motherboard connector one way round. It is therefore impossible to

All the case metal work including the PSU is connected to the earth wire of the mains supply via the mains cable. This also connects to a common 0 V plane inside the PSU and the motherboard, where it helps to screen out interference. This earth plane is often referred to as *earth*, *ground*, *chassis*, *common* or *0 V*, they all mean the same thing. The positive and negative DC supply voltages from the PSU are all with respect to this common earth plane.[2]

[2] Strictly speaking the words 'earth' and 'ground' should only be used when referring to equipment that is properly earthed.

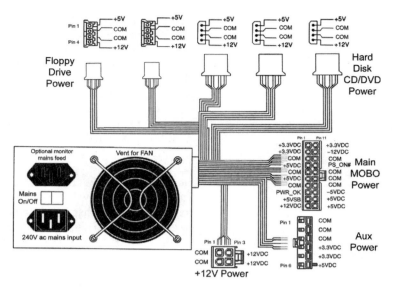

**Figure 2.8**  *ATX PSU and connector configuration*

connect the PSU to them the wrong way round using moderate insertion force.

The main features incorporated in the ATX PSU are as follows:

The older AT-style PSU has two identical motherboard power connectors and these can quite easily be fitted incorrectly. To prevent this situation the golden rule is to connect them so that all the black 0 V wires align in the middle.

- software power-on/power-off;
- standby power;
- well-regulated, stable DC supplies of +3.3, ±5 and ±12 V;
- extra cabling for new and future technologies.

### *Software power-on and power-off (PS_ON#)*

To switch an AT PSU on–off required a fairly heavy duty mains on–off switch located on the PSU itself or on the front panel. The latter method meant that live mains cables were present inside the system-unit. This is not particularly desirable even though the switch terminals are usually heavily insulated. The ATX design uses a far neater and safer solution using a PS_ON# input signal. PS_ON# allows the motherboard to remotely control the PSU on–off function via software or via a light-action switch. The dangerous voltages are therefore contained inside the PSU case.

When the PS_ON# line is held to a low voltage between 0 and 0.8 V the power supply turns on the five main DC supply rails and the PC powers up.

When PS_ON# is held to a voltage between 2 and 5 V the DC output rails shut down to 0 V.

This basic power on/off function is implemented via a simple front panel switch connected across the motherboard's front panel PWR and GND pins or via software features such as Wake-on-local area network or Wake-on-modem.

The PWR_OK line is almost self-explanatory. It was labelled the Power Good (P.G.) signal on the XT and AT-style PSU. It is held high (i.e. held at a few volts positive) by the PSU to signal the motherboard that the main outputs have reached a stable level.

*Standby power (5 V SB)*

Modern operating systems (OSs) such as Windows 98/2000/XP together with a supporting BIOS have the ability to support automatic shutdown and standby, it was therefore a logical step to include facilities for this in the ATX PSU specification. In standby mode the +3.3, ±5 and ±12 V supplies are turned off but the 5 V SB feed from the PSU remains energized. This supplies an auxiliary ± 5 V to the motherboard circuits to keep the essential soft-power control circuitry operational.

*3.3 V output*

Intel introduced the low-voltage +3.3 V outputs in the ATX PSU design to suit the new lower-voltage CPUs, memory chips and motherboard chipsets. Earlier integrated circuits (ICs) used a standard +5 V system but modern high-speed devices consume too much power at this voltage – lower-operating voltages mean lower-power consumption.

*Extra cables*

The ATX 2.1 specification includes details of additional power feeds and connectors for systems that consume DC currents in excess of the maximum connector rating. The main motherboard power connector has a maximum current carrying capacity of 6 amp per pin. These extra feeds provide 0, +3.3 and +5 V on a six-pin in-line 'auxiliary power connector' and 0 and +12 V on a separate four-pin +12 V connector. This effectively spreads the total current load over more pins.

# The motherboard

The motherboard is the main PCB in the system-unit. It acts as the central connection point for the whole system and carries out all the major data processing and control functions. The motherboard contains several major system components including the following items:

- CPU and FAN assembly,
- the system ROM,
- CMOS setup,
- RAM modules (primary storage),
- expansion buses,
- I/O interfaces,
- disk drive interface circuits.

You do not need to be familiar with these components at this point as each one will be described in detail, later in this section.

## Motherboard form factors

There are several motherboard form factors used on the PC. These specify the PCB board size and hole fixing positions, as well as recommended strategic positions for major components. The main ones used on today's PCs are as follows:

- full size AT (obsolete),
- Baby-AT,
- ATX,
- NLX,
- ITX.

## Full size AT

The original International Business Machines (IBM) PC-AT used a full size motherboard, measuring 13.8 in. by 12 in. (35.02 cm × 30.48 cm). This is considerably larger than the early PC-XT motherboard and is no longer in production.

## Baby-AT

The Baby-AT board has a width of 8.57 in. (21.77 cm) the same as the original XT motherboard but the length varies considerably from one board to another. It was a popular alternative to the full size AT form factor as it gave the PC manufacturer more scope for designing compact and therefore more marketable systems than the full size offerings. The full size AT motherboards therefore soon faded into obsolescence.

Today, even though the Baby-AT form factor is no longer used for new systems, motherboards conforming to the Baby-AT form factor are manufactured for upgrade purposes (Figure 2.9). They are often used as a convenient way to upgrade the CPU/RAM and BIOS without having to renew the case and power supply, as you would be upgrading to a more modern form factor.

The Baby-AT motherboard was smaller than the original full size 286 AT board but the positioning of the CPU and *single in-line memory modules* (SIMMs) and the lay of the ribbon cables from the separate I/O card made the task of upgrading quite awkward. In

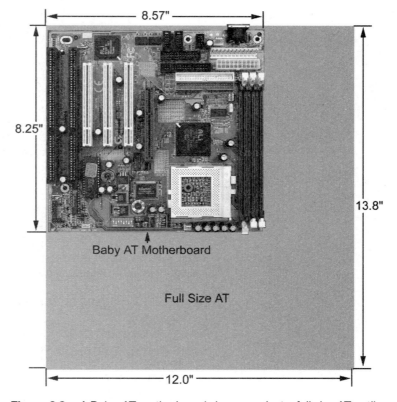

**Figure 2.9** *A Baby-AT motherboard shown against a full size AT outline*

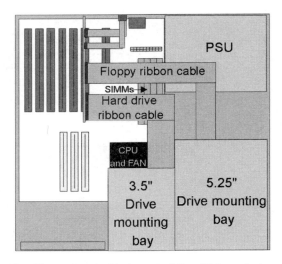

**Figure 2.10**  *The rather untidy layout of the AT form factor*

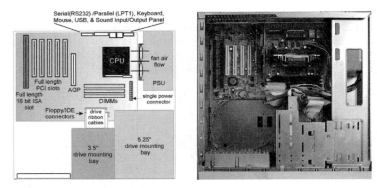

**Figure 2.11**  *The ATX form factor allows for a far less cluttered interior*

some systems the drive mounting bay and/or PSU had to be removed temporarily to perform RAM or CPU upgrades (Figure 2.10).

A new form factor was desperately awaited to sort out these issues.

### ATX motherboard

The ATX form factor was introduced by Intel in 1995 to overcome the shortcomings of the Baby-AT design (Figure 2.11). This major change to the basic PC architecture involved physical changes to the PSU, the motherboard and the case. The main motherboard embellishments involved are as follows:

- Repositioning the SIMM and CPU sockets to make them more accessible.
- Integrating the standard serial/parallel, universal serial bus (USB), sound keyboard and mouse ports into a neat array at the rear of the motherboard. This simplifies installation and removes the cable clutter at the rear of the case.
- Fitting a single 20-way motherboard power connector instead of the two connectors used on AT boards. This allows extra feeds

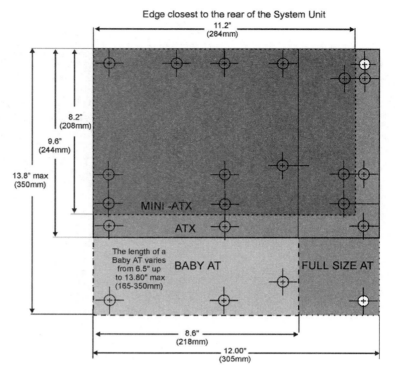

**Figure 2.12**  *ATX and AT board size comparison (dimensions are only approximate)*

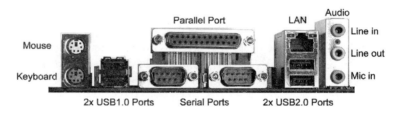

**Figure 2.13**  *Typical motherboard ATX 2.1 I/O panel layout*

to be accommodated including a +3.3 V supply and software-controlled power on/off and standby.

The ATX board is based on a Baby-AT board rotated through 90° (Figures 2.12 and 2.13), so the long side fits along the rear of the case. The maximum recommended dimensions are 12 in. by 9.6 in. (30.48 cm × 24.38 cm). Smaller mini-ATX boards are also available with maximum dimensions of 11.2 in. by 8.2 in. (28.45 cm × 20.83 cm).

The ATX motherboard/case system is a neat evolution of the old AT design and a boon for new system builders. As the ATX system is a complete rethink of the PC layout it is not downwards compatible to the old AT standard. This highlights a situation often met by the systems support engineer. Here is a typical real-life scenario where you can imagine you are the systems support person.

Your customer Mrs Malone has a young family attending junior school:
    You have just completed a fault diagnosis of their stand-alone non-networked AT system. The family uses this for internet access and general purpose applications including games. You have diagnosed a faulty CPU and PSU. The CPU is obsolete but you have a second-hand one in an old working motherboard. You can obtain a replacement AT PSU from your main supplier. Before you commence any repair work, you politely state your diagnosis, remedy and the cost of the repair to the customer. At this point Mrs Malone asks you if the repair will be cost effective and whether a better option would be to upgrade the system with a new CPU and motherboard able to accept modern add-ons like a digital camera. However at present they cannot afford a completely new system, but she is thinking of buying her husband a digital camera for his birthday.
    To upgrade the old AT machine to an ATX system, the case, PSU, motherboard, CPU and RAM would have to be replaced. At the very best, the only salvageable items from the old system-unit are the graphics card, sound card, hard drive, compact disc read only memory (CD-ROM) drive and floppy drive. But when these are compared to today's accepted specifications it is questionable whether these items are worth reusing. The existing monitor, keyboard, mouse and modem can be reused.

**SAQ 2.1** How would you advise the customer and what would you suggest?
    *Hint:* You could start by listing all the options available, their relative cost and then use positive, open questioning and recommendations to reach a satisfactory mutually acceptable outcome.

## NLX

Low-profile PCs have been around for a long time but there has never been a satisfactory standard for the board size and interconnection system. In all low-profile systems, the low case height is achieved by placing the bus expansion slots on a riser board instead of on the motherboard. This allows the adapter cards to lay in the horizontal plane thereby taking up less vertical space.

The NLX standard was jointly developed by Intel, IBM and Digital Equipment as a low-profile motherboard form factor for small desktop solutions. It supports current and future processor technologies and accepts standard SIMM and *dual inline memory modules* (DIMM) devices. A major feature of the system is the easy removal and fitting of the motherboard and riser board.

The main board holds the CPU, BIOS, main memory, cache memory and controller ICs and the riser board holds the bus expansion slots. The riser board supports both the *Industry Standard Architecture* (ISA) and *Peripheral Component Interconnect* (PCI) buses, and slots onto one edge of the NLX main board it also bolts onto the main chassis.

Compared to previous low-profile designs the NLX specification improves the accessibility of upgradeable components, such as the CPU, RAM and the motherboard itself by making more efficient use of the space inside the case. The motherboard, which slides into an edge contact on the riser board, can easily be installed and removed, by simply removing a side panel.

Just like the ATX form factor the NLX standard is supported by a broad range of manufacturers, original equipment manufacturers (OEMs) and third party vendors (Figures 2.14 and 2.15).

**Figure 2.14**  *NLX board arrangement*

**Figure 2.15**  *Conventional low-profile layout*

### Micro-ATX, ITX and mini-ITX

In a move away from traditional PC enclosures, several manufacturers have produced miniature PC systems small enough to fit in the base of a liquid crystal display (LCD) monitor. The motherboards used in these systems provide very little expansion capability but they do cram a lot of power in a small space. The boards in decreasing size are known as micro-ATX (244 mm × 244 mm), ITX (215 mm × 191 mm) and mini-ITX (170 mm × 170 mm). An example mini-ITX board by Virtual Interface Architecture (VIA) featured a built-in low-temperature CPU, on-board sound and graphics and even a spare PCI slot for future expansion (Figure 2.16).

## The motherboard – a closer look

Imagine your embarrassment turning up to service a customers PC only to find that it has a problem with a device or technology that you know nothing about! It is therefore imperative for everyone in the business of supporting PC systems to be aware of the wide variety of different styles, layouts and specifications in use today. These range from legacy systems that have seen better days, to brand new

**Figure 2.16** *A VIA mini-ITX motherboard*

state-of-the-art systems. Gaining this knowledge is easier said than done. Even while this book is being published, new technology and new standards are evolving. What the author describes as 'state-of-the-art' may well be 'run-of-the-mill' by the time the book finally appears in print.

So how do you keep abreast of the latest technology? The answer is by doing what all good systems support engineers do – read technical journals and browse manufacturer's web sites and FAQs, regularly![3] It is also important to be aware of the older established technologies.

Going back to the later half of the 1990s for a moment: most Pentium CPU-based PC systems available used a motherboard with a ZIF socket known as 'Socket 7'. Socket 7 motherboards were very popular as they worked satisfactorily on a range of Pentium-style CPUs including Intel P54/P55, advanced micro-devices (AMD) K5/K6 and the CYRIX 6 $\times$ 86/MII range. However as the demand for faster CPUs at lower prices forced the manufacturers into more competitive practices the standard Socket 7 format was gradually abandoned. The first major watershed in the evolution of Pentium-style motherboards occurred when Intel introduced their radical Slot 1 CPU (Figure 2.17), the Pentium II (PII). The Slot 1 design required a special motherboard with a slot for the CPU instead of Socket 7 (Figure 2.18). AMD followed suit by introducing their own slot design called Slot A. Both designs were incompatible with one another and CPUs of different manufactures could not be interchanged on the same motherboard. Unfortunately this situation remains unchanged today. If you build a system based on a current Intel CPU, for example a Pentium (PIV), you must use a motherboard specifically designed for that CPU range. The same applies to AMD CPUs.

Soon after the introduction of slot CPUs, both Intel and AMD reverted to ZIF socket CPU designs.[4] Again there was no standardization. Intel used a 370-pin socket (Socket 370), and AMD decided on a 462-pin socket (Socket A, alternatively known as (a.k.a.) Socket 462). One of the latest CPUs from Intel, the Pentium IV, requires a 478-pin socket (Socket 478) (Figure 2.19).

[3] A list of technical web sites can be obtained from the Elsevier website at http://books.elsevier.com/ companions/0750660740
[4] See CPU section for more information.

**Figure 2.17** *Intel Slot 1 motherboard*

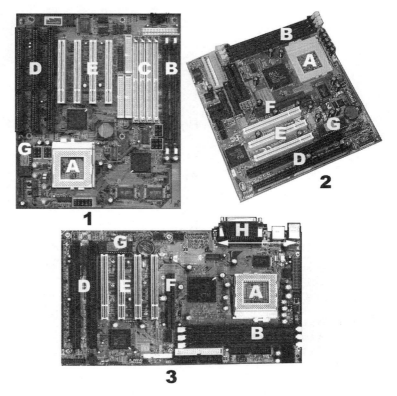

**Figure 2.18** *Typical Socket 7 motherboards: (1) Baby-AT, (2) Baby-AT with AGP and (3) ATX. Key of the major system components: (A) CPU socket, (B) RAM (DIMMs), (C) RAM (older SIMMs), (D) ISA slots, (E) PCI slots, (F) AGP, (G) system ROM and (H) rear I/O panel*

As well as the differences in CPU design, motherboards for Intel and AMD CPUs now tend to support different RAM technologies. Intel favours a proprietary RAM technology called Read Direct RAM (RDRAM) and AMD prefers the open Data Direct DRAM (DDR-DRAM). This disparity is in direct contrast to the multi-vendor Socket 7 design.

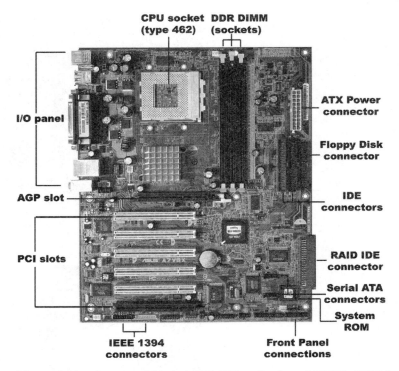

**Figure 2.19**    *A top-end Socket 462 ATX motherboard (ASUS A7V8X)*

A typical top-end motherboard is equipped with some impressive technology. For example the motherboard shown above incorporates several of the latest bus systems and I/O ports including USB2.0, IEEE1394, Serial and Parallel ATA, Accelerated Graphics Port (AGP) 3.0 (8X). It also accommodates the latest AMD CPUs and DDR RAM devices.[5] The same motherboard manufacturer also markets a similar board for the latest Intel CPUs and memory devices.

As this is a book intended for the beginner as well as the more familiar computer user, we will start our investigation of the motherboard by taking a look at the basic CPU/RAM/I/O architecture.

## The chipset

PC motherboard technology has come a long way since the launch of the AT machine. No longer are the CPU and RAM the only dominant items determining the system power. The motherboard chipset also plays a very important role. In terms of electronic complexity it could be argued that the chipset is the most important feature. So what is the chipset?

In a nutshell the chipset controls data flow throughout the motherboard, linking the CPU and RAM to everything else in the system. The chipset is analogous to a traffic control centre of a large city. The traffic flows at different speeds and volumes depending on the location. City traffic is the slowest whereas traffic on the outer ring roads and motorway links flows at a higher rate. The control centre manages the traffic system using traffic lights and signs to direct the traffic to various destinations.

It is unfortunate that the two major CPU manufacturers now support different architectures and pinouts. At one time an AMD CPU would work quite happily in place of an Intel CPU and vice versa. Now the situation has changed completely and one is not compatible with the other. This means that the two makes of CPU require a different chipset, CPU socket and hence a different motherboard.

[5] These are discussed in detail later in this unit.

On a PC the chipset manages the data flow between the various motherboard devices using a series of carriageways analogous to the roads around a city.

There are slow-, medium- and high-speed routes and these are assigned to a device accordingly. Devices such as the keyboard, mouse, floppy drive and legacy I/O ports transfer data quite slowly so they only require a narrow carriageway. In direct contrast, the CPU and RAM handle an enormous amount of very fast data so they require a very wide carriageway.

In a computer, a measure of the ability to handle fast data is termed the bandwidth. The greater the bandwidth of the data-handling system the greater the data throughput. The bandwidth is determined by three parameters:

1. the width of the carriageway,
2. clock frequency,
3. bus architecture.

The data carriageways of a computer are known as buses. The more conductors making up a bus the greater its bandwidth. The bus width is literally the number of conductors used. Each conductor carries 1 bit at a time so the bus width is stated in bits. On a Pentium-based system and above, the CPU/RAM data bus is 64-bit wide.

The clock frequency determines how quickly bits are transferred along the bus. This depends on the chipset, CPU and type of RAM in use. The more up to date these components, the higher the clock speed and greater the bandwidth. For example, a 64-bit bus operating at a clock frequency of 66 million pulses per second (66 MHz) can transfer 528 million bytes per second (MB/s) – assuming one block of 64 bits are transferred per clock cycle.[6]

With a touch of electronic wizardry the bandwidth can be doubled or quadrupled by transferring more than one block of data per clock cycle. We will discuss clocks and clock frequencies presently.

### North bridge and south bridge

The chipset is usually divided into two separate chips. One for the high-speed CPU/RAM interface and one for medium- to low-speed I/O devices. These are known as the north bridge (NB) and south bridge (SB), respectively – from their conventional position in block diagrams and from the fact that they bridge several highways (buses) together (Figure 2.20).

### Frontside bus, memory bus and AGP bus

Conventionally the NB chip is used to handle the high-CPU/RAM data rate but recent chipset implementations also use the NB for the latest AGP bus standard – known as AGP 3.0. Four buses are involved:

1. 64-bit CPU data bus (*frontside bus*),
2. 64-bit RAM data bus (*memory bus*),
3. 32-bit AGP 3.0 bus,
4. SB to NB link.

Some vendors use different names for the NB and SB chips. For example, Intel has recently used the name *memory controller hub* for the NB and *I/O controller hub* for the SB.

The operation of the NB chip is complex but basically it organizes the flow of data across the bridge by storing data from each bus until the recipient device is ready. This is known as *bus buffering*.

[6] A 64-bit bus is 8 bytes wide. So at 66 MHz, the bandwidth is $8 \times 66$ MB/s.

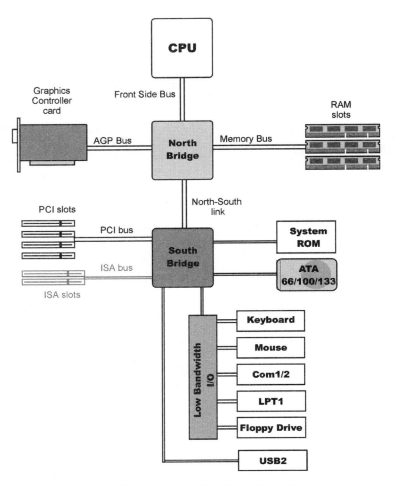

**Figure 2.20**   *North bridge and south bridge chipset bus arrangement*

The most accessed device in the system is RAM. Both the CPU, AGP card and SB are continually vying for memory access. So the greater the bandwidth of the memory bus the more quickly it is available for another device. As mentioned previously, the bandwidth is determined by the chipset architecture, bus width and clock frequency.

## Clocks

Every device on the motherboard involved in data transfer is controlled either directly or indirectly by a clock and there are several on the motherboard (Figure 2.21).

A clock, in electronics/computer terms, is a regular series of voltage pulses each with a fast rising and falling edge and a high and low steady-state portion. The number of pulses that occur in 1 s is termed the *clock frequency*. The time interval between two consecutive rising or falling edges is termed the *period* of the clock.

The period ($T$) of the clock is the reciprocal of the clock frequency ($f$), that is:

$$T = 1/f$$

For example, a clock frequency of 100 MHz has a period of 1/10 MHz (i.e. 10 ns).

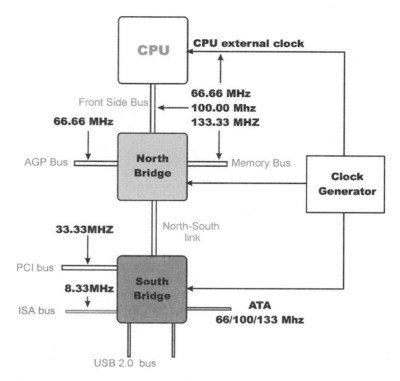

**Figure 2.21** *Chipset clocks*

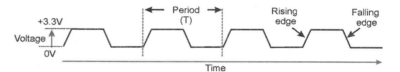

**Figure 2.22** *Typical 3.3 V clock waveform*

**Figure 2.23** *Quartz-crystal clock generator*

The period defines one cycle of the clock and the number of cycles occurring in 1 s is the frequency measured in Hertz (Figure 2.22).

To maintain data integrity the timing of all data transfers in the system must be extremely precise. So the clocks must remain very stable over several hundred clock pulses. This task is carried out by a quartz-crystal-controlled clock generator IC (Figure 2.23). Usually the quartz crystal and sometimes the generator IC itself is housed in a small metal can. There are several clock generators on the motherboard.

Conventionally data is transferred on either the rising or falling edge of each clock cycle. However the latest CPU/chipset combinations can transfer data twice in each clock cycle by transferring data on both edges. Some chipset/CPUs are able to transfer data four times in each clock cycle using special techniques.

### The CPU clocks

The CPU uses two clocks, an *external clock* also called the *system clock* and an internal *core clock* that runs at a multiple of the system clock frequency. The system clock usually operates at 66.66, 100.00 or 133.33 MHz. For example, an AMD Athlon XP 2000 CPU operates at a core clock frequency of 1667 MHz, so with the system clock set at 133 MHz, the core clock multiplier factor is 12.5×.

### Frontside bus clock

The FSB is clocked by the *system clock* (i.e. 66.66, 100.00 or 133.33 MHz), but later CPU/chipset combinations are able to transfer data at double or even quadruple the system clock frequency. For example the Intel Pentium IV CPU with a system clock of 133.33 MHz runs the FSB at 533 MHz.

### Backside bus clock

A few years ago motherboard CPU/chipset systems used an additional bus called the *backside bus*, which clocked at a frequency close to the CPU core frequency. Its purpose was to provide a fast data link between the CPU and the Level 2 cache on the motherboard.[7] Recent CPUs form Intel and AMD have Level 2 cache built onto the CPU so the backside bus is effectively implemented in silicon.

### Memory bus clock

The memory bus usually clocks at the same frequency as the system clock (i.e. 66.66, 100.00 or 133.33 MHz), but again using special techniques, chipset/RAM combinations can transfer more than one data block per clock cycle.

### AGP bus clock

The AGP bus connects to a single AGP slot on the motherboard providing a fast data highway to the graphics accelerator card. There are several AGP implementations. The original AGP1.0 specification uses a 66 MHz clock and a 32-bit data bus and transfers one block of data per clock cycle. This provides at best, a bandwidth of 264 MB/s. Later 2X and 4X versions increased the bandwidth to a maximum of 528 MB/s and 1.1 GB/s by transferring two and four blocks of data per clock cycle, respectively. The recent AGP3.0 specification transfers eight blocks of data per clock cycle giving a bandwidth of 2.1 GB/s.

### Alternative bus names

The chipset buses and bridges are often described in technical documents by different names as shown in Table 2.1.

The following section covers the range of CPUs the support technician is likely to meet when maintaining PC systems in the field. This is quite detailed in places so it is best to use this section for reference rather than reading through it chronologically.

[7]Cache memory is discussed later in this unit.

**Table 2.1** *Alternative bus names*

| Generic name | Intel variant name | AMD variant name |
|---|---|---|
| FSB | System bus | System bus |
| CPU external clock | System clock | System clock |
| CPU internal clock | Core clock | Core clock |
| NB | Memory controller hub | System controller |
| SB | I/O controller hub | Peripheral bus controller a.k.a. HyperTransport™ I/O hub |

# CPU

The CPU, as its name suggests, is the main data processing chip on the motherboard. With the support of the chipset and RAM it forms the 'nerve centre' of the whole system. Back in 1980 when the design of the first IBM PC was under way the IBM engineers decided to use an Intel 8088 CPU – a cut-down version of the 8086. This had an 8-bit external data bus, a 20-bit address bus and a 16-bit internal data bus. The core of the CPU ran at 4.77 MHz, over 200 times slower than the clock speed of today's designs. A range of more powerful Intel CPUs followed over the years, including the 80286, 80386, 80386SX, 80486DX, 80486SX, Pentium and Pentium equivalents. Each new processor sported a more advanced architecture and instruction set than the previous model. Sensibly, each new design contained an instruction subset of the previous model to maintain downwards compatibility. This practice is still adhered to today.

### Maths co-processors – floating-point units

Up to the launch of the 80486 CPU, there was no built in floating-point maths capability in any of the foregoing CPUs. So maths intensive applications ran very slowly. To improve this situation, a separate maths co-processor could be linked to the main CPU. To accommodate this option, motherboards designed for any CPU prior to the 80486DX had a separate socket alongside the CPU socket for an optional maths co-Processor IC. Unfortunately these ICs often cost more than the CPU and motherboard combined so they were not a popular upgrade option.

Thankfully a separate maths-co-processor chip is no longer required as all CPUs from the 804086DX up now incorporate a built-in maths processor called the *floating-point unit* (FPU). The FPU takes over when floating-point arithmetic operations, are being handled leaving the main CPU to get on with its normal processing duties.

Table 2.2 shows a list of old CPUs and their corresponding co-processors.

Several of today's applications including games, feature advanced 2D and 3D graphic rendering effects that are very maths intensive. For example some 3D appliances use floating-point routines to calculate points in the Z plane (image depth). So there is often a great performance boost when a high-performance FPU is used. Several of the earlier Pentium-equivalent CPUs had inferior FPUs, compared to Intel's engine, so these otherwise fine CPUs, performed poorly when

**Table 2.2**   *Old CPUs and their corresponding co-processors*

| Main CPU | Floating-point maths unit |
| --- | --- |
| 8088/86 | Separate 8087 co-processor required |
| 80286 | Separate 80287 co-processor required |
| 80386SX | Separate 80387SX co-processor required |
| 80386DX | Separate 80387DX co-processor required |
| 80486SX | 80486DX (used in place of old CPU) |
| 80486DX, Pentium and beyond | Built into the CPU |

confronted with such tasks. Today's CPUs feature advanced pipelined FPUs, so the performance differences are less pronounced.

### Pentium (fifth-generation CPUs)

Intel launched their fifth-generation CPU in 1993. Instead of naming it the 80586, in line with the previous series of processors, Intel registered it under the name 'Pentium' to prevent clone CPU manufacturers from using the same name.

The Pentium represented a revolutionary advance in PC processor architecture but it still offered full downwards compatibility with the entire installed base of PC applications software.

Some of the significant enhancements were:

- superscalar architecture,
- dynamic branch prediction,
- pipelined FPU,
- two internal 8 KB caches,
- 64-bit data bus.

### Superscalar architecture

The Pentium features a superscalar architecture, which means it can execute two instructions at once. It does this by implementing two instruction pipelines called the 'U' and 'V' pipes. The U-pipe, also known as the primary pipe, can execute all integer and floatingpoint instructions. The V-pipe or secondary pipe, can only execute integer and a limited range of floating-point instructions. However in many situations this technology enables two instructions to be executed in one clock cycle. At best the 80486 could only achieve an average of one instruction every two clock cycles.

### Dynamic branch prediction

As mentioned in Chapter 1, all generations of the Intel CPU from the early 8088 to the present, use the pipeline technique. CPUs based on the instruction pipeline principle, fetch subsequent instructions into the pipe before the current instruction has been completely executed. However there is a slight disadvantage with this system that dynamic branch prediction can improve.

### Pipelined FPU

The FPU of the Pentium used faster algorithms to produce results up to 10 times faster than that of the 80486DX FPU.

If the current instruction is a branch-type command, then the next instruction in the pipeline could be from the wrong location. So the command must be fetched from memory instead of the pipe. *Branch prediction* is a technique that attempts to infer the next instruction address, from the current one, using a small storage area called the *branch target buffer* (BTB). The BTB tries to predict where the next instruction after a branch instruction is located. If successful, it fetches it into a prefetch buffer. If a branch mis-prediction occurs, instruction speculation ceases, the buffers are flushed and the correct instruction is fetched from the slower main memory in the normal way, this is known as *branch recovery*.

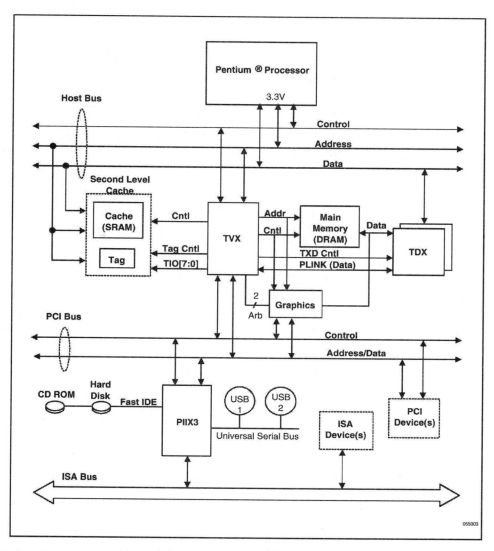

**Figure 2.24**   *The complex internal organization of the Pentium Processor. Courtesy Intel*

### Internal cache memory

There are two internal caches on the Pentium, each 8 kB in size. One acts as a cache for data and the other for code.[8]

### 64-Bit data bus/32-bit address bus

To improve the data transfer rate the Pentium has a 64-bit wide internal/external data bus (the 80486 data bus was only 32-bit wide). The address bus and internal registers however are 32-bits wide – the same as the 80486 (Figure 2.24).

The first Pentiums released in 1993 ran at 60 and 66 MHz. Initially the Pentiums received a fairly cold welcome by PC vendors – not surprising as they were relatively expensive and their lack-lustre performance with the software of the day did not help matters. Despite the fact that the first Pentiums were inherently more powerful than 80486 CPUs, the 80486DX2 66 MHz offered a better

---

[8] Cache memory is covered later in this unit.

performance/price ratio. Another disadvantage with the 60–66 MHz Pentiums was their relatively high-heat dissipation.

In 1994 Intel announced the second-generation Pentium CPUs using 0.6 μm technology. These were available in 90 and 100 MHz versions and were altogether a better device. They ran faster, consumed less power and used a more compact pin arrangement. The corresponding socket, *Socket 7*, became the industry standard for a whole generation of Pentiums and Pentium alternatives.

The last of these popular fifth-generation Intel Pentiums, known as the P54C series, were supplied in speeds up to 200 MHz. Due to the fabrication technology used, the power consumption of the Pentium 200 MHz, was lower than the earlier 66 MHz Pentium. And to improve heat dissipation even further, the P54C was also manufactured in 0.35 μm – what was then the latest fabrication technology,[9] for use in laptop PCs where low heat dissipation and power consumption is paramount.

### Fifth-generation Pentium alternatives

Ever since the launch of the early Intel 8086 CPU, several chip manufacturers have produced clones of Intel processors. Direct copies were often produced under licence from Intel. For example AMD produced versions of the 80386 and 80486 CPU. Since the advent of Intel's Pentium however, alternative chip manufacturers have strived to design and produce advanced Pentium equivalents to equal or better the performance of their rival's product. From these 'chip wars', two main Intel alternative CPU manufacturers emerged. One was Cyrix who later amalgamated with VIA to produce inexpensive low-power CPUs for laptops, etc. and the other was AMD. The AMD have remained in direct competition with Intel ever since.

Importantly, in those days the three main manufacturers kept to the Socket 7 specification so their CPUs were fully compatible with any Socket 7 motherboard.

One of AMD's successful alternatives to the Pentium P54C was their 'K5' Socket 7 CPU. This chip featured six parallel execution units (EUs) arranged in a five-stage pipeline. It also contained a 16 kB instruction cache and an 8 kB data cache. To make it an attractive alternative to the Intel Pentium it was priced very competitively.

Another popular Pentium P54C alternative was the CYRIX 6 × 86 processor. This clever CPU featured two superpipelined integer units, an FPU and some innovative techniques.

The most notable of these were:

- out-of-order processing,
- data-dependency removal,
- multi-branch prediction and speculative execution.

*Out-of-order processing* allows a simpler instruction to complete in one pipeline ahead of a previous more complex instruction in the other pipeline.

*Data-dependency removal* effectively unravels the serial nature of some of the complex 8086 instructions to achieve a form of parallel processing and hence faster execution.

*Branch prediction and speculative execution* is based on the same principle described for the Intel Pentium P54C (Figure 2.25).

[9] The lower this factor the smaller the chip die and the heat dissipation for a given clock speed.

**Figure 2.25** *Socket 7 Pentium and Pentium equivalents circa 1996*

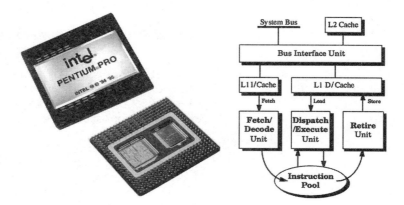

**Figure 2.26** *Block diagram of the three core structure of the P6. Courtesy: Intel*

The 6 × 86 however used a 256 entry, four-way set associative BTB to store possible branch addresses and an eight-entry return address stack. A four-state branch prediction algorithm is used to achieve a claimed 90 per cent hit rate.

### Sixth-generation CPUs

The first sixth-generation CPU from Intel was launched in 1995. It was named the 'Pentium Pro', it was also given the code name 'P6'. It was available in speeds from 150 to beyond 200 MHz and still maintained downward compatibility with the instruction codes of the earlier generations of Intel processors, right down to the 8086/8.

The P6 featured a complicated internal structure based on a 12-stage, superpipelined architecture (Figure 2.26). Instead of the usual pipelined fetch and execute system used on previous processors it uses an *instruction pool in association* with a *fetch/decode unit*, a *dispatch/execute unit* and a *retire unit*.

This new architecture enables instructions to be started in any order but essentially completed in the original program order. The processor also has the ability to 'look ahead' at the instructions in the instruction pool and carry out those that can be executed immediately. Data is fetched for those instructions needing extra information, while the simpler instructions are executing. The results are then stored back into the pool until they can be placed in the correct sequence in the retire unit.

By executing instructions 'out of order' in this way, the processor neatly eliminates the time delay while data is being fetched from main memory.

The Pentium Pro (P6) was a major departure from the conventional run of PC-type processors. Rather than using just a single processor core, it integrates several system components into one package. It accommodates an Advanced Programmable Interrupt controller (PIC), a Level 2 cache and cache controller, as well as the CPU and its Level 1 cache, into a single package that fits into a special CPU socket known as Socket 8.

One innovative feature of P6 was its ability to operate with up to three other P6s in a multiprocessing mode. This and its high price indicated that it was mainly aimed at the professional market but it signalled that even better CPUs designs were around the corner.

### Processors with MultiMedia eXtension technology

To ensure continuing success in the commercial CPU market, Intel launched MultiMedia eXtensions (MMX) in January 1997 as an upgrade to the standard Pentium, called the 'Pentium MMX', code named 'P55C'. The P55C was based on a souped up P54C with 57 extra instructions specially geared to make multimedia software run more efficiently. It was pin-to-pin compatible with a standard P54C so it fits a standard Socket 7 motherboard.

After research by Intel and several leading software developers into the most computationally intensive routines. It was found that many routines used multiple repeating loops that only operated on small data quantities, but consumed a large portion of the processors time. MMX technology was devised to improve this situation by adding the following features:

- SIMD technology (single instruction, multiple data);
- 57 additional instructions;
- eight 64-bit wide MMX registers, each overlaying two existing 32-bit registers;
- four new packed data types.

The new instructions allow several single byte data items to be acted upon in parallel. For example in many graphics applications, '$n$' blocks of data are handled slowly 1 byte at a time by repeating a particular instruction '$n$' times. Using the MMX SIMD technology, 8 bytes of data (in the form of two 32-bit integers) can be handled at once with a single instruction. Thereby increasing the execution speed eight-fold.

Parallel operations can be performed on single byte, word (2 bytes), double words (4 bytes) and quad words (8 bytes).

To utilize MMX enhancements no additional motherboard hardware or BIOS amendments are required, all the enhancements are built into the processor. However, performance increases are only noticeable on software using MMX.

In addition to MMX, the P55Cs were upgraded with a 16 kB data cache and a 16 kB instruction cache – double the size of the P54C cache. Also branch prediction, pipelining and write buffer performance were improved. The new features allowed more instructions to be executed in one clock cycle thereby increasing the processor performance even in the absence of MMX compatible software.

AMD and Cyrix quickly followed suit adding MMX technology to their new range of Pentium P55C equivalent CPUs. The AMD K6, and the CYRIX 6 × 86MMX were clearly excellent alternatives to the Intel Pentium P55C range of processors.

**Figure 2.27** *Pentium II slot processor from Intel*

With such a wide choice of CPUs, competition between the major manufacturers increased and prices fell quite dramatically. In 1999 Cyrix decided to pull out of the top-end CPU race and concentrate more on the low-power laptop CPU market.

MMX still forms part of the instruction set of the latest CPUs.

### Pentium II and Pentium III

The first Intel Pentium II code name 'Klamath', was launched in May 1997 and represented yet another radical change in processor design. Early versions were based on the P6 architecture and used commercially available Level 2 cache chips and a processor die similar to the P6 but with MMX technology.

The revolutionary feature of this processor was the device packaging. The processor and cache chip dies are mounted on a PCB substrate with a double-sided single edge contact (SEC) cartridge and an aluminium thermal plate. This is then encased in a Lexan cover, with mounting latches.

The SEC system uses 242 copper edge finger contacts on the PCB substrate and this connects to a double-sided double-layer edge connector known as Slot 1 (Figure 2.27).

When it was launched Intel suggested that this form of packaging would be the next logical step forward for this and future high-speed processors, as it allows the Level 2 cache to be placed right next to the CPU for maximum operating speed, and also improves the overall handling protection, and ease of installation. As it happens this was not to be the case. For one thing Intel patented the Slot 1 design which prevented rival CPU manufacturers from using it for their new CPU designs. The second factor was cost and the third factor relates to the fabrication technology itself. Chip fabrication techniques were developing rapidly over this period. We will come back to this point in a moment (Figures 2.28 and 2.29).

The original Pentium II was available in core clock speeds from 233 to beyond 500 MHz. The FSB was clocked at 66 MHz. It also featured a 32 kB Level 2 cache – 16 kB each for the data and instruction cache.

The Pentium II did not incorporate the Level 2 cache chip on a silicon die like the Pentium Pro, instead it is soldered onto the PCB substrate using commercially available burst pipelined synchronous

**Figure 2.28** *Side A of the Pentium II PCB showing the CPU (centre) and two L2 cache chips*

**Figure 2.29** *Side B showing a controller IC and two more BSRAM ICs*

Static RAM (BSRAM). This meant that it could only reliably operate at 50 per cent of the core clock frequency. Nevertheless the Pentium II performed well in comparison to rival socket CPUs at the time.

In 1998/99 a succession of more powerful versions of the Pentium II slot-style CPU were announced by Intel including the Pentium III (PIII). Improvements included larger Level 2 caches and higher clock speeds on the FSB and CPU Core. The first release of the Pentium III in 1999 code named 'Katmai' had a 100 MHz FSB and 70 new instructions given the unlikely name 'streaming SIMD extensions' (SSE). Basically the use of SSE improves the performance of some repeating single-precision floating-point maths operations, in much the same way as MMX improves the performance of repeating integer operations. These new commands are only beneficial when OSs and applications make use of them. Windows 98/ME/2000/XP does.

Controversially Intel added a processor serial number (PSN) tag to the Pentium III that allowed the serial number to be read directly off the CPU electronically. Many critics regarded this as an invasion of privacy as the PSN could be read via the Internet. However some companies saw it as useful facility for keeping track of the installed CPU based on networked PCs – particularly as top-end Pentium IIIs at launch, cost several thousand pounds each. Around this time CPU and RAM theft reached epidemic proportions and weight for weight they were worth more than gold.

In 1998 a lower-cost alternative of the Pentium II called the 'Celeron' was launched for the entry level market. It was basically a

**Figure 2.30** *AMD's seventh-generation Athlon Slot A CPU*

slot-style Pentium II with the Level 2 cache removed. Unfortunately the performance did not match up to the current MMX Socket 7 CPUs from AMD and Cyrix so it was not hugely successful.

### Seventh-generation CPUs

The Intel Pentium II, Pentium III and AMD K6-II and K6-III are all variants of sixth-generation CPU architectures. The Intel Pentium II and Pentium III CPUs are based on the P6 and the AMD K6-II and K6-III are enhanced versions of the K6. AMD was the first to launch a seventh-generation CPU in 1999 called the Athlon (Figure 2.30).

The Athlon was originally fabricated in 0.25 μm technology and was available in several core clock versions from 500 to 650 MHz. Later when 0.18 μm fabrication technology became a reality, higher speed versions were released.

The architecture of the Athlon was a radical leap in 80 × 86 compatible processor design. It incorporated some 22 million transistors compared to the 9.1 million transistors used in the Pentium III CPU. A large portion was used for the large Level 1 cache and additional EUs. The Level 1 cache consisted of a data cache and an instruction cache, both had a capacity of 64 kB compared to the 16 kB caches used on the Pentium III. It also sported nine EUs – the Pentium III had six. Three of the EUs are dedicated to floating-point instructions and offered a performance that surpassed the Pentium III FPU at comparable clock frequencies. The Athlon was a major improvement on previous AMD CPUs which under-performed during FPU intensive calculations compared to equivalent Intel CPUs.

The Athlon added another 19 instructions to the 3D Now technology – first introduced in their K6-II and K6-III series – and also added five special instructions for digital signal processing (DSP) applications. The net result was a noticeable improvement in graphics and multimedia performance, compared to sixth-generation CPUs of the same clock speed.

A radical feature of the Athlon was its 200 MHz FSB. This was twice the maximum speed of the FSB of previous CPU generations (Figure 2.31). The bus technology was licensed from Digital's Alpha EV6 interface protocol. One major feature allowed two devices to share the bus at the full clock speed (200 MHz) using packet

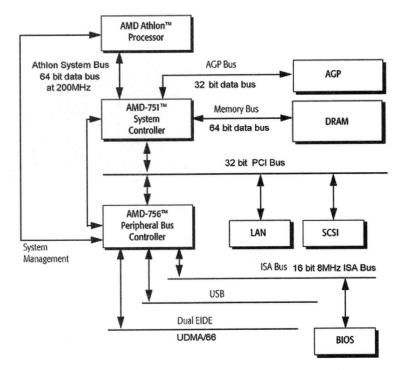

**Figure 2.31**   *The system-unit block diagram of an Athlon PC*

**Figure 2.32**   *Fitting a slot-style CPU module*

switching technology. For example an AGP card could use memory at the same bandwidth as the CPU. The bus also supported the then prevalent 100 MHz DRAM technology.

The construction of the original Athlon was based on a slot design mechanically similar to the Slot 1 design used on the Pentium II, Penitium III and Celeron. However the pin designations are totally different, so the two-rival slot CPU ranges are not interchangeable. An AMD Athlon can only operate on a purposely designed Athlon Slot A motherboards and Intel slot CPUs can only operate on specific Slot 1/ Slot 2 motherboards.

Power dissipation is always a problem with high-speed CPUs and the slot-style CPUs are no exception. To provide adequate cooling a substantial heat sink and fan are essential.

After assembling the CPU SEC cartridge to the heat sink, the whole assembly slides into a framework built around motherboard slot. The device is then pushed down until it clicks in place in the slot.

The slot CPUs and motherboards were regarded by many as an unwelcome deviation from the standard Socket 7 system that had prevailed since the launch of the Pentium in 1993 (Figure 2.32).

# From socket to slot and back again

Ignoring marketing factors, the main technical reason why Intel and AMD moved from socket CPUs to the radically different slot design, was to incorporate the Level 2 cache chips as close to the CPU die as possible. This is necessary to maximize the operating speed of the CPU core/cache link (i.e. the backside bus). Even over distances of just a few centimetres, the longer the backside bus, the slower the data transfer rate. This is because the bus wires act as delay-lines and seriously reduce the operating speed.

The delay-line effect is caused by electrical parameters such as resistance, inductance and capacitance. But even if we ignore these factors voltage pulses can only propagate along a wire at two-thirds the speed of light. So signals take at least 0.5 ns to travel 10 cm along a perfect conductor. A CPU core operating at 1 GHz for example, has a clock period of 1 ns (i.e. the delay is of the same order as the clock period!).

In the older Socket 7 system, the Level 2 cache is placed on the motherboard so the backside bus is quite long and propagation delays are at a maximum. This was not a major issue however because Socket 7 CPUs have relatively low-core clock frequencies. In modern systems the CPU core frequencies often extend well beyond 1.5 GHz so the backside bus must be kept as short as possible if the cache is to run at the same frequency. The obvious solution was to put the cache on the same silicon slice as the CPU itself. But around the time slot CPUs were being developed, chip fabrication technology had not evolved sufficiently to achieve this. The slot CPUs system was therefore a workaround allowing fast L2 cache ICs to be mounted close to the CPU using the SEC cartridge PCB system.

From the outset, Intel's Slot 1 and AMD's subsequent Slot A design appeared to many people to be 'a solution looking for a problem'. The production costs of such a design seemed to outweigh the small speed advantage it offered.

It was therefore not surprising when Intel introduced a lower-cost version of the Celeron by repackaging the CPU die into a 370-pin socket named Socket 370.

Technology permitting, it was almost inevitable that both Intel and AMD would eventually go back to producing socket versions of their latest CPUs. The turning point came when chip fabrication technology finally advanced to resolutions of 0.18 μm and below. At last it was possible to reduce the area of silicon used by the CPU and Level 2 cache sufficiently to fit them both onto the same silicon slice. This heralded a major speed-versus-cost advantage.

However, nothing is straightforward in the chip industry. When the chip area is reduced, the conductors linking the various sub-circuits are thinner and their ability to conduct electrical signals is impaired. If the reduction process is taken too far, the circuits fail to function properly and excessive heat generation becomes a problem. Up to this point the chip conductors were made of aluminium which has a poor conductance compared to copper or silver. It is used because it is easier to fabricate on silicon.

The next leap in chip fabrication technology took place when copper conductors were successfully implemented. With copper technology, even though the conductors are considerably thinner at 0.18 µm technology and below, their conductance is about the same as aluminium at 0.25 µm. Intel aptly called their new chip fabrication process 'Coppermine'.

## The current socket CPUs

### Intel Pentium III and Celeron

Due to the new 18 µm chip fabrication advance, in April 2000, the Intel Pentium III was launched as a socket CPU with full-speed Level 2 caches of 256 kB on the chip. Also a new version of the Celeron with Level 2 cache was announced. The new Celeron was based on the Pentium III die but with only half of the Level 2 cache enabled (i.e. 128 kB).

Intel's new socket CPUs are quite different to Socket 7, there are more pins and the chip is placed on the top of the substrate. This allows the heat sink to make intimate contact with the face of the chip thereby transferring heat from the CPU more efficiently.

### Intel Pentium IV

In November 2000 Intel released the Pentium IV range featuring their new NetBurst micro-architecture. The first Pentium IVs were shipped on a 423-pin socket (Socket 423) but later versions from the 1.9 GHz Pentium IV and up, use Socket 478. The NetBurst micro-architecture featured a new hyper-pipelined technology which is basically 20 pipelines working in parallel. This is double the pipeline depth of previous Intel CPUs and this alone significantly increases the processor performance. To handle the extra throughput of the pipelines, it incorporates two arithmetic logic units operating at twice the core frequency. The FSB also runs at a much higher frequency than previous designs by transferring data at four times the system clock frequency. For example with a 133.33 MHz system clock the FSB runs at 533.3 MHz. The Pentium IV (Figure 2.33) also incorporates an extra 144 instructions and SIMD extensions 2 (SSE2) which improves MMX™ and SSE1.

(a)                     (b)                     (c)

**Figure 2.33**  *Typical modern CPUs; (a) Intel Pentium III, (b) Intel Pentium IV and (c) AMD Athlon XP*

### AMD Athlon (thunderbird), Athlon XP and Duron

The current socket CPUs require decent sized heat-sink/fan assemblies and must not be operated, even for a few seconds, without a heat sink. Even if you are just testing the CPU, make sure the correct fan and heat sink are fitted. Unlike previous CPU versions, these new CPUs with their combined on die Level 2 cache, will quickly overheat and fail if allowed to run without a heat sink.
In case the fan should fail, the BIOS used with these CPUs supports temperature monitoring and an audible warning system and automatic shutdown options (Figures 2.34 and 2.35).

Just after the launch of the Intel new socket CPUs, in June 2000 AMD launched their socket equivalent of the Slot A Athlon, code named 'Thunderbird'. It was virtually a full Athlon on a single silicon die with several performance enhancements including a Level 2 cache running at the full core clock frequency. AMD also launched a low-cost version socket version of the Athlon named the 'Duron'. The Duron has a CPU core based on the Athlon architecture but with a smaller purpose built 64 kB Level 2 cache. Owing to its lower price and slightly better performance, it proved to be a good competitor to Intel's Celeron.

In competition with the Pentium IV, AMD released the Athlon 'Palomino' in May 2001. The first Palominos were sold as top-end notebook CPUs and featured a new Athlon-based core (fabricated using copper). Several architectural improvements were also incorporated including Intel's SSE instruction set. The new CPUs consumed significantly less power and thus ran cooler than the Thunderbird.

The desktop version of the Palomino the 'Athlon XP', was launched in October 2001 and managed to perform more instructions per clock pulse than an Athlon Thunderbird running at the same clock speed. So in a masterful marketing ploy, AMD used their own form of benchmarking and processor speed branding. For example an Athlon XP 2000 is so named because even though it only operates with a core frequency of 1667 MHz, it performs like a Thunderbird running at 2 GHz.

The exposed silicon chip on these new CPUs means that careful handling is essential when fitting the heat sink/fan combination. Careless fitting of the heat sink/fan can easily result in a cracked chip. AMD's Athlon/Duron CPUs are particularly susceptible to this form of damage. The face of the heat sink/fan assembly must be positioned in the same plane as the chip surface before the retaining clamps are secured. To achieve this there are four high-density foam spacers on the CPU substrate.

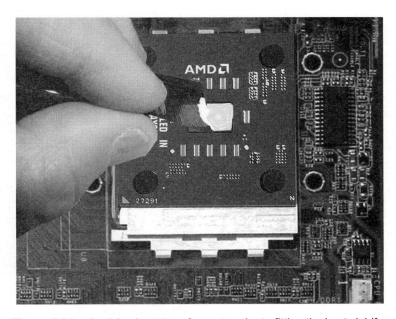

**Figure 2.34**  *Applying heat-transfer paste prior to fitting the heat sink/fan*

## Pentium CPU sockets

With all the socket changes that have taken place since the launch of the Pentium CPU it is fitting to list the sockets found on today's installed PC base (Table 2.3). *Note*: All CPU sockets from Sockets 1 to 6 are used on obsolete CPUs and will not be discussed further.

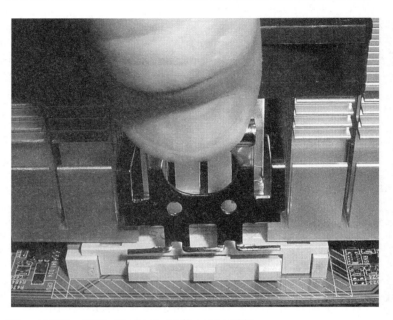

**Figure 2.35**   *Fitting the heat sink/fan assembly*

**Table 2.3**   *Sockets and slots from the 80486 to present-day CPUs*

| Socket name | Contacts | Pin layout | Processors supported |
|---|---|---|---|
| 1 | 169 | PGA 17 × 17 | 486SX, DX, DX2 |
| 2 | 238 | PGA 19 × 19 | 486SX, DX, DX2, P/Overdrive |
| 3 | 237 | PGA 19 × 19 | 486SX, DX, DXn, P/Overdrive |
| 4 | 273 | PGA 21 × 21 | Pentium 60 and 66, P/Overdrive |
| 5 | 320 | PGA 37 × 37 staggered | Pentium 90 and 100 |
| 6 | 235 | PGA19 × 19 | 486DX, DXn, P/Overdrive |
| 7 | 321 | PGA 37 × 37 | Pentium 75 & up inc. clones |
| 8 | 387 | PGA modified | Pentium Pro (P6) + P6 Overdrive |
| Slot 1 | 242 | SEC | PII, some PIIIs, Celeron |
| Slot A | 242 | SEC | AMD Athlon 500–1 GHz |
| Socket 370 | 370 | PGA | Intel Celeron and PIII |
| Socket A | 462 | PGA | AMD Thunderbird, Duron and XP |
| Socket 478 | 478 | PGA | Intel PIV |

*Key:* PGA, pin grid array; PGA (staggered/modified), plastic pin grid array; SEC, single edge contact.

### Socket 7

Socket 7 is an upgrade to Socket 5 with one extra pin and was used on the popular Pentium P54 and P55 range of CPUs and equivalents.

### Socket 8

Socket 8 was originally destined to be the last CPU socket according to Intel. It has 387 pins in a staggered pin grid array (PGA). This socket was specially designed for the Pentium Pro (P6).

### Slot 1

Slot 1 was the first of a new CPU connector series by Intel. Introduced in 1997, it was designed for the Pentium II, Celeron and future PC-type processors but the production cost of this style of CPU proved to be its downfall.

### Slot A

Slot A was AMD's venture into the CPU slot arena. It was physically similar to Slot 1 but with a completely different pin configuration.

### Socket 370

Socket 370 signalled the return to socket CPUs. It was originally introduced as a cheaper alternative to the Slot 1 system for the early range of Intel Celerons. Later the Pentium III used the same physical socket but the pinouts are different from that of the Celeron.

### Socket A/Socket 462

The AMD Duron, Thunderbird and XP use a 462-pin socket. This was named *Socket A* by AMD but it is often referred to as Socket 462.

### Socket 478

The Intel P4 range of CPUs use a 478-pin socket (Figure 2.36).

## CPU performance ratings

In order to provide the consumer with a clear assessment of the relative performance of different CPUs, several organizations have produced CPU benchmark criteria and software. Producing a good benchmark program is quite a difficult task as it must take into account the strengths and weaknesses in the architecture of all the current range of CPUs. It must then run tests simulating today's applications to exemplify these differences.

Just comparing the performance of a CPU based purely on clock speeds is not a good indicator of processing power, particularly when used across different architectures. However, everything else being equal, it stands to reason that CPU architecture X running at a core frequency of 2.0 GHz will definitely execute more instructions per second than the same architecture X running at 1.0 GHz. How this extra power is perceived depends on the user and on the applications in use. For example, while typing a document in a word processor the user would not detect any difference as the CPU is virtually idle most of the time waiting for the next key press. However in processor intensive applications such as 3D imaging the difference in performance would be very noticeable.

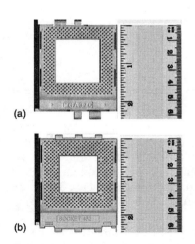

(a)

(b)

**Figure 2.36** *(a) Intel's Socket 370 and (b) AMD's Socket A/Socket 462*

### Benchmarking complete systems

The CPU is obviously an important factor in determining the relative power of one system versus another but there are many other important hardware factors that should be taken into account too. The RAM, chipset, graphics card and hard-disk drive also contribute significantly to the perceived system performance. In an attempt to access the whole rather than part, the Ziff Davis media group PC Magazine, assigned VeriTest to produce Winstone and WinBench two notable benchmark programs.

### Winstone

*Business Winstone* is a system-level, application-based benchmark test that measures the overall performance of a PC system when running eight popular 32-bit Windows 95/98 and Windows NT, business applications. It generates a weighted overall comparison of system performance based on the execution times of real Windows 32-bit business applications. The tests are categorized into a series of specific applications of the following types: *database, publishing* and *word processing/spreadsheet*. The results from each category as well as an overall performance score is calculated, and displayed as a bar chart.

### WinBench

WinBench tests is a subsystem-level benchmark, that is, it measures the performance of a PC's graphics, disk, processor, video and CD-ROM subsystems in the 32-bit Windows environment.

Another company that has recently issued some well-respected benchmark programs is Futuremark Corporation. Their products include 3DMark2001 and PCMark2002. Lite versions of their benchmarks can be downloaded from http://www.futuremark.com

The 'Win' benchmarks can be downloaded from http://www.veritest.com

## Real-time clock

Since the introduction of the AT computer, all PC motherboards including the latest ATX versions have a real-time clock circuit (RTC) incorporating battery-backed CMOS RAM. The RAM has a large enough capacity – a couple of 100 bytes – to accommodate the date, time and CMOS setup, information, so when the machine is powered down, the battery keeps the information intact. The date, time and setup information can be updated by the user by entering CMOS setup.

CMOS memory circuits consume extremely minute amounts of power when operated at low-clock frequencies. The RTC circuit operates at a frequency of just 32,768 Hz. This particular frequency is used as it is exactly divisible by powers of two to give 0.01, 0.1 and 1 s intervals for the time and date function.

The RTC circuit – which is usually part of the SB chip – is powered continuously for many years by a small lithium cell about the size of a 1 p piece. Lithium cells are used because they provide a relatively high voltage (approximately 3 Vper cell) and are remarkably stable, providing a constant voltage for a decade or so without leaking.

# The system (BIOS) ROM

A read only memory (ROM) chip, as the name suggests, is a memory device that can be read but not edited or erased under normal circumstances. The information stored on ROM therefore remains intact even after the power is turned off. Such memory is referred to as *non-volatile* or *permanent* memory. The programs stored on ROM chips are generically termed *firmware* to distinguish them from disk-based *software*.

ROM chips are used extensively in computer equipment and are available in several different forms. The types commonly used today are known as *Flash ROMs*. They take this name because the contents of ROM can be updated quickly and in situ by the support technician. When updating a Flash ROM no additional erasure/programming equipment is required, unlike previous types. While the whole point of ROM is that its contents *cannot* be changed, on a few occasions being able to update the ROM is extremely useful.

The System ROM contains a suite of machine code programs (routines) that must be instantly available at power up. These programs are, the *power on self-test* (*POST*), *basic input output system* (*BIOS*), *disk boot loader* and *CMOS setup*. The BIOS subsystem is the largest of these routines so some folk refer to the system ROM as the BIOS ROM. In this book we prefer the term system ROM to describe the actual chip and its contents.

When a PC is first switched on, the CPU uses the firmware on the system ROM to test the hardware and prepare the system to run the OS. The operation of the firmware is described briefly below. A more detailed explanation is given in Unit 7.

## POST

Each time the PC is powered up, the first program the CPU executes is the *POST*. This runs automatically as soon as the PSU establishes a stable supply. The POST is a sequence of short test routines that check the integrity of the hardware subsystems prior to running the OS. The items tested include, the motherboard chipset, subsystems, graphics adapter, keyboard, mouse and disk drives. If a fault is detected on either of these, the POST issues an error code in the form of audible beeps and/or an on-screen report. If a serious fault is detected before the sound and graphics subsystems have been tested, the POST issues an 8-bit error code to output port 10000000 (80h). This allows the support technician to read the error code using a POST diagnostics card.[10]

## CMOS setup

If the hardware tests out okay, the POST scans the keyboard buffer to see if the user has made a request to enter *CMOS setup*. This is usually initiated when the user holds down a certain key, for example *Del* or combination of keys during the POST. CMOS setup is a convenient menu-driven system that allows the support technician to set system parameters and options via the keyboard. Normally it is used when a new PC is being configured or when the technician is carrying out maintenance checks. The main CMOS setup program is stored permanently in the system ROM but the parameters entered by the technician must be stored in read and writeable memory

On some of the latest motherboards POST and even PRE-POST faults can be reported via a special *Speech Controller* chip. For example: the ASUS A78XV motherboard featured in Unit 5, uses a Winbond™ speech controller chip and the ASUS POST Reporter™. This clever system sends a spoken message to the sound card or system speaker immediately reporting the faulty section. This has enough intelligence built in to even indicate a faulty CPU – in the conventional POST system the CPU must be partially working otherwise the POST itself cannot run.

[10] POST error codes and diagnostics cards are discussed in Unit 6.

(i.e. RAM). However it must not lose its data when the system is powered down like ordinary RAM. To achieve this a special battery-backed CMOS RAM circuit is used. The CMOS RAM is also used by the RTC circuit to store time and data information.

### Disk boot loader

After completing the POST tests, if a request to enter CMOS setup was not made and no errors were detected, the POST hands control over to the *DBL*. This little program scans the disk drives for a bootable operating system (OS). When a bootable disk is found, it loads the boot sector of the disk into memory and executes it, thereby handing control from the system ROM to the OS. The process of loading the OS and running it automatically are known as *boot strap loading* and *booting*, respectively.

### BIOS

Once the OS has booted, the *BIOS* is used by the OS to link user commands and application programs to the system hardware. The BIOS is a suite of machine code programs each with a specific group of I/O tasks to perform via the system hardware. These small programs are called *interrupt service routines* (*ISRs*). This name derives form the fact that the OS calls specific software interrupts to perform various tasks relating to hardware. The BIOS performs the important run-time communication between the OS and the hardware. In the computer command hierarchy, the BIOS forms the lowest-program layer. The next layer up is the OS, followed by *applications*. The *user* sits at the top of the hierarchy (Figure 2.37).

## RAM

The primary storage area of the PC is made up from banks of RAM modules that slot into the motherboard. There are several package styles and technologies currently in use. The main package styles used on desktop PCs are SIMMs, DIMMs and *RamBus*© *in-line memory modules* (RIMMs). Each memory module is composed of a

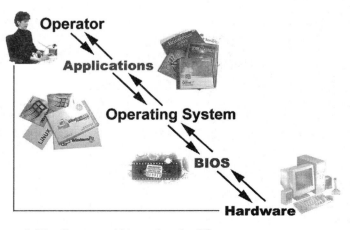

**Figure 2.37**   *Command hierarchy of a PC*

small rectangular PCB substrate with groups of RAM chips soldered to one or both sides of the board. This arrangement results in quite a small size-to-capacity ratio (Figure 2.38).

## Memory address bus width

The smallest unit of addressable memory is one memory location composed of 8 bits (1 byte). Each location can be individually addressed by the CPU using the address bus. Socket 7 CPUs have a 32-bit wide address bus so up to 4 GB ($2^{32}$ bytes) of memory can be accessed. The latest CPUs and those on the horizon offer even higher addressing capabilities taking the maximum possible memory capacity well into the terabyte realm and beyond. Current desktop motherboard technology and chipsets however limit this to 32 bits (i.e. 4 GB).

## Memory data bus width

All Pentium-style CPUs including the latest AMD XP and Intel P4 range have a 64-bit data bus. So although a 1 byte of memory is addressable by the CPU it can accommodate data from eight memory locations at once. The memory banks must therefore provide a 64-bit wide data path to match. Some legacy Socket 7 motherboards use

**Figure 2.38** *RAM modules currently in use: (a) 72-pin SIMM 32-bit wide SDRAM used on older Socket 7 motherboards (must be used in pairs); (b) 168-pin DIMM, 64-bit wide SDRAM used on Sockets 7, 370 and A and Slots 1, 2 and A motherboards; (c) 184-pin DIMM 64-bit wide DDR-SDRAM the most popular type used on current motherboards; (d) 184-pin RIMM 64-bit wide RDRAM (RamBus© memory) used on several Intel platforms*

72-pin SIMMs and these are only 32-bits wide so they have to be used in pairs to make up 64 bits. DIMMs support the full data bus width of the CPU so they can be used singly.

So RAM, however it is implemented, is as an array of 8-bit memory locations each with a unique address and each able to connect to the data bus under the control of the CPU. The individual bits making up the contents of each location are stored in eight memory cells as a specific voltage or electrical charge. There are two main types of storage cell. The first type is based on an electronic switch called a *bistable* or *flip-flop* and the second type uses a tiny capacitor to store the presence or absence of a charge.

RAM built from flip-flop memory is called static RAM (SRAM) because once a cell has 'flipped' or 'flopped' to a certain logic state it remains in that state indefinitely until a different state is set or until the power is removed.

Capacitor-based memory is called dynamic RAM (DRAM). The name derives from the fact each memory cell must be continually topped up with charge to prevent it leaking away.

The two types will now be described in more detail.

### SRAM

An SRAM cell is constructed from a circuit called a flip-flop or bistable. In a typical SRAM circuit each flip-flop is composed of six metal oxide semiconductor field effect transistors (MOSFETs) (the strange sandwich-like symbols in Figures 2.39 and 2.40). One of these circuits stores just one binary digit. The binary levels are simply two different voltage levels – a low voltage for '0' (e.g. 0.2 V) and a higher voltage for '1' (e.g. 2.5 V).

The flip-flop circuit will reliably store binary '1' by placing '1' on the SET input and '0' on the RESET input while the row select line is active. On deactivating the row select line the logic state of the bistable is retained. To store binary '0' the '0' is placed on the SET input and '1' on the RESET input while the row select line is active. Again on deactivating the row select the logic state of the bistable is retained. Reading the contents of the bistable is achieved by activating the row select line when no input data is present.

In an actual SRAM device there are millions of flip-flop circuits fabricated on a tiny slice of silicon not much bigger than an orange pip in diameter. Due to the large number of MOSFETs required for each cell, the price of SRAM is high compared to alternative RAM types at a given capacity.

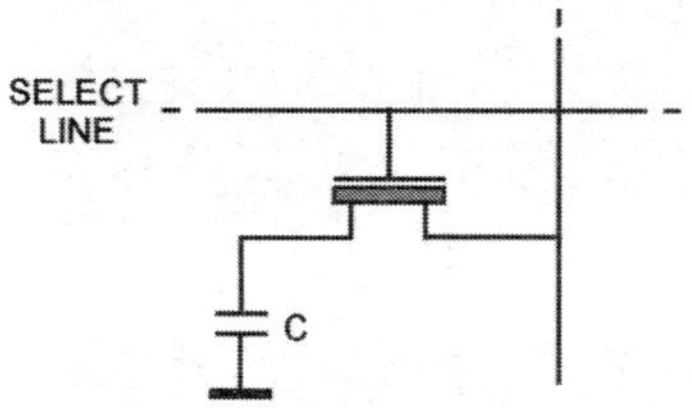

The major advantage or SRAM is its fast operating speed which is due to the use of MOSFETS as switches. SRAM is so efficient in speed terms, it is used for Cache memory – discussed later. The cost of SRAM, prohibits is use as the main RAM in a PC.

### DRAM

The circuit (Figure 2.40) shows a simplified DRAM (DRAM) cell. It is immediately evident that DRAM uses far fewer components than SRAM. This means that more cells can be accommodated on a given chip area.

It relies on the principle of holding a charge on a tiny capacitor. The presence of absence of charge represents the two binary states '0' and '1' or vice versa. The capacitor is formed by filling tiny trenches in the

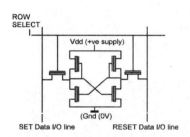

**Figure 2.39**   *A single SRAM cell uses six transistors (MOSFETs)*

**Figure 2.40**   *A DRAM cell using one MOSFET*

silicon chip with a dielectric material. In its simplest implementation, only one MOSFET (transistor) and a corresponding trench capacitor is required per storage cell. This allows a very high storage capacity to be obtained. At the upper end of the fabrication scale, one silicon chip can hold in excess of 200 million bits.

The trench capacitor is very tiny so the stored charge must be *refreshed* by a new write or a read/write operation every millisecond or so. This makes the supporting circuit more complex but the overriding advantage is that many more cells can be crammed onto the silicon slice, compared to a SRAM device. This means that cheaper and larger memory capacities are possible with DRAM.

### CPU and memory

The CPU works in close collaboration with electronic memory, namely ROM and RAM. ROM is *permanent memory* which means it is retained even when the computer is switched off. The CPU cannot write to ROM it can only read what is already permanently stored in its memory.

RAM on the other hand is *volatile* or temporary memory as its contents are lost forever when the computer is switched off or reset. It is the main memory used by the CPU to store programs and data while the computer is running. However, any useful data in RAM that the user wishes to keep, must be saved to disk before the program is closed. You may wonder why RAM is used at all if it is only temporary. Well for one thing, it is very much faster than disk storage and for another, it leaves a clean slate for the next session after powering down the computer. Memory is perhaps the most important commodity when the comparing the performance of two similar PCs. If there is insufficient RAM capacity even the most powerful CPU will struggle when presented with large processing tasks.

Electronic memory is composed of a matrix of 8-bit storage devices, made up from the basic single bit cells discussed earlier. These are connected in rows and columns and addressed via a row and column decoder built into the memory device. To transfer data to a particular memory location it has a unique address based on a 32-bit binary number.

Imagine that the CPU at a particular instant in time is about to send data to memory location 18. It places this number in binary, on the address bus and this activates just one unique memory register out of the millions available, to accept the data about to be placed on the data bus.

If we focus on a particular memory location, most of the time it will be disconnected from the data bus. However, on the odd occasion when the CPU is specifically reading or writing data to this location, it will be automatically connected to the data bus to facilitate the data transfer and then disconnected when the transaction is complete.

The illustration below shows the basic architecture of a RAM device using a cut-down model. This is based on just 32 memory locations so it only needs five address wires. The address wires are split up into two groups. One group, address wires A0 and A1, are connected to a column decoder and the other group, address wires A2 to A3, are connected to a row decoder (Figure 2.41).

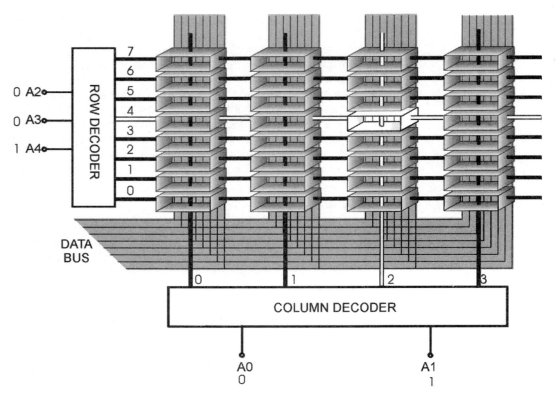

**Figure 2.41**   *A simple block representation of RAM device*

For example, to access the register at location $10010_B$ (18 in decimal), the 3 row bits activate one of the eight select wires of the row decoder and the 2 column bits activate one of the four select wires of the column decoder. So the register at row 4, column 2 is activated and connected to the data bus. In a real memory chip there are many more address wires arranged in a similar matrix arrangement thereby providing millions of memory locations.

The advantages gained from arranging the address into rows and columns are two-fold:

- All locations are accessed quickly and in the same amount of time, regardless of their location, for example location $11111_B$ can be accessed just as quickly as location $00000_B$. This type of addressing is known as random access addressing. Both ROM and RAM chips use this form of addressing.
- The row and column matrix arrangement greatly reduces the number of internal conductors required in the chip fabrication process.

### Synchronous dynamic random access memory

In the late 1990s, synchronous dynamic random access memory (SDRAM) became the most widely used DRAM device for the main memory of the PC. It has also gained popularity for use in a wide range of applications requiring fast read/write memory. Its ability to synchronize to the PC's system clock has maintained its popularity particularly as prices have fallen substantially over the years.

SDRAM contains two independent array banks that can be accessed individually or interleaved. This allows enhanced performance compared to conventional DRAM. The fact that it synchronizes to a common clock also adds to the performance. All data is written or read in burst fashion. Given a single starting address, SDRAM internally accesses a sequence of locations from that address. The length of the burst sequence can be programmed to 1, 2, 4, 8 or 256 locations. This makes it ideal for reading and writing in applications that use large blocks of data in consecutive memory locations (e.g. graphics intensive programs).

### Double data rate SDRAM

Today double data rate SDRAM (DDR-SDRAM) is the most popular main memory for new systems. It is inexpensive and it offers excellent performance. DDR-SDRAM features all the benefits of SDRAM while transferring data twice in each clock cycle on the rising and falling edges of the clock. There are several categories of DDR-SDRAM *buffered*, *unbuffered*, *error correcting code* (*ECC*) and *non-ECC*.

### Buffered DDR-SDRAM
Buffered or registered DDR-SDRAM has a set of additional registers and logic gates built-in that increase the strength of the signals received from the motherboard DIMM sockets. This allows very large memory capacities to be accommodated. Without buffering there is a limit to the maximum memory that can be accommodated. There is a slight downside in that the extra circuitry slows down the access time slightly compared to the non-buffered variety.

### Unbuffered DDR-SDRAM
Unbuffered DDR-SDRAMs are currently the most popular type because they are less expensive than the buffered variety. They are fine for most systems that do not require huge memory capacities beyond 2–3 GB.

### ECC and non-ECC DDR-SDRAM
Top-end systems and servers use DDR-SDRAM with ECC capability. ECC allows the system controller on the motherboard chipset to perform error detection and correction. Currently the most popular type is the non-ECC variety.

### DDR-SDRAM speed ratings
DDR-SDRAM is supplied with several different speed ratings to suit a range of system clock frequencies. The naming system may seem rather strange but the number portion, relates to the approximate maximum data transfer rate of the device in megabytes per second. This is arrived at by multiplying the device operating speed in megahertz by its bus with in bytes.

For example (Table 2.4):

1. A 266 MHz DDR-SDRAM device is labelled as PC2100 (i.e. 266 MHz $\times$ 8 bytes = 2128 MB/s (i.e. approximately 2100 MB/s).
2. A 400 MHz DDR-SDRAM is labelled as PC3200 (i.e. 400 MHz $\times$ 8 bytes = 3200 MB/s).

**Table 2.4**   *DDR-SDRAM speed ratings*

| Type | System bus speed (MHz) | 2 × operating speed (MHz) | Data bus |
|------|------------------------|---------------------------|----------|
| PC1600 | 100 | 200 | 64 bits |
| PC2100 | 133 | 266 | 64 bits |
| PC2700 | 166 | 333 | 64 bits |
| PC3200 | 200 | 400 | 64 bits |

The motherboard manufacturer usually specifies the memory type, and range of speed ratings it can accommodate. Some manufacturers also specify recommended makes of DDR-SDRAM that have been certified as compatible with their product.

### RDRAM

A revolutionary high-performance type of DRAM called RDRAM is supported on some motherboards and in particular Pentium IV systems. This type of ram uses a high-speed bus called the RamBus channel operating at 800 MHz (400 MHz clock transferring data on the rising and falling edges). The system uses protocol-based data transfer and offers a scalable architecture for future memory demands. Unlike DDR-SDRAM which is an open standard, RDRAM and its patents are registered to RamBus Inc.

## Cache memory

Cache memory (pronounced 'cash'), is a block of RAM, separate from the main memory, where data can be stored temporarily. The function of a cache is to speed up data transfer. For example, *a hard-disk controller cache* is a block of RAM that speeds up the apparent writing and reading speed of a hard disk. Instead of sending and receiving data directly to and from the relatively slow disk, it is directed to a cache of memory chips on the hard-disk controller board, where the data can be accessed quickly. The contents of the cache are then transferred to and from the hard disk at appropriate intervals.

In a similar way, the CPU benefits from having a cache. With Pentium-based systems the cache memory is composed of fast SRAM which is much faster than the DRAM used for the main system memory. The idea of the cache is to eliminate the time wasted by the CPU as it waits around for the RAM to catch up with it. Data is transferred in blocks from the slower main memory to the faster cache RAM. When filled with the correct data, the cache RAM is able to keep up with the CPU's demand for data.

It may seem silly to mess around with a cache system, when fast SRAM could easily be used for main memory. The reason, as always, is down to cost. SRAM is considerably more expensive than DRAM, but in the small quantities used by the cache – 256 kB to 1 MB – the extra cost is negligible, especially when compared to the overall system cost.

All CPUs from the 80486 up have cache memory fabricated on the CPU chip. This internal cache is known as the *primary cache* or

Level 1 cache. Today's CPUs also have a block of *secondary cache* or *Level 2 cache* fabricated on the CPU. Cache memory is composed of fast SRAM with a very short access time compared to normal DRAM.

A cache relies on two probabilities:

1. When the CPU accesses a certain memory location, it is likely that it will access the same location in the near future. This is called temporal locality.
2. When the CPU accesses a certain memory location it is also likely that it will access a nearby location in the near future. This is called spatial locality.

The cache effectively stores a copy of the memory contents, that are most likely to be accessed in the near future.

The Level 2 cache works basically as follows:

The CPU transfers data in four-block bursts, where each block is the width of the data bus. A Pentium has a burst length of 32 bytes. The data in the cache is also arranged in 32 byte groups to match the CPU. Each group is called *a line*. The number of lines stored in the cache is determined by the size of the cache.

During a memory read operation, the CPU addresses a certain location in RAM and the cache checks to see if this location is present in the cache. If it is, then the cache provides the data and the read operation is completed very quickly. This is called a *hit*.

If the location is not present in the cache, then an entire line starting at that location, is copied from RAM into the cache. This is called a 'miss' and several extra clock cycles are used up as the main memory is activated. The good thing about this however is that there is a high probability that the next CPU address location will be in the cache.

The cache uses an extra SRAM device just to store the address of each line held in the cache. This is called TAG RAM as it is analogous in use to the address tags used on parcels, etc.

When the CPU addresses a certain memory location, the tag bits of the address are compared with all the tag addresses stored in the cache. If the address matches, it is a hit, and the data is extracted from the relevant line in the cache and sent to the CPU.

If a miss occurs, the data including its tag address is fetched from RAM and stored in a spare line location in the cache. If a spare location is unavailable, then the cache controller will allow older lines in the cache to be overwritten.

## Serial data transmission

All PCs have a generic serial communication system that conforms to Reference Standard 232 revision C (RS232C and CCITT V24). This standard has been implemented on PCs ever since their launch in the early 1980s – decades before the launch of advanced serial bus systems: USB and Firewire™. The RS232 ports on the PC are known as *asynchronous communication* ports and are abbreviated to COM1 and COM2. The ports are accessed via two 9-pin male 'D' connectors at the rear of the system-unit. Old XT PCs used 25 way male 'D' connectors.

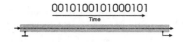

0010100101000101

Time

**Figure 2.42**  *In serial transmission 1 bit at a time is send down the cable*

A serial communication interface is essentially based on transferring one bit of data at a time down a cable. As only one data wire and ground is used, there is no cross talk between the wires, so longer cables can be used. However in practice, the maximum cable length is limited by other factors such as the bit rate and the voltage levels available. The penalty of serial data transmission compared to parallel transmission – everything else being equal – is speed. Data is sent 1 bit at a time as a train of pulses (Figure 2.42), instead of several bits simultaneously, as in parallel transmission.

### Serial data transmission/reception

Serial transmission and reception via a computer host requires a certain amount of data conversion. It must be changed from an inherently parallel form at the CPU's data bus, to a serial form at the I/O port. Parallel-to-serial and serial-to-parallel conversion is carried out using an electronic circuit called a *shift register*. Shift registers convert data from parallel to serial and vice versa. There are two types of shift register used to perform the conversion process. These are known as parallel in serial out (PISO) registers for sending data from the computer to the serial line and serial in parallel out (SIPO) registers, for receiving data in from the serial line to the computer.

### Universal asynchronous receiver transmitter

To provide the two-way serial-to-parallel and parallel-to-serial data conversion required by a computer, a special circuit called a universal asynchronous receiver transmitter (UART) has evolved. When data is transmitted serially, each bit occupies a certain time slot, the length of which depends on the clock frequency set for the UART, which in turn determines the data rate of the serial bit stream. This is measured in *bits per second* (bps) and is known as the *bit rate*. The range of bit rates that can be set on the COM ports of the PC are 300, 1200, 2400, 4800 and 9600 bps. The default bit rate on a PC is 9600 bps.

The data from the PC is in the form of 8-bit wide 'characters'. So to send data via the serial port, each character must be converted to serial form by temporarily storing each character in the UART and then shifting out each bit of the character at a predetermined rate as mentioned previously.

To maintain the integrity of the character before transmission, a *start* bit and one or two *stop* bits are added to the character. These enable the receiving device to distinguish one character in the stream of bits, from another. In addition, an option is available to include an odd or even *parity bit* to the data frame to provide a small degree of error detection.

The RS232 system relies on the fact that the sending and receiving UART clocks are stable enough to keep in step with one another over the short period of a complete character frame. At the receiver, the UART records the logic state at its input at regular intervals determined by the local clock frequency. As long as both ends of the line have approximately the same clock frequency, the start bit ensures that the data bits that follow are read at a valid point in time. The precise details of this process are described in several electronics textbooks on the RS232 standard. For the support technician: the important practical point to grasp is that a serial RS232 link will only work correctly when the same, UART parameters are set on both devices.

The default parameters set on the PC's serial ports 'COM1' and 'COM2' are: Bit rate 9600 bps, 1 stop bit, 8 data bits, no parity.

You can check the serial port in Windows 95/98/ME by clicking: | **Start | Settings | Control Panel | System | Device Manager | Ports (COM and LPT)** | Then click **COM1** or **COM2**.

Windows XP: | **Start | Control Panel | System | Hardware | Device Manager | Ports (COM and LPT)** | Then click **COM1** or **COM2** (Figure 2.43).

If a reliable serial communication link is to be established using this system, a means must be found to inform the sending computer when an error has occurred. This can be done either by using software or via separate *handshake lines*. Either method can be set up on the PC.

The RS232-C specification defines two types of equipment that can be connected to a serial link, these are:

- Data terminal equipment (DTE), which includes such items as printers, computers and terminals (*a terminal is a single operator site, usually a visual display unit (VDU) and keyboard connected to a network or a mini- or mainframe computer*).
- Data communication equipment (DCE), which is designed to communicate over very long distance cables such as the telephone line. Modems are DCE devices.

As well as the main data signal, the RS232 standard specifies a set of signals to control the flow of data across the serial communication

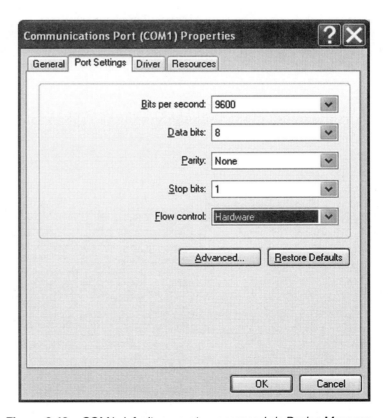

**Figure 2.43** *COM1 default parameters accessed via Device Manager*

link. These are known as handshake signals. They are used to control the flow of data, when the receiving device is busy or awaiting new data. Unfortunately extra wires (lines) are needed for the handshake signals in addition to the transmit, receive and ground wires needed for the data itself.

There are many different ways of wiring up the handshake lines on a serial lead and these details are usually included in the user manual of the peripheral being installed. However as long as the type of connection is known (i.e. 'DTE to DTE' or 'DTE to DCE'), then one of the connection systems shown below will usually work satisfactorily. Sometimes however the peripheral uses a strange connection system which makes it imperative to obtain the correct information or buy the correct cable from the manufacturer.

A computer–printer serial RS232 connection uses a DTE to DTE link and a computer–modem connection uses a DTE to DCE link. See the connection tables (Tables 2.5–2.8) below.

### Simplex, half-duplex and duplex transmission

Data sent in one direction only, from transmitter to receiver, it is known as '*simplex*' transmission. For example radio and television broadcasts are essentially *simplex*. In the Semaphore example shown

**Table 2.5**   *DTE–DTE printer lead*

| Legacy 25-way female D connector (computer end of lead) | | 25-way male D connector (printer end of lead) | |
|---|---|---|---|
| Pin | Function | Pin | Function |
| 2 | TD | 3 | RD |
| 3 | RD | 2 | TD |
| 4 | RTS | 5 | CTS |
| 5 | CTS | 4 | RTS |
| 6, 8 | DSR-CD | 20 | DTR |
| 7 | Ground | 7 | Ground |
| 20 | DTR | 6, 8 | DSR to CD |

**Table 2.6**   *DTE–DCE modem lead*

| Legacy 25-way female D connector (computer end of lead) | | 25-way male D connector (modem end of lead) | |
|---|---|---|---|
| Pin | Function | Pin | Function |
| 2 | TD | 2 | TD |
| 3 | RD | 3 | RD |
| 4 | RTS | 4 | RTS |
| 5 | CTS | 5 | CTS |
| 6 | DSR | 6 | DSR |
| 7 | Ground | 7 | Ground |
| 8 | CD | 8 | CD |
| 20 | DTR | 20 | DTR |

in the picture below, data can be sent in both directions but only one way at a time, this is termed '*half-duplex*' transmission.

Data sent and received in both directions simultaneously is known as '*duplex*' transmission. A normal telephone link uses *duplex* as it is possible to both talk and listen at the same time (Figure 2.44).

### RS232 COM1/COM2 pin assignment and signal descriptions
For more details refer to Table 2.9.

**Table 2.7**   *DTE–DTE printer lead*

| 9-way female D connector (computer end of lead) | | 25-way male D connector (printer end) | |
|---|---|---|---|
| Pin | Function | Pin | Function |
| 3 | TD | 3 | RD |
| 2 | RD | 2 | TD |
| 7 | RTS | 5 | CTS |
| 8 | CTS | 4 | RTS |
| 6, 1 | DSR to CD | 20 | DTR |
| 5 | Ground | 7 | Ground |
| 4 | DTR | 6, 8 | DSR to CD |

**Table 2.8**   *DTE–DCE modem lead*

| 9-way female D connector (computer end of lead) | | 25-way male D connector (modem end) | |
|---|---|---|---|
| Pin | Function | Pin | Function |
| 3 | TD | 2 | TD |
| 2 | RD | 3 | RD |
| 7 | RTS | 4 | RTS |
| 8 | CTS | 5 | CTS |
| 6 | DSR | 6 | DSR |
| 5 | Ground | 7 | Ground |
| 1 | CD | 8 | CD |
| 4 | DTR | 20 | DTR |

Duplex          Simplex          Half Duplex

**Figure 2.44**   *Communication channel operating modes*

**Table 2.9**   *RS232 COM1/COM2 pin assignment and signal descriptions*

| Signal pin no. | Signal name | Direction from/ to printer | Description |
|---|---|---|---|
| 1 | FG | – | Frame ground: the 0 V line which usually connects the computer frame (chassis) with the printer frame |
| 2 | TXD | Output | Transmit data: data from the printer to the computer Connected to the RXD data line of the computer connector Logic '0' = +3 to +25 V Logic '1' = −3 to −25 V |
| 3 | RXD | Input | Receive data: data from the computer to the printer Connected to the TXD line of the computer connector Logic '0' = +3 to +25 V Logic '1' = −3 to −25 V |
| 4 | RTS | Output | Request to send: a signal from the printer to the computer Usually set permanently to Logic '0' (+3 to +25 V), and cross connected with the computer's CTS connection |
| 5 | CTS | Input | Clear to send: input from the computer's RTS to the printer When the signal line is Logic '0' (+3 to +25 V), the computer is ready to send data Sometimes left unconnected in which case it is permanently enabled to Logic '0' |
| 6 | DSR | Input | Data set ready: a Logic '0' input from the computer to the printer to say that the computer is ready to operate Often cross connected with the computer's DTR connection Sometimes left unconnected in which case it is permanently enabled to Logic '0' |
| 7 | SG | – | Signal ground, connected with the internal ground line in the plotter |
| 8 | DCD | Input | Data carrier detect: input from the computer to the plotter Same as for CTS and DSR when unconnected |
| 20 | DTR | Output | Data terminal ready: a Logic '0' signal to say that the printer is connected and ready to receive data Often cross connected with the computer's DSR connection Sometimes left unconnected in which case it is permanently enabled to logic '0' |

## Serial data transmission down long cables

Although RS232 is ideal for serial transmission over reasonably long cable runs up to 20 m or so, it is no good for very long cable runs. This is because the electrical capacitance, resistance and inductance of the cable, attenuates the signal very severely particularly on the rapidly changing parts of the signal. So the rising and falling edges of the signal become less distinct and distortion and signal corruption occurs.

The Internet allows computers all over the world to communicate with one another using the public switched telephone network (PSTN). The PSTN naturally uses extremely long cables. However, the maximum range of audio frequencies that can be reliably transmitted on the PSTN extends from 300 to just over 3000 Hz (3 kHz). This is fine for intelligible speech but it is pretty useless as it stands for conveying digital data. To allow the PSTN to convey data, the binary signals at the computer must first be converted to a series of modulated tones. The simplest possible modulation system uses a single tone for each digit (i.e. one tone for binary '0' and a different tone for binary '1'). For example, the first 300 Baud modem (the Bell 103 duplex standard) used two modulated channels one for sending and one for receiving data. One channel used a 1270 Hz tone for binary '1' and a 1070 Hz tone for binary '0'. The other used tones of 2225 and 2025 Hz, respectively. This technique is known as '*frequency shift keying*' (FSK). The tones were switched at a rate of 300 Hz.

At the receiving end, the modulated signal is then demodulated to recover the original data. The electronic device that performs the modulation and demodulation process is known as a modem, from modulator demodulator.

### *Faster modem communication*

In this technological age, there is an increasing need to transfer larger amounts of data more quickly, at less cost. The telephone network (PSTN) is an obvious choice for data communication, due to the sheer number of lines and distances covered. The PSTN was originally designed for voice communication, but the low-signal bandwidth of the system imposes an upper limit on the data rate achievable. Using the simple analogue FSK modulation system, data can only be sent down a telephone line at a few hundred bits per second. Solutions were therefore devised to increase the apparent data rate.

The most drastic solution, and the best in the long term, is to install completely new telephone lines based on fibre optics. A fibre-optic link has a bandwidth several million times that of the old wired system!

The other solution is to utilize the existing lines and evolve more elaborate methods of modulation, error detection and error correction. Instead of just using a separate tone for each binary state of the signal, as in FSK, the phase, frequency and amplitude of the carrier can be changed. This greatly increases the amount of data that can be squeezed onto the line, but it inevitably increases the relative cost of the modem. The initial high equipment cost however, is outweighed by the potential saving in telephone bills. High-performance modems can produce bit rates beyond 56,000 bps which is well over 100 times faster than the maximum rate possible with a simple tone system.

### Baud rate and bit rate

In the pioneering days of telegraphy, the rate of sending 'marks and spaces' on punched ticker tape, was known as the *baud rate* – pronounced 'board', named after J. Baudot the French inventor. Unfortunately the baud rate is still used as a measure of transmission rate on serial transmission lines and this causes some confusion when used on modern transmission systems.

With all modern modems, the fundamental *baud rate* of the clock or carrier is always a lot lower than the achievable *bit rate*, so it is important to distinguish between the two.

The baud rate is the basic *signalling rate* and the data rate is the effective number of bits transferred in 1 s. For example, in the old Bell 103 modem mentioned earlier, the two tones in each channel are switched at 300 Hz. This is the baud rate. Now as only 1 bit is transmitted for each tone, the bit rate is the same as the baud rate (i.e. 300 bps). However on more elaborate modems with advanced modulation techniques a basic baud rate of 300 could give rise to data rates of way beyond this, depending on the modulation method used.

We will explain how this works using a simple example.

If the binary bits entering the modem are taken in pairs, then 4-bit patterns are possible; 00, 01, 10, 11. Each of these could be represented by a change in the amplitude, phase or frequency of the carrier (tone). To ease the explanation, four changes in frequency will be used – in a real system, the phase, frequency and amplitude of the carrier are changed.

The frequency $f_0$ could be assigned for 00, $f_1$ for 01, $f_2$ for 10 and $f_3$ for 11. So the bit pattern 0110110100 – representing a complete character frame including a stop and start bit – would be sent in the frequency sequence, $f_1$, $f_2$, $f_3$, $f_1$, $f_0$. If the frequencies are switched at a signalling rate of 300 Hz (the baud rate), then the data rate is 600 bps, because 2 bits are transferred each period of the signalling frequency.

If the bits where transferred in blocks of 4 bits, then 16 different state changes of the carrier would be needed (i.e. 16 different frequencies). The data transfer rate would then be four times the baud rate (i.e. a baud rate of 300 would give a bit rate of 1200 bps). Today, advanced modulation techniques and error correction codes have pushed the maximum data rate on the PSTN to 56,000 bps.

## The Centronics parallel port standard

The parallel port LPT1, name derived from Line PrinTer 1, on the PC is 8-bits wide so that a whole byte of data can be transferred at once. It is simpler to implement than a serial port as no voltage level shifting is required and no start, stop and parity bits are used. However it is still necessary to incorporate *handshake* signals to control the flow of data between the computer and the peripheral.

The PC uses the Centronics communication standard. This was specifically designed for one way, 8-bit data transfer, from a computer to a peripheral and it has remained the standard for parallel printers for over 20 years. The pinouts and signal descriptions for the Centronics specification are shown in Table 2.10.

There have been several enhancements to the parallel port over the years. The first and most common type used on all PCs was the unidirectional Centronics port. This was used on the original IBM XT LPT1 port. LPT1 was designed purely to send data from the PC to a printer, with *busy*, *acknowledge* and *strobe* lines for the handshake signals. Even on a modern PC this is the default setting, but more advanced port configurations can be selected via CMOS setup.

**Table 2.10**   *Parallel and serial I/O port default settings*

| Port description | Port addresses | Interrupt |
|---|---|---|
| Parallel ports | | |
| LPT1 | 378h–37Fh | IRQ7 |
| LPT2 | 278h–27Fh | IRQ5 |
| Serial ports | | |
| COM1 | 3F8h–3FFh | IRQ4 |
| COM2 | 2F8h–2FFh | IRQ3 |

Some handshake signals travel from the printer back to the computer but such signals do not convey user data, they are essentially control signals. The data corresponding to printable characters and printer control go in one direction only – from the computer to printer. So the original standard LPT1 port is a simplex system.

Today's PCs incorporate an advanced parallel port specification, downwards compatible to the original IBM LPT1 port. It supports Intel's *enhanced parallel port (EPP)* specification also called the *fast mode* parallel port. This allows intelligent bi-directional communication (duplex) between the PC and an appropriately equipped peripheral. The new LPT1 port also supports the enhanced capabilities port (ECP) specification which offers a similar performance to the EPP specification (i.e. both enhancements offer bi-directional communication at speeds at least 10 times that of the original IBM Centronics specification.

Table 2.10 shows the default I/O port address and interrupt settings for the both LPT1 and additional LPT2 and LPT3 ports.

### Centronics interface

Until the rise in popularity of USB, the Centronics parallel data communication specification was the most widely used printer interface. It still is the main connection system for office and top-end printers. The main features are:

- Parallel data transmission, with synchronization based on a logic low $\overline{\text{STROBE}}$ pulse that initializes the transfer of data from the computer to the printer.
- Handshaking via a logic low acknowledge $\overline{\text{ACK}}$ signal and a logic high BUSY signal, from the printer to computer. Also special signal lines such as paper out, off-line and error are provided.
- All interface, control and data signals are transistor/transistor logic (TTL) compatible.

The generic Centronics interface connection used on printers is a 36-way IEEE 488 socket. The LPT1 port connection at the back of the PC is provided by a 25-way female 'D' connector. The 36-way IEEE-488 connector is often referred to as a 'Centronics' connector (Figure 2.45 and Table 2.11).

**Figure 2.45** *Centronics IBM PC compatible/printer lead*

**Table 2.11**    *Centronics port pin connections (LPT1/LPT2)*

| Signal pin no. IEEE488 | Return pin no. IEEE488 | Signal name | Direction from/to printer | Description |
|---|---|---|---|---|
| 1 | 19 | STROBE | Input | STROBE pulse to read data into the printer. Normally high. Data is read in when it goes low. The pulse must stay low for about 1 μs |
| 2 | 20 | Data bit 1 | Input | These eight signals form 1 byte of data from the computer. Each bit is a TTL level, high for Logic '1' (2.4–5 V) and low for logic '0' (0 to 0.8V). This data must be present for 0.5 μs before and after the strobe pulse |
| 3 | 21 | Data bit 2 | Input | |
| 4 | 22 | Data bit 3 | Input | |
| 5 | 23 | Data bit 4 | Input | |
| 6 | 24 | Data bit 5 | Input | |
| 7 | 25 | Data bit 6 | Input | |
| 8 | 26 | Data bit 7 | Input | |
| 9 | 27 | Data bit 8 | Input | |
| 10 | 28 | ACK (short for acknowledge) | Output | The printer sends this signal automatically after the strobe line has changed from high to low and after the printer is switched on-line. The line is normally high going low for about 12 μs to indicate that the printer is ready to receive another block of data |
| 11 | 29 | BUSY | Output | This signal is high when the printer is busy and cannot receive data. It goes high when the printer buffer is full, when the printer is processing data and when the printer is off-line or in an error condition. |
| 12 | 30 | PO (PE) | Output | A Logic '1' signal indicating 'paper out' |
| 13 | – | SLCT | Output | Indicates the on-line or off-line state of the printer. '1' on-line, '0' off-line |
| 14 | – | Autofeed XT | Input | When this signal is 'low' the paper is automatically advanced one line after a carriage return command. Effectively a 'line feed' command is added after a 'carriage return' command |
| 15 | – | No connection | – | – |
| 16 | – | 0 V | – | Signal ground ( TTL 0 V) |
| 17 | – | Chassis GND | – | The metal chassis of the printer |
| 18 | – | No connection | – | – |
| 19–30 | – | GND | | Twisted pair ground returns |
| 31 | – | INIT | Input | This signal is used to initialize the printer and is normally high going low for less than 50 μs to reset the printer |
| 32 | – | ERROR | Output | This line is normally 'high' but goes 'low' when an error condition occurs. This can be caused by: (i) a paper out condition, (ii) printer is off-line, (iii) cover is open, (iv) overloaded condition |
| 33 | – | GND | – | Twisted pair ground return |
| 34 | – | No connection | – | – |
| 35 | – | +5 V | Output | pulled up to +5 V via a resistance usually 2–3.3 kB |
| 36 | – | SLCT IN | Input | Normally 'low' set by an internal dip-switch. Used by 'DC1' and 'DC3' control codes to enable '0' disable '1' the printer |

Note: The bar over some signals indicates they are active in the logic low state.

**For the inquisitive reader**
When data is transferred from the PC to the printer a precise handshaking procedure is adopted. This is depicted in the timing diagram (Figure 2.46). This shows the timing between the three main printer control signals, STROBE, ACK and BUSY.

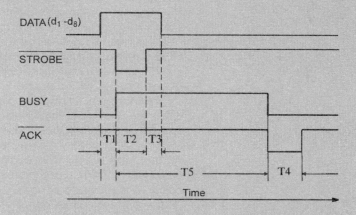

**Figure 2.46** *T1 = 0.5 μs (minimum); T2 = 1 μs (minimum); T3 = 0.5 μs (minimum); T4 = 5 μs (maximum); T5 = 1 ms or less when buffer not full, but typically 1 s or less when buffer is full*

In the Centronics specification, the PC places a byte of data on the eight data wires, D0 to D7, at least 0.5 μs before the application of a logic low pulse on the STROBE. On receipt of the strobe pulse, the printer reads the data into its internal memory (the printer buffer) and pulls the BUSY line high. If the buffer is partially empty, it remains in this state for up to 1 ms. If the buffer is full, the busy line can remain high for several hundred milliseconds while the data in the buffer is being utilized. The ACK signal then goes 'low' for up to 5 μs, from the instant the busy signal changes from 'high' to 'low', signalling to the computer that the printer is now ready to receive more data.

The Centronics interface is an excellent and extremely reliable interface ideal for medium to slow peripherals like printers.

You can check the parallel LPT1 port in Windows 95/98/ME by clicking:

- | **Start** | **Settings** | **Control Panel** | **System** | **Device Manager** | **Ports (COM and LPT)** | Then click **LPT1**.
- Windows XP: | **Start** | **Control Panel** | **System** | **Hardware** | **Device Manager** | **Ports (COM and LPT)** | Then click **LPT1**.

# The system buses

IBM introduced the first PC in 1981 as an open architecture so that new features could be added to the system as they became available. This naturally opened the doors for other manufacturers to offer their own PC upgrades, and a huge new market in PC upgrades was born. The massive success of the PC was due in part to the build quality and prestige of IBM and its upgradability thanks to the expansion bus system.

The original IBM PC used the INTEL 8088 CPU, this has a 16-bit internal data bus and an 8-bit external data bus. So IBM designed the PC expansion bus to match. The bus originally ran at 4.77 MHz, the same clock frequency as the processor itself.

Later when IBM designed the AT machine with its 16-bit 80286 CPU, they wisely decided to retain the original 8-bit slots and include a 16-bit slot as an extension. This allowed full downward compatibility with the older 8-bit system.

Manufacturers of PC compatible machines began to introduce enhanced versions of the AT machine by increasing the CPU's clock speed and therefore the expansion bus speed. However problems soon started to appear, as expansion boards that worked fine on IBM AT machines, did not work correctly on the faster machines. To overcome this problem, several manufacturers formed a group to create a better AT bus standard known as the *industry standard architecture* or *ISA* bus for short. Among the recommendations the bus clock frequency was set at 8 MHz.

When the even more powerful Intel 32-bit 80386DX CPU was introduced, a need again arose for a wider, faster bus standard. IBM responded to this by introducing their PS range of computers using a new 32-bit bus. The bus was known as the micro-channel architecture (MCA). Unfortunately with hindsight this was a bad move for IBM as the new design was a total departure from the ISA standard and offered no downward compatibility. It was therefore not widely accepted by the PC fraternity.

Around the time of the PS/2 launch, a group of manufacturers introduced a new 32-bit standard but one that was downward compatible with the ISA bus. They named it the *extended industry standard architecture* (EISA) bus. This bus used deep slots, with a row of ISA contacts and a row of EISA contacts, placed one above the other in a two-tier arrangement. These cards were far more expensive than a comparable ISA card. To the average PC user, the EISA bus did not offer a sufficiently superior performance, to warrant the extra cost. The public preferred to stay with the cheap and cheerful 16-bit ISA standard which is still supported today on some motherboards.

As the demand for better graphics and faster processing speeds grew – brought on by the increasing sophistication of software and falling CPU prices – the old ISA bus became a major bottleneck. A wider, faster BUS standard – one that would be inexpensive to implement – was deemed essential.

## Video Electronics Standards Association *local bus*

To go some way in meeting the specification for a wider, faster bus, the *Video Electronics Standards Association* (VESA) – a group of video card/motherboard manufacturers, originally formed to set realizable standards for video graphics adapters – proposed a new standard called the VESA local bus (VL bus).

The bus had to achieve goals that were missed on previous attempts at a new bus standard. It had to be cheap, so that the buying public could afford it, and it had to offer significant technical advantages over other bus standards. The features of the VL bus standard are listed below:

- low cost;
- significantly higher performance than the 16-bit ISA bus, to handle the bus speeds of modern processors and the high bus traffic rate from modern peripherals like multimedia systems and advanced high-resolution graphics cards;
- open architecture so that anyone can use it;
- independent of software drivers;
- built in expandability to allow it to keep up with future technology.

The VL bus provided direct access to the system memory at the same speed as the CPU itself, which was then around 50 MHz. This allows peripherals such as disk controllers and graphics cards to transfer data relatively quickly along a 32-bit wide bus.

The bus was a direct branch off the main CPU BUS and as such, limited the maximum number of cards that could be on the bus simultaneously. At best two cards could occupy the bus at 33 MHz. This dropped to one card at 50 MHz.

### PCI bus

In 1992, Intel in collaboration with several motherboard manufacturers, formed the *Peripheral Component Interconnect* (PCI) *Special Interest Group* (SIG), to promote and develop the PCI BUS to meet the requirement of existing and future generation systems. SIG now has over 500 members world wide and the PCI bus has become popular not only as a PC bus but also as an international bus standard for industrial applications.

The PCI bus is linked to the CPU bus by a bridging circuit inside the motherboard chipset. This makes it independent of the CPU and therefore more easily adapted to other processors. The PCI bus on a PC is sometimes referred to as a *mezzanine* bus, because it forms another layer between the CPU bus and the ISA bus. As the PCI bus is not directly connected to the microprocessor, it overcomes the loading limitations found on the VESA bus. It also accepts several PCI adapter cards, with minimal drop in performance.

PCI was primarily intended for fifth-generation CPUs and beyond but chipsets were available to suit 486 systems. Like VESA, the PCI can improve the performance of graphics and other hardware devices, including network, hard drive, I/O and multimedia cards.

The PCI 2.1 specification sets a maximum data bus-width of 64 bits and a maximum clock frequency of 66 MHz, giving a maximum theoretical data transfer rate of 524 MB/s!

On a most AT/ATX motherboards, the PCI bus runs at 33MHz, half the speed of the system clock, with a data bus-width of 32 bits. This gives a theoretical maximum data transfer rate of 132 MB/s. In practice this drops to about 100 MB/s in a real system. This is still approximately 20 times faster than the old ISA bus.

Most motherboard chipsets today support PCI 2.1. This is called Concurrent PCI, because it is an improved architecture that overcomes some of the latencies present in the early implementations. The main improvement was made by minimizing the time wasting that occurs when the controller is waiting to perform an ISA transaction.

In the earlier PCI implementation this held the PCI bus inactive, until ISA responded. In the new implementations, an ISA transaction can be continually attempted while the other buses are being served, thereby allowing simultaneous activity on the CPU, PCI and ISA buses.

### Personal Computer Memory Card International Association bus

From the late 1980s onwards there has been an explosion of PC compatible laptop computers onto the market. Which inevitably gave birth to a whole new industry manufacturing add-on products, from physically tiny hard drives to advanced communication adapters.

In 1989, in an effort to build in compatibility and expandability to laptop systems, an association of several hundred manufacturers known as the *Personal Computer Memory Card International Association* (*PCMCIA*) was formed. The PCMCIA specifies a bus standard that allows extra memory and peripherals – such as fax modems, network interfaces and disk drive adapters – to be conveniently slotted into a laptop computer. These peripherals come in the form of thin cards, known as PCMCIA cards. In 1996 the PCMCIA decided to rename the rather cumbersome '*PCMCIA*' *card* to simply '*PC*' *card*.

The PCMCIA bus is intended for use on laptop and the more recent notebook computers, where small size and low weight is of paramount importance. The bus is such an important development in personal computing, that it is now supported on several desktop PCs. This allows the user to connect a hard disk, or flash card, for example, from a notebook, directly to a desktop PC, to facilitate fast data transfer between the two machines.

The PC card standard is based on a 68-pin parallel bus interconnection system, supporting 26 address bits, 16 data bits and various control and power lines. Since 1990 there have been three major revisions of the 68-pin PCMCIA plug-in card, all with the standard width and length of 54 and 85.6 mm, respectively. The thickness however has been changed to support the extra requirements of each revision:

- Type I cards are 3.3 mm thick and are mainly used for memory expansion.
- Type II cards are 5 mm thick and offer a versatile range of plug-in options such as modems, network cards and sound cards.
- Type III cards are 10.5 mm thick and allow miniature hard-disk drives to be plugged in directly into the PCMCIA socket on the side of the notebook.

### Universal Serial Bus (USB)

USB is a relatively new open standard bus specification that supports Intel's plug-and-play (PnP) scheme. It also supports 'hot plugging', allowing peripherals to be added quickly and easily without the need to switch off the system. It was developed by a group of leading PC and telecommunication companies with the aim of making the process of adding and removing peripherals easy enough for a operator to do.

The first USB implementation USB 1.0, can transfer data at a maximum rate of 12 million bps (12 Mb/s), which is 1.5 MB/s. This is fast enough for slow to medium speed devices such as joysticks, mice, scanners and digital cameras. It has plenty of expansion capability as the specification allows up to 127 different devices to be connected to the bus at once.

The USB has eliminated the need to remove the case cover or set jumpers to install a new peripheral – a task that the novice often finds rather daunting. It is designed to allow peripherals to be hot plugged/unplugged at will, without the need to switch off and reboot the system, as is the case with conventional PC upgrades. It can detect when devices are added or removed from the bus and can also automatically determine the resources needed for each device present on the bus. Like current ISA, PCI and AGP devices, it makes full use of Windows PnP so there are no manual IRQ, direct memory access (DMA) and port address settings to worry about.

The USB cable has four wires, +5 V DC, ground (0 V), data− and data+. The 5 V and ground wires supply power to each peripheral thereby removing the need for separate power adapters. The data wires provide a differential logic signal (i.e. the wanted signal is extracted as the voltage difference between the two data wires). The cable is constructed using two multi-strand wires for the power feed and a screened twisted pair for the data feed (Figure 2.47).

The differential data signals provide a fair degree of noise immunity as stray magnetic fields and radio frequency (RF) radiation from the monitor induce the same noise–voltage in the twisted data wires, so the interference is almost completely cancelled out at the receiving end where the D+ and D− signals are subtracted from one another. The screen also provides additional interference protection.

If we consider the USB port on the PC as the uppermost port (the host), then all other connections on the bus can be considered as downstream of the host.

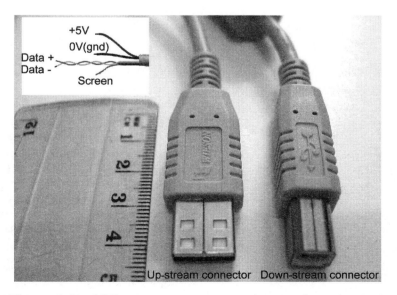

**Figure 2.47** *USB connectors, type-a (upstream) and type-b (downstream)*

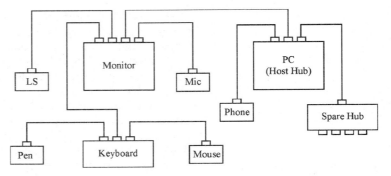

**Figure 2.48**   *Black diagram of a typical USB hub arrangement*

The cable from the host can either go directly to a peripheral or to a hub where several devices can be attached. Hubs are the key elements of the USB architecture. They improve connectivity by providing a robust and easy connection point for several devices, including other hubs. The hub acts in a slightly analogous way to a miniature telephone exchange, controlling the transfer of data between the downstream devices and the host. It contains a repeater and controller circuit.

The repeater is a software-controlled switch connected between the upstream port and the downstream ports and the controller handles communication between the hub and the host. Commands from the host configure the hub and allow it to monitor and control its ports. This allows low-speed and high-speed devices to connect to the same hub without data conflicts, and enables cables to be removed and refitted while the system is powered (Figure 2.48).

The drawing above depicts a well-populated USB system with hubs on the keyboard, monitor and PC. These items provide a stable and sensibly located site for the additional devices. For example, the keyboard hub is ideal for the mouse and touch pen and the monitor hub is the ideal site for the amplified loudspeakers and microphone.

When the USB was first introduced there was only a small range of USB devices to support it. Today the situation has changed quite dramatically. All manner of peripherals including printers, scanners, pointing devices, digital cameras, keyboards, data pads, sketch pads, joysticks, game pads and modems are now supplied with USB connections as standard.

## USB 2.0

The USB 2.0 standard launched in 2001 supports a much higher maximum data transfer rate of 480 Mb/s. This is 40 times the maximum data transfer rate of USB 1.0. The USB 2.0 standard is fully downwards-compatible with USB 1.0. This means that a USB 2.0 interface will quite happily support USB1.0 devices as well as USB 2.0 devices. The high data transfer rate of USB 2.0 allows wide-bandwidth devices like portable hard-disk drives and digital video cameras to connected and disconnected at will.

## Firewire

A similar but superior serial bus system known as 'FireWire™' was developed by Apple before the launch of USB 1.0. FireWire supports wide-bandwidth data transfer up to 400 Mb/s and allows devices to transfer data point-to-point directly on the FireWire bus without the PC acting as a go between. The next version of FireWire will support even higher data transfer rates.

Serial bus systems in all flavours, look set to go through even more evolutionary changes in the next few years. This is because serial bus systems, although inherently slower than parallel bus systems, offer a simpler connection system (i.e. compare the wide ribbon cables and connectors used on internal disk drives to the simple USB cable system).

## Input/Output (I/O)

Like all computers a PC must have a means of attaching peripherals so that data from the outside world can be accepted and processed in binary. The resulting data must then be converted from binary to a form that can be assimilated by the operator. To do this the CPU must have the ability to transfer data to and from a selected peripheral when needed. This is achieved in much the same way as memory access – by assigning an address to each peripheral. However, this cannot be done directly. Instead interface circuits are used to link the data bus to each peripheral. These interfaces are known as I/O ports and each one has a unique address in much the same way as memory locations. In a nutshell an I/O port is just an extension of the data bus via an addressable interface. The CPU is able to distinguish between an I/O addresses and a memory address by the type of instruction it is executing but they both function in a similar way.

There are several I/O ports on a PC and these can be categorized into *dedicated ports* as used by the keyboard and mouse and general purpose ports. The *general purpose ports* available on a standard IBM PC are limited to two serial ports – *com1 and com2* – and one 8-bit parallel port – *lpt1*. The serial ports conform to the *RS232C* standard and the parallel port to the *Centronics* standard.

Today's PCs also support additional high-performance general purpose I/O interfaces in the form of external bus systems. These include two serial bus systems: the *USB* and *firewire bus* and two parallel bus systems; the *small computer systems interface (SCSI)* and *personal computer management computer interface adapter (PCMCIA)*. Unlike the traditional IBM general purpose ports, these external buses allow several devices to connect at once thus greatly expanding the I/O capability of the PC. Also the inherent data transfer rate of these buses is several orders of magnitude greater than the traditional buses.

## Interrupts

I/O data is continually vying for processor time even while other programs are running. For example, while an application program is running, the user is frequently using the mouse to access menu functions. The signal from the mouse interrupts the application program and forces the CPU to run the mouse control program. After the interrupt has been serviced, the CPU continues from where it left off, prior to the interrupt.

An interrupt is therefore an interruption of the current work being carried out by the CPU. The CPU is diverted to do work elsewhere and then it is returned to carry on exactly where it left off.

The PC handles two categories of interrupt known as *hardware* (*external*) *interrupts* and *software* (*internal*) *interrupts*. The former category is the only type of interest to the support technician. In this section we will refer to them as hardware interrupts.

### Hardware interrupts

Hardware interrupts are so called because they are sent to the CPU as electronic signals from the I/O ports and other hardware devices including adapter cards and drives. There are two types: *non-maskable* and *maskable*.

### Non-maskable interrupts

Non-maskable interrupts are applied to a pin on the CPU labelled NMI and these are only used to signal catastrophic events such as a sudden drop in power or memory failure.

### Maskable interrupts

Maskable interrupts are applied to the interrupt request line (INTR line) pin of the CPU via a special sub circuit on the motherboard called the PIC. The IBM XT PC uses a single PIC chip – the Intel 8259A – to carry out this function. Unfortunately the 8259A provides just eight hardware interrupts and this severely limited the expansion capability of the XT. To improve on this situation the IBM AT PC was equipped with two 8259A PIC chips wired in series (cascaded), thereby providing 16 hardware interrupt lines. Modern AT and ATX PCs still use a similar PIC system – equivalent to two 8259As – but the circuits are embedded in the motherboard chipset.

To accommodate two PICs, the output from the second PIC is routed through IRQ2 of the first PIC (Figure 2.49). So out of the 16 interrupt inputs available, the IRQ2 pin is lost for the privilege. To retain the normal function of IRQ2 – as it must be for XT compatibility – IRQ9 is wired to the same destination as IRQ2 on an XT.[11]

Hardware interrupts are prioritized in the PIC depending on the IRQ number. The lowest numbers have the highest priority. When an interrupt or group of interrupts arrive, the PIC signals the CPU via the INTR line), that an interrupt has been received. When the CPU is ready to respond, it completes the current instruction of the program it is running and saves the address of the next instruction so that it can subsequently return to it. The CPU then sends an *interrupt acknowledge* (INTA) signal to the PIC via the control bus. The PIC responds by placing an 8-bit code onto the data bus. This points to the start address of an *ISR*. This is a special program stored in RAM, specifically written to serve the needs of the interrupting device. The CPU executes the ISR (e.g. the mouse-handling routine) and at the end of the routine it fetches the previously saved address of the original program so that it can continue where it left off.

[11] IRQ2 on an IBM XT is wired to the B4 pin on the 8-bit expansion bus therefore AT compatible PCs connect IRQ9 to the same place on the 8-bit portion off the ISA bus.

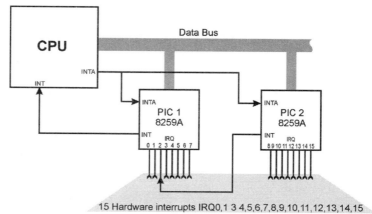

**Figure 2.49** *Basic block diagram of the PC-AT's cascaded PIC arrangement*

**Table 2.12** *Hardware interrupts*

| Hardware interrupt | | Expansion slot |
|---|---|---|
| IRQ0 | Timer tick | Not available |
| IRQ1 | Keyboard | Not available |
| IRQ2 | Used to link PIC2 | Not available |
| IRQ3 | COM2 serial port | 8-bit ISA |
| IRQ4 | COM1 serial port | 8-bit ISA |
| IRQ5 | LPT2 parallel port | 8-bit ISA |
| IRQ6 | Floppy disk | 8-bit ISA |
| IRQ7 | LPT1 parallel port | 8-bit ISA |
| IRQ8 | Real-time clock | Not available |
| IRQ9 | Acts as IRQ2 | 8-bit ISA |
| IRQ10 | Spare | 16-bit ISA |
| IRQ11 | Spare | 16-bit ISA |
| IRQ12 | Mouse | 16-bit ISA |
| IRQ13 | Maths co-processor | Not available |
| IRQ14 | Hard disk | 16-bit ISA |
| IRQ15 | Spare | 16-bit ISA |

Of the 15 hardware interrupts available on the PC-AT and ATX, several are pre-assigned to specific devices as shown in Table 2.12.

### Software interrupts

Software interrupts on the PC are signalled to the CPU internally and are initiated with the machine code INT instruction. Software interrupt routines form the heart of the BIOS and OS services. All BIOS and OS calls are made using the INT instruction. For example, when an application program needs to execute a disk-operating system (DOS) function it calls the DOS INT 21 service.

The PC can use up to 256 software interrupts, 00H to FFH. With so many interrupt sources available on the PC, some form of prioritizing is essential. Software interrupts are given the highest priority, followed by non-maskable then maskable interrupts.

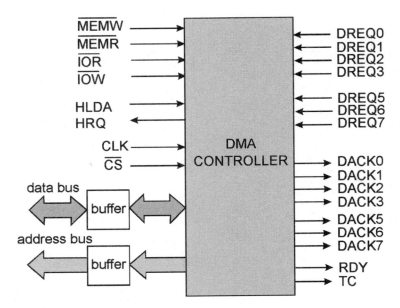

**Figure 2.50**   *Basic circuit of the PC-AT DMA controller. In reality the tracks and sectors are invisible; they are drawn here to show the layout of the magnetic boundaries*

# Direct Memory Access (DMA)

Another way to input and output data on a PC, is for the external device to borrow the system's address and data bus temporarily and transport the data between the device and memory directly, without using the CPU. This type of data transfer is known as Direct Memory Access (DMA).

To implement DMA the peripheral device must be capable of performing fast data transfers – so a high-speed, intelligent interface is essential. Hard-disk controllers contain such a device, they have a built in processor that can perform read and write operations independent of the main CPU.

It basically works as follows: the peripheral device requests the use of the system address and data buses by sending a *direct memory access request* (DREQ) signal to a special part of the motherboard chipset known as the DMA controller. On receiving DREQ, the DMA controller sends a HOLD signal to the main CPU. The CPU then completes the current instruction and places its address and data bus pins into the floating state – effectively disconnecting itself from the buses. It then sends a *hold acknowledge signal* (HLDA) to the DMA controller which in turn sends an acknowledge signal (DACK) to the I/O device to begin the data transfer. While the HOLD signal is high, the CPU is disconnected from the bus, and the peripheral has access. As soon as the hold line goes low again, the CPU regains control of the bus. In the PC, the DMA process is highly sophisticated, with DMA transfers taking place during intervals when the buses are idle.

A simplified version of the DMA circuit used on the PC is shown in Figure 2.50. The control signals are provided by a bus controller circuit which is part of the chipset.

**Table 2.13** *DMA channel assignments*

| DMA channel | | Function | Expansion slot |
|---|---|---|---|
| PC-XT | 0 | DRAM refresh | Not available |
| | 1 | Spare | 8-bit ISA |
| | 2 | Floppy-disk controller | 8-bit ISA |
| | 3 | Hard-disk controller | 8-bit ISA |
| PC-AT | 0 | Spare | 8-bit ISA |
| | 1 | Spare | 8-bit ISA |
| | 2 | Floppy-disk controller | 8-bit ISA |
| | 3 | Spare | 8-bit ISA |
| | 5 | Spare | 16-bit ISA |
| | 6 | Spare | 16-bit ISA |
| | 7 | Spare | 16-bit ISA |

The reason for performing DMA rather than normal I/O port transfers, is to gain a speed advantage. When the I/O device is faster than the computer's CPU it can improve the data throughput. However with modern CPUs, DMA transfers can be slower than direct I/O port transfers.

The DMA controller system used on the original PC-XT had four channels– DMA0 to DMA3. The PC-AT and ATX has seven DMA channels– DMA0 to DMA3 and DMA5 to DMA7 – it is composed of two controllers similar to the design used on the PC-XT. DMA4 is used internally to switch between the two controller circuits. On a modern PC system the DMA controller circuits are incorporated into the motherboard chipset (SB).

On a PC-AT/ATX the DMA channels 0 to 3 are used for 8-bit data transfers and are therefore available on the 8-bit portion of the ISA expansion slots. Channels 5–7 are used for 16-bit transfers and are therefore available on the 16-bit extension part of the ISA slots (Table 2.13). The PCI slots do not use DMA requests.

Add-on expansion cards often use both DMA channels and IRQ lines, for example, some sound cards require up to three DMA channels and one IRQ line.

# The OS

However powerful the computer's CPU, it is useless without a program telling it what to do. For example, to make the CPU carry out 'PC-like' tasks, the CPU must be programmed to do the right things. Part of this special program is permanently stored in the system ROM and part is temporarily stored in RAM. The ROM-based part of the program is called the BIOS and the RAM-based part is called the Disk Operating System or Operating System. The OS is a set of programs, that allows the user and application programs to utilize the system hardware in an efficient and controlled manner. It does this by enabling and managing communication between the user end of the system and the BIOS.

In modern PC systems, the OS, for example 'Windows' is loaded from hard disk to RAM every time the user starts the computer. The advantage of such a disk-based OS is that it is easy for the user to install upgrades, which in turn ensures that the software keeps pace with advancements in hardware technology. This prolongs the useful life of the OS and ensures a profitable return for the manufacturers.

Each time a PC is powered up, the CPU automatically runs a special system-test program that is built into the same ROM as the BIOS. This is called the *Power On Self Test (POST)*. The POST runs a series of tests to check the integrity of the various parts of the system hardware and then passes control to a 'boot-loader' program (routine) which is also stored in the ROM. The boot-loader's job is to search for an OS start-up program on the floppy disk or the hard disk, copy into RAM, and then run it, thereby allowing the OS to take the helm.

Once the OS is loaded and active, the user is effectively put in control, and the OS sits in the background waiting for an input from the keyboard, mouse or other input device. The user is now able to interact directly with the OS using a set of keyboard commands or by using the mouse and clicking on buttons displayed on the VDU.

When an input device – like the keyboard or mouse – is activated by the user, it sends an interrupt request to the CPU which activates a device-handling routine stored in the BIOS ROM. This in turn invokes other routines in the BIOS or OS, depending on the particular key press or mouse action.

The command hierarchy in a PC system is illustrated in Figure 2.37. The user is at the top using *high-level* English-like commands or mouse actions to communicate with the application program. The application program communicates with the OS which in turn communicates with the BIOS, using *low-level* commands – *machine code* – to control the hardware.

Computer programs are sequences of 8-bit binary numbers, temporarily stored in RAM or permanently stored on disk, that mean something to the CPU and usually provide some useful function. Each byte or group of bytes in the program instructs the CPU to carry out specific tasks on data stored at specific locations in the system. The greater the amount of RAM available to the CPU, the larger and more powerful a program can be. Large amounts of RAM accommodate full programs and data, so program execution can take place quickly, without the need to access the slower disk drives. Once the data has been processed it can then be saved to a more permanent form of storage using magnetic or optical disks. Disk- or tape-based memory is known as auxiliary storage.

## Summary

We have seen that a digital computer contains a CPU, primary and secondary memory, and an I/O unit. Even the simplest computer can input and output data and perform arithmetic, logic, sorting and data transfer operations. A non-dedicated computer like the PC is capable of carrying out an almost limitless amount of useful tasks, by following a specific *Application Program* loaded into memory by the operator. The Application Program communicates with the computers OS, which in turn communicates with the BIOS. The BIOS then configures the hardware to suit the action to be performed. At any particular instant in time, the CPU may be executing an instruction from an Application Program, the OS or the BIOS or even from an externally connected device. The system ROM contains the BIOS, boot-loader and POST routines.

# Auxiliary storage devices

### *Floppy-disk drives*

The floppy-disk drive was invented at IBM by Alan Shugart and his design team in 1967. They were called 'floppy' disks because the storage media was housed in a large flexible 8-in. diameter flat plastic envelope. The whole thing literally felt 'floppy' compared to the heavy and very rigid Winchester disks used on the mainframe computers of the day. These true 'floppy' disks now seem unbelievably large and unwieldy compared to today's 3.5 in. variety.

In 1974 a 5.25-in. diameter 360 kB floppy-disk drive was introduced. This became the standard drive for many computers of the day, including the IBM PC-XT. It remained the standard until IBM introduced their PC-AT in 1984. These new PCs were fitted with a 5.25-in. 1.2 MB drive and therefore offered a capacity four times that of an XT drive.

By 1986 several manufacturers were producing floppy-disk drives and a more compact 3.5-in. 720 kB disk drive evolved. Then in 1987, IBM announced the PS/2 range of computers sporting a new 3.5-in. 1.44 MB drive. Needless to say, it was not long before PC manufacturers started to incorporated these new high-capacity drives in their AT machines. The 1.44 MB drive quickly became the most popular format due to its compact size and high capacity. This format is now the standard floppy-disk drive for PC-AT and other PCs including laptops.

Although floppy-disk drives are now available with capacities far greater than the 3.5-in. 1.44 MB floppy disk, it is still the most popular system for small scale file storage and backup. A few years ago, most application software was supplied on 3.5-in. floppy disks that the user then installed onto hard disk. Nowadays most software is provided on CD-ROM. For example, a popular office application software suite used to be supplied on 30 numbers of 1.44 MB floppy disks!, the same suite is now provided on one CD-ROM. The 3.5-in. disk is becoming less and less used for this purpose and will soon become obsolete like its predecessors.

Both floppy-disk drives and hard-disk drives store information as minute changes of magnetization on the surface of a spinning magnetic disk. A floppy disk is made from plastic film such as Mylar. First the disk is coated with a magnetic compound and then with a protective layer of polymer. The completed disk is encased in a protective sleeve that incorporates a low-friction lining. This helps to keep the surface of the disk clean and allows the disk to rotate freely in its sleeve.

Information is stored at precise locations on the upper and lower surfaces of the disk in a series of concentric circles known as 'tracks'. As both surfaces of the disk are used, a pair of circular tracks – one on each side – forms an imaginary 'cylinder' and is named the same. Each track is subdivided into a number of 'sectors'.

A 1.44 MB 3.5-in. floppy disk has 80 cylinders and 18 sectors-per-track and each sector holds 512 bytes of information. The capacity of the disk is found simply by multiplying these together (i.e. 2 sides $\times$ 80 tracks $\times$ 18 sectors $\times$ 512 bytes per sector) gives a capacity of 1,474,560 bytes, incorrectly dubbed 1.44 MB.

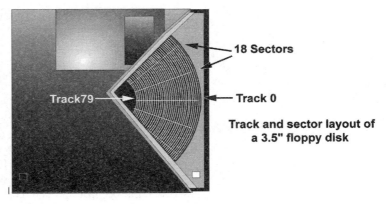

Track and sector layout of
a 3.5" floppy disk

Figure 2.51

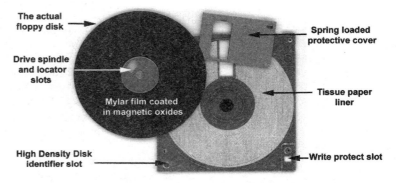

**Figure 2.52**   *Exploded view of a 3.5-in. 1.44 MB high-density floppy disk*

The old format 5.25-in. 360 kB floppy disks have 40 tracks and 9 sectors with 512 bytes per sector (Figures 2.51–2.54).

*Note:* There are $2^{10}$ (1024) bytes in a kilobyte, and $2^{20}$ (1,048,576) bytes in a megabyte. When referring to disk drive capacity however, a megabyte is often defined by some manufacturers as 1000 kB (1000 × 1024) not 1024 kB (1024 × 1024) as it should be.

When a disk is pushed into the drive's slot it is clamped, quite firmly, between the two read/write heads and there it stays motionless, until a read or write request is received from the computer – or the user ejects the disk. A sensor inside the drive, signals that a disk is present and on receiving a read or write request, a *motor-on* signal spins up the disk. At this point, as the disk starts to rotate, it suddenly locates in the drive spindle slot. This can be heard as a faint thud or click as the magnetic spindle plate locates into the metal drive slot. It may seem odd to use a magnetic method of holding the disk to the spindle, particularly as the disk coating is sensitive to magnetic fields (Figure 2.55). This is not a problem however as the nearest tracks are a couple of centimetres away from the magnetic centre spindle so the field is too weak to erase the data. It is wise however to always remove a disk from the drive when it is not in use. Leaving the disk in the drive for long periods will warp the disk and may corrupt the data.

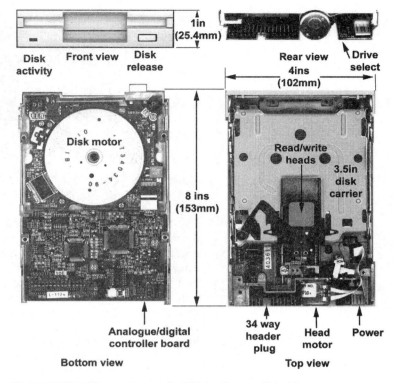

**Figure 2.53** *The anatomy of a 3.5-in. floppy-disk drive*

**Figure 2.55** *The drive motor mounting plate inside a 3.5-in. disk drive*

**Figure 2.54** *A low-profile (0.5-in. high), 3.5-in. 1.44 MB floppy drive*

## How a floppy-disk drive works

A floppy-disk drive has two read/write heads – one for each side of the disk. These are mounted on a moving arm that moves tangentially to the tracks on the disk. The whole head assembly is moved across the disk using a motor-driven system known as the head actuator. The most common type in use is a rack and pinion system based on a motor-driven Archimedes-screw.

The lower head is fixed to the head rack assembly on a diaphragm-spring allowing the head to mould to the disk's bottom surface.

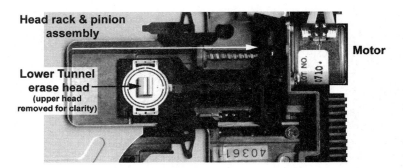

**Figure 2.56**  *The head actuator and bottom head assembly, the top head has been unscrewed from the assembly for clarity*

The top head is fixed to a plastic arm in a similar fashion, using a diaphragm-spring. This is then screwed to the rack assembly. Both heads therefore move in and out together under the control of the head actuator motor (Figure 2.56).

The top head arm is hinged to allow the disk to be inserted and removed without restriction. When the disk is pushed fully into the drive, the arm springs down and both heads grip the disk under gentle pressure. On pressing the disk release button, the top head arm is hinged up, out of the way, by the disk cover plate.

The heads actually make intimate contact with the disk surface, so head wear and surface abrasion of the disk can be a problem – especially when you consider that the disk revolves at 300 rpm. To minimize wear, the disk surface is coated in a thin but tough layer of plastic and the head blocks are made from a tough hard-wearing material such as ceramic.

The heads themselves are manufactured to an extremely fine tolerance and consist of a read/write head and a 'tunnel-erase' head. The pole pieces of each head are embedded in a tough plastic resin with tiny gaps only a few micrometers wide.

The recording process is designed to place thin, well-defined magnetic tracks onto the disk's surface without interaction between tracks. This is achieved using *tunnel-erase recording*. In this form of recording, as the disk spins in a clockwise direction. First the read/write head records the data onto the track and then the tunnel-erase head erases the outer extremes of the track – effectively trimming the track to a thin, well-defined band (Figure 2.57).

All floppy-disk drives have a built-in PCB to control the heads, motors and various sensors in the system. It must also provide an interface to the computers floppy-disk controller circuitry. The system conforms to a standard set out by Shugart Associates (SA) known as the SA-400 interface. The interface/controller board of a typical 3.5-in. drive is shown in Figure 2.58. The disk motor and associated electronics are on a separate PCB similar to that shown in Figure 2.53. In addition to the built in electronics, the drive connects via a 34-way ribbon cable to a floppy-disk controller card inside the PC.

The electronics on a modern floppy-disk drive, at first sight, looks simple, as all the control, driver and interface circuitry is taken care of by one or two LSI chips. Most of the basic electronic components like resistors, capacitors, inductors and diodes are surface mounted on the opposite side of the board to the chips, adding more to the deceptively simple appearance.

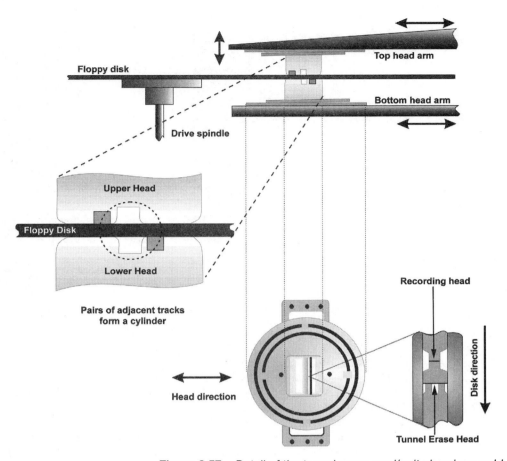

**Figure 2.57**  *Detail of the tunnel-erase read/write head assembly*

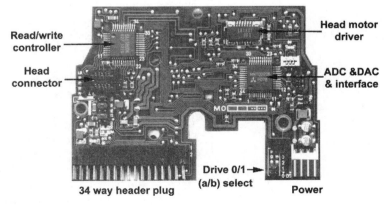

**Figure 2.58**  *The main interface/controller board built into a 3.5-in. floppy-disk drive*

The disk drive circuitry performs the following basic functions:

- Interprets and acts on commands sent from the PC's floppy-disk controller card.
- Moves the read/write heads to any desired track.

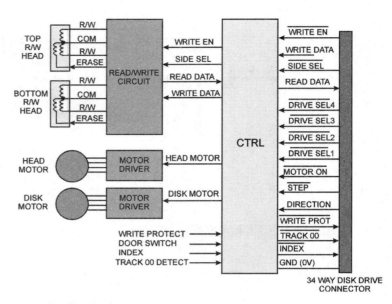

**Figure 2.59** *Block diagram showing the signals present on a 3.5-in. disk drive*

- Reads and writes the disk's magnetic data.
- Maintains a constant disk velocity.
- Monitors and acts on the various sensors inside the drive.

### Data read operation

With reference to the simple block diagram in Figure 2.59: Data from the disk read/write head– in the form of a weak current, is passed to the read/write block where it is amplified to a suitable level, filtered and then converted to a digital signal. The stream of data is then passed to the control block (ctrl). Here the signal is converted to the correct serial data format for transmission to the floppy-disk controller. The signal is sent via the *read data* line of the 34-way ribbon connector.

### Data write operation

The write data from the computer – received via a wire on the 34-way ribbon cable as a serial bit stream – is applied to the control block and then on to the read/write block. Here it is converted to a current signal and applied to one of the read/write heads. The particular head in question is selected by the SIDE-SEL signal via the 34-way cable. A disk write however can only occur if signals from the write protect detector and write enable pin are at appropriate logic levels. The read write logic is contained in the read/write block.

### Track 00 detection

The track 00 sensor is located at the front edge of the drive and detects the head arm when it is at the bottom of its travel, (i.e. when the read/write heads are exactly positioned over track 0) the outermost track of the disk. At power up, the head arm automatically returns to the track 0 position, this sends a track 00 signal to the control block which is now able to determine the position of any subsequent tracks via the STEP and DIRECTION signals.

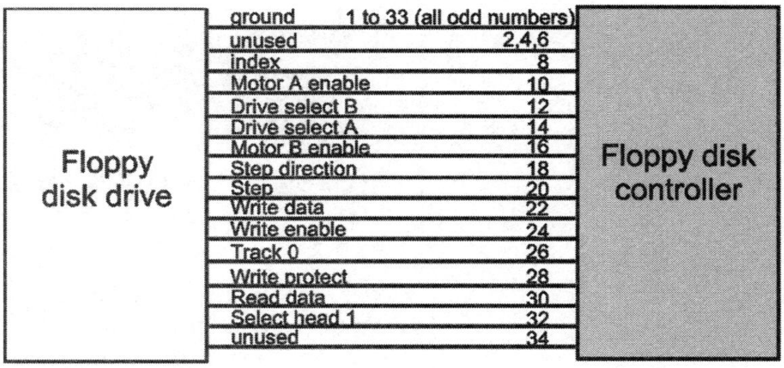

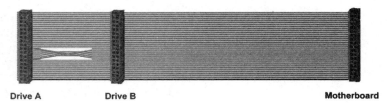

| Floppy disk drive | | Floppy disk controller |
|---|---|---|
| | ground 1 to 33 (all odd numbers) | |
| | unused 2,4,6 | |
| | index 8 | |
| | Motor A enable 10 | |
| | Drive select B 12 | |
| | Drive select A 14 | |
| | Motor B enable 16 | |
| | Step direction 18 | |
| | Step 20 | |
| | Write data 22 | |
| | Write enable 24 | |
| | Track 0 26 | |
| | Write protect 28 | |
| | Read data 30 | |
| | Select head 1 32 | |
| | unused 34 | |

**Figure 2.60**  *The floppy-disk-to-FDC ribbon cable connections.*

**Drive A**          **Drive B**                                                **Motherboard**

**Figure 2.61**  *The PC floppy-disk drive ribbon cable, note the seven twisted wires*

The $\overline{\text{STEP}}$ pulse moves the head one track at a time in a direction determined by the direction signal. When direction signal is low, the STEP pulse moves the head towards the disk centre and when it is high, it moves it away from the disk centre.

To make use of the floppy-disk drive, an interface is required between the disk drive and the computer bus. On a PC this is provided using either a plug-in adapter card or using one built in to the motherboard. The trend nowadays is to incorporate all the I/O functions including the drive interfaces onto the motherboard. The interface used for the floppy-disk drive is known as *floppy-disk controller* (FDC). It provides an intelligent link between the computer and the disk drive and essentially converts the serial bit stream at the disk drive into parallel data at the data bus. The FDC circuit is based on a single VLSI IC known as the FDC chip. In essence it is a CPU with a SIPO and PISO converter built-in. The FDC provides all the control signals for the floppy-disk drive and also decodes signals into a form suitable for the main CPU. If an adapter card is used, the FDC circuit is usually integrated with other I/O functions such as serial, parallel, games and hard-disk ports. The card is known as the multi-I/O adapter (see Figure 2.60).

### Identifying disk drive ribbon connectors

Floppy-disk drives connect to the FDC circuit using a 34-way ribbon cable (see Figure 2.61). On 3.5-in. drives, both the controller and drive end of the cable, use a female DIN 41651 IDC connector. 5.25-in. drive ribbon cables have a 34-way edge connector for the drive and a female DIN 41651 IDC connector for the controller.

To allow two floppy drives to be fitted, some ribbon cables are usually supplied with two sets of connectors. These cables have the 'drive select', 'motor enable' and associated ground wires (10–16) crossed

**Figure 2.62**   *The rear view of a 3.5-in. floppy-disk drive*

over, causing a twist in the cable. The connectors on the twisted end of the cable go to the A: drive and those on the non-twisted end go to the B: drive. The picture below shows a universal ribbon cable suitable for both 3.5- and 5.25-in. drives.

Each floppy drive is allocated a unique drive number selected by a 'jumper' on the circuit board of each drive. The jumpers are usually labelled DS0, DS1, DS2, DS3 with drive A: as 'DS0' and drive B: as 'DS1'. However, using a cable with twisted wires eliminates the need to set the A: and B: 'drive select jumpers'. The twisted wires swap jumpers DS0 and DS1 around on the drive connected on the end of the cable, so both drives can be left in the default 'DS1' position. To reduce costs some manufacturers no longer fit jumpers.

The signals present on the floppy-disk cable are essentially TTL compatible. With logic '0' defined by a signal voltage between 0 and 0.4 V and logic '1' as a signal voltage between 2.4 and 5 V. It is worth noting that all outputs are 'open collector' so connecting it the wrong way round, should not cause any damage (Figure 2.62).

### Disk drive power connectors

There are two different sized power connectors used on disk drives. The larger type is used on all 5.25-in. floppy-disk drives, hard drives and CD-ROM drives. The smaller type is used mainly on 3.5-in. floppy-disk drives. The connector supplies +12V and +5 V DC. The +5 V supply powers the electronics and the +12 V supply feeds the head and disk motors. On some modern 3.5-in. floppy-disk drives the +12 V wire is not used (Figure 2.63).

### *Hard disks*

The operation of both floppy and hard-disk drives is similar in many respects. Hard disks are made of highly polished aluminium, coated on both faces with a thin magnetic compound of iron and cobalt. The disks are both lightweight and rigid, so they can be stacked one above the other to increase the overall capacity. During operation, the rigid structure prevents excessive flexing, so the read/write head can float above the disk's surface without actually making contact. The disk is therefore able to spin at a very high speed with virtually no wear to the head.

Each head is mounted on the end of a lightly sprung nib like finger which allows the head to move perpendicularly to the disk surface on a cushion of air. The heads are mounted on the end of a rigid aluminium moulding, known as the pivot arm. As the arm moves, the heads move in unison across the disk. but never in contact with the spinning surface. As the disk spins up from rest, the heads, starting from a safe resting position, lift up on a cushion of air and float a few microns above the surface.

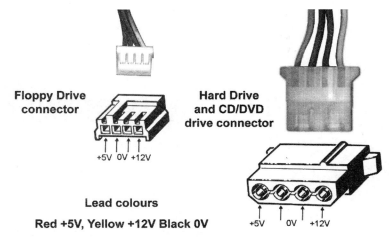

**Figure 2.63** *Floppy-disk drive power connectors*

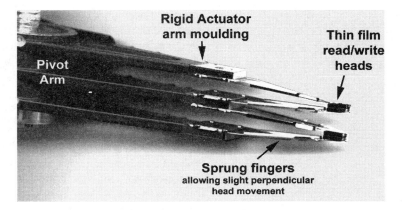

**Figure 2.64** *A head stack assembly showing four TFI heads*

## The read–write head

The head is purposely made small and lightweight to maximize, the number of tracks, and increase the data access time (Figure 2.64).

Head design is a crucial factor in achieving high-areal disk densities. The smaller the head gap, the finer the width of the recorded track, so more tracks can be squeezed onto the disk. Several improvements in head design have occurred since the first inductive heads were invented. A modern inductive head uses thin film technology (TFT) similar to that used for semiconductor chip manufacture. In the TFT process, layers of conductive and non-conductive material are spluttered onto the head substrate, thus forming an extremely fine head gap. Nickel–iron pole pieces and a copper coil are then electroplated onto the substrate.

A more recent advancement in head technology uses the principle of *magneto-resistance* (MR).

Such a substance exhibits a change in resistance when subjected to changes in magnetic field strength. An MR head, employs two heads closely spaced on the same substrate. The heads are fabricated using TFT. A normal inductive head element is used for recording data and the MR head element is used to read the data. The MR head is far more sensitive to magnetic flux changes than a corresponding inductive

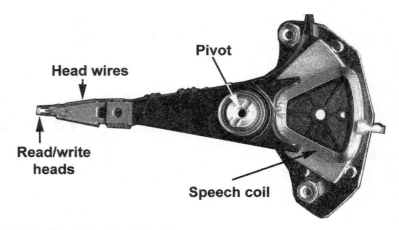

**Figure 2.65**   *Speech-coil head actuator arm*

head, so fewer turns are required on the write head. This has the advantage of reducing the head mass and increasing the maximum operating frequency, allowing faster data transfer rates and increased areal densities. The use of two heads also reduces the problems of noise at these very low magnetic field strengths.

### Head actuator assembly

The arm that moves the head across the disk's surface is designed to operate at a high average speed–often in excess of 50 mile/h on a modern drive (Figure 2.65). This high speed is achieved by using a *speech-coil* head actuator, which is much more responsive than the motor-driven head assemblies used on floppy disk drives and some older hard-disk drives. The name 'speech coil' is derived from the 'moving coil loudspeaker' which uses a coil attached to a paper cone suspended in a magnetic field. The current in the coil, varying in sympathy with an audio signal, sets up a corresponding varying magnetic field across the coil. This interacts with a fixed magnetic field from a permanent magnet and vibrates the coil and cone, producing an audible signal. The speech-coil head actuator uses the same principle. By applying a current in proportional to the distance the head needs to move, the arm can be positioned with a high degree of accuracy.

### Disk motor

The rotational speed of a 3.5-in. hard disk can be over 7200 rpm on some top-end drives–about 24 times the speed of a floppy disk. The disks must rotate at these high speeds with virtually zero perpendicular movement, so high-precision low-friction motor bearings are essential. The stack of disks is directly attached to the drive shaft of the motor via a spacer and clamp assembly. The motor is a special type known as a *stepper motor*. This type of motor uses pulses of current to rotate the drive spindle. Each pulse rotates the spindle a few degrees, so applying a series of pulses at a precise frequency, the speed of the disk is maintained at a constant speed (Figure 2.66).

The high head and disk speeds of a hard-disk drive dramatically reduce the data search and seek times. The performance of a modern hard disk is compared with a typical 3.5-in. floppy disk in Table 2.14.

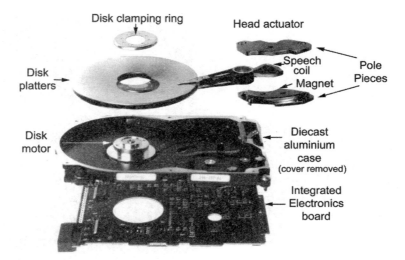

**Figure 2.66** *Exploded view of a hard-disk drive*

**Table 2.14** *Floppy disk/hard disk comparison*

| Parameters | Floppy disk | Hard disk |
|---|---|---|
| Capacity | 1.44 MB | 80 GB |
| Maximum data transfer rate | 45 kB/s | 100 MB/s |
| Average seek time (ms) | 94 | 7 |
| Track density (tpi) | 135 | 6461 |
| Rotational speed (rpm) | 300 | 7200 |
| Head floating height | n.a. (*touches surface*) | 5 μ in. |

*Note:* (i) 1 μ in. = 1 millionth of an inch. (ii) 1 GB = $2^{30}$ bytes = 1,073,741,824 bytes. (iii) tpi = tracks per inch. (iv) ms = milliseconds or 1/1000 s.

The inside of the hard disk must be kept dust-free. At the high head-surface speed and the tiny distances involved, even a tiny particle of dust can damage the surface of the disk or even the head. The disk is assembled in a clean-room environment with an allowable dust particle size many hundred times smaller than that of a modern operating theatre. The inside of the disk case is kept in a clean air environment using air filters, so the cover should never be taken off except by an authorized service centre equipped with a clean-room and associated equipment.

## Hard-disk types

Several hard-disk types and control/interface systems, have been used on PC systems since the launch of the XT PC, the main ones are modified frequency modulation (MFM), enhanced small disk interface (ESDI), SCSI and AT attachment integrated drive electronics (ATA IDE).

MFM and ESDI types are obsolete, so these systems will not be discussed further, except to compare cable differences. Currently, (ATA IDE) and SCSI are the most widely used standards in use.

Speech coil head actuator

Stack of coated metal disk platters

Upper read/write head

**Figure 2.67**   *A typical 3.5-in. ATA IDE hard disk with the top cover removed, showing details of the head arm and the upper disk (platter)*

### ATA IDE drives

Figure 2.67 shows a typical 3.5-in. ATA IDE hard disk with the top cover removed and details of the head arm and the upper disk (platter).

The disk platters are totally enclosed in a dust-proof case in a filtered air environment to keep disk surfaces clean and free of dust and dirt particles.

AT IDE was originally designed as a plug-in hard-disk drive option for the 16-bit expansion bus of IBM PC-AT and compatible computers. These drives incorporate all the signal processing, controller and I/O electronics on the drive itself, unlike MFM and ESDI drives which need a separate controller card. The integrated design reduces electrical interference and allows faster data transfer rates. It also has a cost and reliability advantage, as a separate disk controller card is not needed. The only extras required to link the drive to the AT bus, are a 40-way ribbon cable and a simple IDE I/O port. Nowadays this is built into the motherboard. On old pre-Pentium systems the IDE I/O port was usually contained on an ISA I/O adapter card. The IDE drive and host adapter combination is known as the ATA. The salient features of the ATA system are high performance, high reliability and low cost (Figure 2.68).

IDE drives became so popular, that the committee representing the major SCSI drive manufacturers known as ANSI *Common Access Method* (CAM) *Committee*, decided to support the ATA specification and offer performance enhancements on a regular basis. The first CAM ATA specification was introduced in 1989. Followed by ATA-1 (ANSI X3.221-1994). The later specifications, ATA-2, ATA-3, ATA-4, ATA-5, etc., support added features and enhancements.

The CAM ATA standards specify:

- Cables and connectors,
- Electrical and logical signal characteristics,
- Commands and protocols,
- Mechanical details.

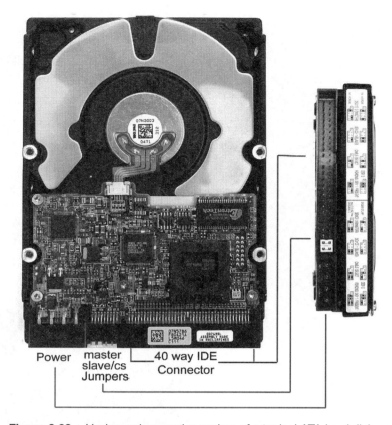

Power   master   40 way IDE
        slave/cs   Connector
        Jumpers

**Figure 2.68**   *Undercarriage and rear view of a typical ATA hard disk*

The enhanced ATA standard known as ATA-2, recommends ways to increase the speed of data transfer and improve the way the drive informs the PC about itself. The latter is particularly important when utilizing Intel's *PnP* system – mentioned later in this section. ATA-3 was announced in 1996 to extend the capabilities of the interface. Modifications to the mechanical and electrical aspects of the previous specification to suit laptop computers and to enhance data transfer rates to match advances in interface electronics. Power saving techniques and password protection are major features of this new specification. ATA-3 also maintains a high degree of compatibility with previous ATA versions. ATA/AT Packet Interface Standard (ATAPI)-4 (1997) includes additional specifications for the ATAPI used on IDE CD-ROM drives and ULTRA-DMA, a new hard-disk protocol for faster data transfer rates. Further revisions will be made on a frequent basis, to keep pace with improvements in disk technology and the PC hardware in general. For example ATA/ATAPI-6 includes specifications for even higher maximum data transfer rates to 100 MB/s, well as upgrades to the ATA-ATAPI-4 and -5.

The main hard-disk manufacturers have implemented the enhanced ATA standards and produced their own advanced software drivers and matching hard disks. Some manufacturers named their system *enhanced integrated drive electronics* (EIDE) and others named *fast ATA*. Both have similar specifications to the recommended ATA standards.

### Logical block addressing

The ATA-2 specification supports logical block addressing (LBA). This is a linear method of addressing the data on each sector of the disk drive. The old addressing method used on low-capacity drives is based on the cylinder, head and sector (CHS) number. LBA is used on high-capacity drives and works as follows.

Logical block address 0 (LBA0) is sector 1 of head 0, cylinder 0 and LBA1 is sector 2 of head 0, cylinder 0 and so on in a linear fashion. The addressing scheme obeys the formula:

$$\text{LBA} = ((\text{cylinder} \times \text{heads-per-cylinder} + \text{head}) \times \text{sectors-per-track}) + \text{sector} - 1$$

*Note:* A cylinder is the number of tracks under the heads at a particular head arm position. For example, in the drive shown below there are 6 heads so each cylinder is 6 tracks deep. See Figure 2.69.

In the ATA standard a certain number of bits are assigned to represent the number of cylinders, heads and sectors as follows.

**Figure 2.69** *(a) One cylinder of a possible maximum of 65536 across the platters (b)*

*Sectors-per-track* are numbered 1 to 255 – using 8 bits. *Heads* are numbered 0 to 15–using 4 bits and *cylinders* are numbered 0 to 65535 using 16 bits.

Each sector holds 512 bytes, so the maximum capacity ATA drive possible, based on this numbering system is as follows: 65536 cylinders $\times$ 255 sectors $\times$ 512 bytes $\times$ 16 heads = 127.5 or 136.9 GB based on the decimal definition where 1 GB = 1,000,000,000 bytes.

*Note:* Some hard-disk manufacturers use 1,000,000,000 bytes to represent a giga byte whereas the binary definition is $2^{30} = 1,073,741,824$ bytes. To avoid any confusion over the correct definition, the rest of this section will use term $GB_d$ to represent the decimal definition and $GB_b$ to represent the binary definition.

Even though the ATA system is capable of addressing a maximum disk capacity of 127.5 $GB_b$, the AT BIOS and DOS layers impose their own capacity restrictions. The maximum drive size limitation of the BIOS and DOS is covered later in this section.

### Faster data transfer

One of the major enhancements introduced in each new ATA specification is a faster data transfer rate between the drive and the system bus. This is done to keep pace with advancements in drive and interface technology.

In an early AT system, the hard drive interface is nothing more than a basic extension to the 8 MHz ISA bus. The extension port is either built-in to the motherboard or provided on a separate adapter card. As the original AT system had a relatively slow CPU compared to today's standards, the ISA bus was fine for most devices.

As faster CPUs, hard drives and graphics cards became the norm, the old 8 MHz ISA bus was a serious data 'bottle neck', with potentially fast devices either side of a slow bus (Figure 2.70).

To improve this situation two high-speed local bus systems emerged, VESA and PCI, but only the PCI bus became the accepted standard for today's systems (Figure 2.71). The VL bus, once widely used on 486 systems, is no longer supported. The PCI bus is now a feature of all Pentium-based machines. On most systems the PCI bus is clocked at a frequency of 33 MHz. It also uses a wide data bus of 32-bits – some top-end systems even use a 64-bit version of the bus. PC bus systems are discussed in Section 2.

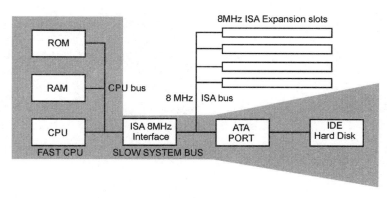

**Figure 2.70** *The ISA 8 MHz bus on a PC-AT compatible*

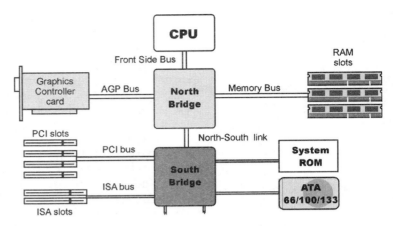

**Figure 2.71** *The PCI 33 MHz bus on a modern PC-AT compatible*

| Table 2.15 | ATA PIO modes | |
|---|---|---|
| PIO mode | Transfer rate (MB/s) | Notes |
| 0 | 3.3 | Old ATA |
| 1 | 5.2 | Old ATA |
| 2 | 8.3 | Old ATA |
| 3 | 11.1 | ATA-2 (using IORDY) |
| 4 | 16.6 | ATA-2 (using IORDY) |

ATA-2 and the later implementation of the standard make full use of the PCI bus to IDE interface, specifying several programmed input output (PIO) modes and DMA modes for increasingly higher data transfer rates.

### PIO modes

PIO data transfer is simply a method of transferring data to and from the drive and the host computer using an I/O port arrangement. In the old ATA PIO modes, 0 to 2, data is transported using 'blind' control. In these modes, data transfer occurs at a fixed rate based on the worst case transfer rate of the system (see Table 2.15).

The ATA-2 standard specifies PIO modes 3 and 4 with a flow-control signal called IORDY (I/O channel ready). This enables a suitable drive to make use of the maximum data transfer rate available on the host system. The drive effectively controls the maximum possible transfer rate by turning the data flow on and OFF with the IORDY control line. Table 2.15 shows the transfer rates for the five PIO modes supported in ATA-3.

### DMA modes

DMA is a system of transferring data directly between the drive and memory without involving the host CPU. Current ATA drives have ultra DMA (UDMA) capability also known as ultra ATA.

There are currently four UDMA standards available, UDMA 33, UDMA 66, UDMA 100 and UDMA 133. The latter three standards require a special cable fitted with standard IDE 40-pin connectors

**Table 2.16** *Hard disk DMA modes (legacy to present day)*

| DMA mode (multiword) | Transfer rate (MB/s) | Notes |
| --- | --- | --- |
| 0 | 4.2 | ATA |
| 1 | 13.3 | ATA-2 |
| 2 | 16.6 | ATA-2 |
| UDMA 33 | 33.3 | Introduced in 1997 |
| UDMA 66 | 66.6 | Introduced in 1998 and implemented in 1999 |
| UDMA 100 | 100.0 | Introduced in 1999 and implemented in 2000 |
| UDMA 133 | 133.3 | Current |

and an 80-way ribbon, so that a ground conductor can be introduced between each signal wire. This maximizes the data transfer rate of the cable. These recent standards support a maximum burst data transfer rate of 33.33, 66.66, 100 and 133.3 MB/s, respectively. Hard drives supporting these standards are labelled *ultra ATA 33, 66, 100 and 133*. These drives will only operate at these data rates on a PC fitted with an up-to-date motherboard and a UDMA compatible BIOS and OS. The DMA transfer rates supported by the leading drive and motherboard manufacturers are shown in Table 2.16.

### SCSI hard-disk drives

SCSI (pronounced 'scuzzi'), a parallel bus system used on many computer types including top-end PCs. The SCSI bus is a device-independent interface designed to support several devices on the bus at one time. Although the words integrated drive electronics (IDE) are used to describe the ATA hard-disk interface. SCSI drives also have the main drive electronics integrated onto the drive. The major difference between SCSI and the ATA system is that SCSI is a fast all singing all dancing bus system supporting a wide variety of devices whereas ATA was designed primarily as a low-cost disk interface.

To implement SCSI on a PC system a SCSI controller card is required, however some motherboards have an SCSI controller built-in. The controller card or 'host adapter' as it is sometimes called, is the heart of the SCSI bus system. It has its own processor and SCSI-BIOS ROM and is thus able to intelligently manage the data flow between the various devices on the bus and the main CPU. The SCSI-BIOS ROM augments the main system ROM, allowing the OS to communicate with the SCSI controller hardware in much the same way as the normal system hardware.

The SCSI bus was used for a wide variety of I/O peripherals, providing a fast convenient connection system. However its popularity as a general bus has declined considerably since the advent of fast serial buses like USB1, USB2 and IEEE-1394 (Firewire™). Also the recent appearance of Serial ATA (SATA) may further add to its decline.

One area where SCSI still holds is own is as a hard-disk interface. A top-end SCSI hard-disk system is generally faster than an equivalent top-end ATA system. Also a major advantage of SCSI is its ability to support a group of hard disks in an RAID array. Raid arrays are discussed in the network section (Section 4).

SCSI is a mature technology that has evolved steadily since the launch of its first standard (SCSI-1) in 1986. The early versions used an 8-bit data bus. Subsequent 'wide' and 'ultra' versions have evolved with 16-bit buses.

There are three different signal types used on SCSI systems, namely single ended (SE), high-voltage differential (HVD) and low-voltage differential (LVD). The older SCSI versions used either SE or HVD. The latest SCSI versions use LVD:

- In the SE system signal bits are send along a single wire of the ribbon cable with an adjacent earth return wire. Only the signal wire carries a voltage.
- LVD and HVD use differential signals. Each bit of the bus still uses two wires but instead of using one wire as an earth return it uses two signal wires per bit (i.e. both wires carry a voltage). The voltages are always of opposite sense in each wire (i.e. while one is +going the other is −going and vice versa). Any interference (noise) picked up on the pair of wires tends to be the same sense in both wires so it can be almost eliminated at the receiving end by simply subtracting the two signals. Lower noise means higher data rates and/or longer cables can be used.

HVD uses relatively high voltages for the data signals and therefore long cable lengths up to 25 m can be used. However modern computer systems and SCSI devices operate at lower-voltage levels so HVD has gone out of favour and LVD is now the dominant SCSI signal standard.

The SCSI bus with its many versions and manufacturer dubbed names has become rather confusing and SCSI devices from one manufacturer are not always compatible with those from another manufacturer. Table 2.17 lists some of the versions introduced since 1986 to the present.

Each device on the SCSI bus is identified by giving it a unique ID number. This is usually achieved by setting a jumper or group of jumpers. The narrow SCSI version controller cards (8-bit buses) can support up to eight devices with ID numbers 0 to 7. The wide (16-bit) controller cards support up to 16 devices with ID numbers 0 to 15.

**Table 2.17**   *SCSI versions (legacy to present day)*

| SCSI version | Bus width (bits) | Bus speed (MB/s) | Maximum bus length (m) | Signal type | Typical internal connector | Maximum number of devices |
|---|---|---|---|---|---|---|
| SCSI-1 | 8 | 5 | 6/25 | SE/HVD | 50-way IDC | 8 |
| Wide SCSI | 16 | 5 | 6/25 | SE/HVD | Mini-D68 | 16 |
| Fast SCSI | 8 | 10 | 3.0/25 | SE/HVD | 50-way IDC | 8 |
| Fast wide SCSI | 16 | 10 | 3.0/25 | SE/HVD | Mini-D68 | 16 |
| Ultra SCSI | 8 | 20 | 3.0/25 | SE/HVD | 50-way IDC | 8 |
| Wide ultra SCSI | 16 | 40 | 1.5/25 | SE/HVD | Mini-D68 | 16 |
| Ultra 2 SCSI | 8 | 40 | 12 | LVD | 50-way IDC | 8 |
| Wide ultra 2 | 16 | 80 | 12 | LVD | Mini-D68 | 16 |
| Ultra 160 SCSI | 16 | 160 | 12 | LVD | Mini-D68 | 16 |
| Ultra 320 SCSI | 16 | 320 | 12 | LVD | Mini-D68 | 16 |

The host adapter is normally assigned to ID 7. There is no set position for a particular device on the bus cable but in case two or more devices try to access the bus simultaneously, each ID is prioritized. The prioritization goes form 7 down to 0. In other words ID 7 has highest priority and ID 0 has the lowest priority. With a 16-bit SCSI bus the priority goes 7–1 then 15–8 (i.e. ID 7 has the highest priority and ID 8 the lowest priority).

Another important consideration when setting up devices on an SCSI bus is *termination*.

Each end of the bus must be terminated with a passive or active terminator. This is essential to stop signal reflections along the bus.

Without a terminator the signal travels along the wire pairs of the ribbon cable to the receiving end where only a small portion of the signal is absorbed. The remainder then reflects back along the cable and is partly absorbed at the sending end where it is re-reflected along the cable again. This process continues until the amplitude of the signal has decayed to zero. The reflections seriously impair the original signal and data corruption occurs.

The most common type of terminator is the passive variety, which takes the form of a series of electrical resistors. The active type is more expensive because it uses electronic voltage regulator circuits to maintain more consistent bus transmission characteristics.

### SCSI versus ATA hard disks

From the foregoing it is obvious that the SCSI bus is capable of high-data transfer rates exceeding that of the fastest ATA UDMA 133 rate. However the differences are not that pronounced. What makes SCSI the winner in the speed stakes is the extra refinement in the hard disk itself. SCSI drives are inherently more expensive than ATA types as they are used more in top-end systems and servers, where performance is more important than cost. Extra performance enhancing features are added which are not normally included in an equivalent ATA drive. Typical enhanced features include: high spindle speeds, large disk caches, short access times and high mean time between failure rates.

The parameters for a typical SCSI and EIDE hard disk are listed in Table 2.18.

From Table 2.18 you can see that the overall performance of the SCSI drives (ignoring reliability factors) is determined more by the spindle speed than by the fact that the external interface is SCSI rather than ATA.

**Table 2.18** *Performance comparison between SCSI and ATA*

| Parameters | SCSI model Seagate Cheetah 15K.3 | SCSI model Seagate Barracuda 36ES2 | ATA model Seagate Barracuda 7200.7 |
|---|---|---|---|
| Interface type | Ultra 160 SCSI | Ultra 160 SCSI | Ultra ATA/100 |
| Transfer rate (MB/s) | 57–86 | 41 | >58 |
| Average seek read/write (ms) | 3.6/4.0 | 7/8 | 8.5 |
| Track-to-track read/write (ms) | 0.2/0.4 | – | – |
| Disk cache memory (MB) | 4 | 2 | 2 |
| Average latency (ms) | 2.0 | 4.17 | 4.16 |
| Disk spindle speed (rpm) | 15,000 | 7,200 | 7,200 |

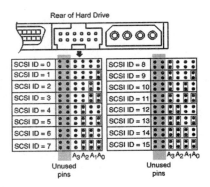

**Figure 2.72**  *ID settings for a typical SCSI hard drive.*

### Setting the ID on an SCSI drive

The ID number of the drive is set by a group of jumpers. On the old SCSI 8-bit drives three jumpers are used in a binary pattern representing ID 0–7, with no jumpers inserted for ID 0 and three jumpers in place for ID 7 (i.e. 000, 001, 010, 011, 100, 101, 110, 111). The latest drives usually have four jumpers providing IDs 0–15. The default setting is usually ID 0. These settings can also be made remotely via a suitable connector and lead assembly and some drives allow the ID to be set via software (Figure 2.72).

## SATA

In 1999 an association of seven PC manufacturers set out to define a new serial interface standard for ATA hard drives and ATAPI drives. The new standard is called Serial ATA (SATA).

The SATA standard has several advantages over the existing parallel ATA standard. The main ones are as follows.

Ease of installation and configuration, as the system utilizes PnP connectivity. No master, slave jumper settings are required.

Neater cabling. The cables are far thinner than the wide ribbon cables used on traditional ATA drives. This not only looks neater inside the system-unit case but it also helps to increases the air flow across the drives.

Architecture is scalable to higher-performance levels to accommodate future advances in hard drive technology. First-generation SATA is the currently available system. It supports data transfer rates of up to 150 MB/s. Second-generation SATA will support data transfer rates of up to 300 MB/s. Third-generation SATA will support data transfer rates of up to 600 MB/s.

The future SATA generations will only really make a noticeable different to hard-disk data transfer rates when a major advancement in the internal disk-to-head transfer rate is achieved.

## Preparing a hard disk

A hard disk must be *formatted* before it can be used to store data. This process is carried out in two stages, with an intermediate stage known as *partitioning*. The two formatting stages are known as *low-level formatting* and *high-level formatting*, respectively. In contrast, when a floppy disk is formatted, both a low- and a high-level format is carried out in one process.

### Low-level formatting (physical formatting)

Low-level formatting is a *physical* process where tracks are laid down on the disk's surface and divided into sectors. During this process, a header and trailer list is recorded on each sector along with a block of dummy data. The inter-track and sector gap positions are also recorded. Today, all hard disks are low-level formatted during manufacture and do not normally need to be low-level formatted again over the life of the drive. However in some instances, a faulty disk that cannot be corrected by high-level diagnostics tools, can sometimes be resurrected by carrying out a low-level format. Nowadays however, it is usually more sensible to fit a new drive.

Some disk drive manufacturers state that their IDE drives should never be low-level formatted. This brought out the myth, much talked about in the PC servicing fraternity, that an IDE drive will be destroyed if it is low-level formatted. With the correct software, low-level formatting can be done, but only if it is absolutely necessary.

Several hard-disk manufacturers provide free disk diagnostics programs and some include low-level formatting options. These can usually be obtained directly from the manufacturer's web site.

### Partitioning

The vast storage area of a hard disk can be utilized as one single area or it can be treated as several smaller areas. These are called partitions. This allows more than one OS to used on the same hard disk. It also allows the disk to be split into several, smaller and therefore more manageable logical drives on one OS.

For example:

1. A PC system could run Linux and Windows XP – two completely different OSs, each on a separate partition.
2. A large capacity hard disk running a particular OS could be split into two or more, logical drives (e.g. C:, D:, E:, etc.)

A single partition may take up all of the available disk space but it can be advantageous to break up a large hard disk into several smaller partitions, as this can ultimately save disk space. This is particularly the case with older systems with smaller capacity hard drives running MS-DOS, Windows 3.1 and Windows 95 release 1. This is due to the way the disk filing system is organized under MS-DOS and early versions of Windows. These OSs use the old *file allocation table* (FAT) system.

The MS-DOS and Win 3.1, FAT supports short, 8-character file names and the Windows 95 FAT variant, known as *VFAT*, supports 256 character file names. The first release of Windows 95, known as Service Release 1 (SR1), had a 16-bit FAT (FAT16). Later versions of Windows 95 (SR2) and Windows 98 support both a 32-bit FAT (FAT32) and FAT16.

The FAT, whatever its pedigree, organizes the disk space into *allocation units* (*allocs*). An allocation unit – formerly known as a cluster – is a group of one or more sectors. The larger the partition size of the disk, the greater the size of an allocation unit, up to a maximum of 64 sectors or 32 kB.

When a file is saved to disk, MS-DOS handles the data in whole allocation units only. It cannot partially use an alloc and keep the rest for another file. So on a large hard disk, with 32 kB allocs, there is a maximum amount of wasted space. For example, if a 48 kB file is saved to the hard disk it will use two allocation units and waste 16 kB. Extrapolate this to a drive containing several thousand files of different sizes, and tens of megabytes of drive space could be wasted. Breaking up the drive into several partitions can save therefore save disk space.

Windows NT, 2000 and XP use an advanced disk filing system called the *new technology file system* (NTFS). NTFS is not hampered by the need to make the system backwards compatible to the older DOS/Windows systems so the aforementioned problems do not arise.

### Partition size limit

The maximum number of allocation units that can be assigned in the FAT and VFAT (FAT16) systems is 65536 and the maximum size of an allocation unit is 32 kB. This unfortunately limits the maximum size of a single partition to $65536 \times 32 \text{ kB} = 2.1 \text{ GB}_d$. So a large hard disk greater than $2.1 \text{ GB}_d$, running DOS or Windows 95 SR1 will have to be split into two or more partitions.

This problem has been overcome in Windows 98 and later versions of Windows 95 (known as OSR2). Both have an updated FAT available, called FAT32. This supports partition sizes up to 2 terabytes (TB) and uses a smaller maximum allocation unit size – so it wastes less space. FAT32 allows you to partition a large hard-disk drive as one single partition. For example an 80 GB drive can be partitioned as one single 80 GB partition or several smaller partitions, each one allocated to a different logical drive.

### MS-DOS/BIOS capacity limit of 528 MB$_d$

The partition size limit was not the only hard-disk size limiting factor. At one time the maximum drive capacity for a PC was a mere 32 MB. This limitation was imposed by all DOS versions up to and including MS-DOS 3.3. Later versions of DOS have an increased address limit of 2 TB. In practice however this was limited to $528 \text{ MB}_d$ by the combined way the heads, cylinders and sectors were addressed by successive data-handling subsystems. In the AT DOS-based system, hard-disk data is handled in a four-layer regime but with no apparent standardization between each layer. The four layers are as follows:

- disk drive,
- controller/interface,
- BIOS,
- DOS.

Each layer imposes a different limit to the maximum number of cylinders, heads, tracks and sectors that can be handled. The overall effect was a considerably reduced maximum hard-disk size. The limits of each layer will now be examined briefly.

## Disk drive (the physical track, sector, head limit )

The data on the disk is divided into tracks and sectors with each sector holding 512 bytes in the same fashion as a floppy-disk drive. However a hard disk usually has several heads and considerably more tracks and sectors than a floppy-disk drive. Also the number of sectors-per-track varies depending on the position of the track. Tracks on the outer edge of the disk are designed to hold more sectors than the inner tracks – this is known as zone recording. As we have seen on the previous pages, an ATA-2/3 drive can have a maximum of 255 sectors-per-track, 16 heads and 65536 cylinders. This gives a maximum possible capacity of 136.9 $GB_d$.

## Controller/interface

Both the ATA-2/3 and SCSI interfaces look at the drive in terms of the sector numbering rather than cylinders heads and sectors (CHS) so the interface imposes no constraint on capacity.

## BIOS limitations

The BIOS provides disk services in the form of ROM-based ISRs. These routines are called via software interrupt 13 (Int 13). Originally intended for floppy-disk service calls, the service routines were modified in later DOS versions to accommodate hard drives. When Int 13 is called, the cylinder number, sector number and head number are stored in the CPUs registers using a set number of bits.

Ten bits were used for the number of cylinders, 6 bits for the number of sectors and 8 bits for the number of heads. This gave $2^{10}$ cylinders, $2^6 - 1$ sectors, and $2^8$ heads which is 1024 cylinders $\times$ 63 sectors $\times$ 256 heads. As each sector holds 512 bytes, this results in a maximum drive size of 8.4 $GB_d$. Unfortunately modern high-capacity drives have far more than 63 sectors and 1024 cylinders but physically they cannot have 256 heads! This causes a few problems particularly as the ATA specification limits the maximum number of physical heads to 16 – quite sensibly as even this number would seriously stretch the capability of the drive mechanics.

## DOS limitations

As mentioned previously, early versions of DOS also imposed a limit to the maximum hard-disk capacity. The hard disk contains a master boot record (MBR), located on cylinder 0, head 0, sector 1. It contains information on the disk's geometry as well as a small boot program. Unfortunately older versions of DOS only allocated 16 bits to hold the total number of sectors for each drive partition. This limits the maximum partition size to 32 $MB_d$ (65535 sectors $\times$ 512 bytes per sector). DOS 5 and higher versions allocated 32 bits to hold the total number of sectors. Increasing the maximum possible drive size DOS can recognize to 2 TB! (Table 2.19).

So a PC system running on DOS 5 or higher has a hard-disk capacity limit imposed by the combined BIOS and ATA limits. The BIOS is the main culprit however, as the Int 13 service routine limits the maximum number of cylinders and sectors. Table2.19 shows the combined effect of the BIOS and ATA limitations.

To overcome the 528 MB limit, most PC motherboards manufactured since the last quarter of 1994, have an enhanced BIOS that translates the drives physical CHS parameters into logical ones. It does this

**Table 2.19** *Factors that limited the maximum partition size*

| CHS parameters | BIOS limit (Int 13) | ATA limit | Combined limit |
|---|---|---|---|
| Maximum number of cylinders | 1024 (0–1023) | 65536 (0–65535) | 1024 |
| Maximum number of heads | 256 (0–255) | 16 (0–15) | 16 |
| Maximum sectors-per-track | 63 (1–63) | 255 (1–255) | 63 |
| Maximum capacity | $8.4\,GB_d$ | $136.9\,GB_d$ | $528\,MB_d$ |

by reducing the drive's physical cylinders by '$n$' powers of two and increasing the number of heads by the same factor.

For example, a drive with a capacity of $2.1\,GB_d$, with actual physical CHS parameters of 8139 cylinders, 8 heads and 63 sectors, would be translated by the BIOS, to 1017 cylinders, 64 heads, and 63 sectors (i.e. in this example the factor '$n$' is 3).

Some older BIOS ROMs can be upgraded to accommodate drives larger than $528\,MB_d$. Alternatively a software solution such as Western Digital's 'EZ-drive' or ONTRACK's 'Disk Manager', can be used.

### Creating partitions

When installing Windows 95/98/2000/XP on new hard disk, partitioning and formatting takes place automatically. The unpartitioned disk is detected by the installation program and partitioning proceeds after prompting the user to choose what disk-filing system to use and the portion of the disk capacity to use for the partition. To manually partition a hard disk, using MS-DOS through to Windows 98/ME, the command FDISK must be used. The *FDISK* menu system has remained almost unchanged since the early versions of MS-DOS right up to the Windows 98 and ME versions.

In Windows 2000 and XP the FDISK command is no longer used. If you want to partition a second hard disk from Windows 2000/XP you use the *Disk Management* utility. To use Disk Management, click *Start*, and then click *Control Panel*. Double click *Administrative Tools*, and then double click *Computer Management* and finally double click *Disk Management*.

In Windows 95/98/ME your start-up floppy disk contains the DISK creation commands including FDISK and FORMAT. If you have mislaid your Start-up disk you can create a new one by launching 'Add/Remove Programs' from Control Panel and selecting the Startup Disk tab.

If you want to look at your partition information in Windows 2000/XP use *Disk Management*. Click *Start*, and then click *Control Panel*. Double click *Administrative Tools*, and then double click *Computer Management* and finally double click *Disk Management* (Figure 2.73).

During partitioning, the OS software writes a boot program and partition table to cylinder 0, sector 1 of the disk. This is known as the

**Warning!**
When a hard disk is partitioned and/or formatted, all data on the disk is destroyed, so only use FDISK on a working PC, when you want to clear everything off the hard disk and start from scratch. However if you want to examine your existing partition information you can do this safely in Windows 95/98 by using the command FDISK STATUS.

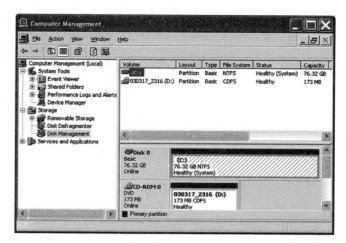

**Figure 2.73**   *Windows 2000/XP Disk Management window*

'*MBR*'. Under MS-DOS and Windows 95/98/2000, the following types of DOS partition can be created on the hard disk:

- primary partition,
- extended partition.

The *primary* partition is the only partition needed to run a single OS. In fact the majority of PCs with a single hard disk are only set up with one partition, allocated as drive C:. This partition is also called the boot partition as it holds the all important bootable files. In MS-DOS and versions of Windows up to ME, the boot partition contains three vital boot files, namely MS-DOS.SYS, IO.SYS and COMMAND.COM. MS-DOS.SYS and IO.SYS are crucial for boot up and so they are hidden from normal directory viewing and set as read only files. This is done to prevent their accidental erasure.

In Windows XP, MS-DOS.SYS, and IO.SYS are present for downward compatibility but the boot partition also includes the Windows 2000 hidden boot files *ntdetect.com*, *ntldr* and *boot.ini*.

An *extended* partition is an optional drive partition that can be used to store any files and compatible software but system boot files will not be recognized (i.e. you cannot boot from an extended partition. In MS-DOS and Windows 95/98 using FAT 16, drives greater than 2.1 GB must be divided into primary and extended partitions, with each one allocated as a logical drive using letters C–Z.

Alternative OSs such as UNIX, and OS2 can also be placed on the hard disk in addition to DOS and Windows, but they must reside in a *non-DOS partition* of the disk.

### High-level format (logical formatting)

After partitioning, it is necessary to high-level format each partition using the OS's FORMAT command. High-level formatting creates a file management structure on the disk. When a new system is being configured – after booting from the OS installation disk – the hard disk is automatically partitioned and high-level formatted. You virtually do nothing!

To manually high-level format a hard disk using versions of Windows up to and including ME, you use the 'FORMAT' command is used. To format the boot partition, you place your MS-DOS boot disk or Win 95/98/ME startup disk, containing the FORMAT.EXE file, into the A: drive and boot up the system. Then enter the command:

FORMAT C:/s

This command formats the primary partition and copies the hidden system files IO.SYS and MS-DOS.SYS and COMMAND.COM from the disk in the A: drive.

Without the '/s' extension a straight format is carried out without transferring the system boot files. This is used to format any extended non-bootable partitions, for example:

FORMAT D:

This command formats the logical drive D: without transferring the system files.

## Hard-disk drive connectors

### ATA (IDE) controller and connections

An ATA or integrated drive electronics (IDE) hard-disk drive has the controller electronics on-board, thus forming one integral unit.

So the motherboard–drive interface only requires a basic I/O port. A single ribbon cable connects the drive to the IDE interface. When building a new PC the ribbon cable is usually supplied with the motherboard.

Today's UDMA 66, 100 and 133 hard-disk drives are designed to be used on special 80-wire ribbon cables (Figure 2.74). These incorporate an extra 40 ground wires between the signal wires. These extra wires act as a screen which greatly improves the signal transmission characteristics of the cable to accommodate the high-data transfer rates of these standards.

These ribbon cables also incorporate the Cable Select (CSEL) system. This allows the drives to be automatically set as master and slave depending on their position on the cable. To implement this feature the drive jumpers must both be set to CSEL. This is an extra jumper in addition to the master and slave jumpers mentioned in the set book. The drive on the far end of the cable is automatically assigned as *master* and the drive on the intermediate connector is automatically assigned as *slave*. In a single drive system with the CSEL jumper in place, the drive must always sit on the far end connector (Figure 2.75).

Master          Slave                          Motherboard
                                               connector

**Figure 2.74**   *A modern UDMA66/100 cable*

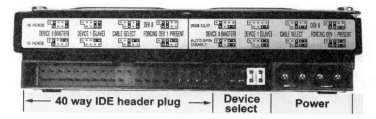

| 16 HEADS | | DEV 0 | | 2GB CLIP | | DEV 0 | |
| DEVICE 0 (MASTER) | DEVICE 1 (SLAVE) | CABLE SELECT | FORCING DEV 1 PRESENT | DEVICE 0 (MASTER) | DEVICE 1 (SLAVE) | CABLE SELECT | FORCING DEV 1 PRESENT |
| 15 HEADS | | | | AUTO SPIN DISABLE | | | |

| ←——— 40 way IDE header plug ———→ | Device select | Power |

**Figure 2.75** *Rear view showing the signal and power connectors*

**Figure 2.76** *A typical SCSI controller card*

## SCSI controller and connections

An SCSI hard disk has the disk controller and I/O interface built into the drive itself in a similar manner to ATA drives but unlike the latter an SCSI controller board is required. The SCSI controller board, or '*host adapter*' as it is sometimes called, is the heart of the SCSI bus system. It has its own processor to generate the SCSI bus control signals and manage data transfer along the bus. The controller board also incorporates a BIOS ROM to augment the main system ROM, allowing the OS to communicate with the SCSI controller hardware (Figure 2.76).

There are three main types of SCSI cables: the standard SCSI 50-way cable (known as the 'A-cable'), the 68-way 'P-cable' (used in 16- and 32-bit wide SCSI) and the 68-way 'Q-cable' (used on 32-bit wide SCSI). The A-cable was the most widely used type up to a few years ago but current SCSI cards and peripherals use the mini-DB 68-way cable and connectors.

## SATA connections

SATA drives have a very neat signal cable system based on a flat 8 mm wide sheathed cable (Figure 2.77). The connectors are identical at each end of the cable and contain seven gold-plated contacts.

Unlike the usual disk drive power connector, an SATA power connector uses a series of small contacts. This spreads the load current over several contacts rather than a few large contacts and this reduces the insertion and extraction force.

**Figure 2.77** *The neat SATA drive connections.*

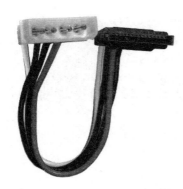

**Figure 2.78** *An SATA power cable adapter lead.*

As the existing range of PC PSUs only support the traditional disk drive power connectors a special adapter lead must be used (Figure 2.78).

## Hard-disk terminology

**Actuator**    The mechanism used to move and position the heads across the disk surfaces. Most modern hard-disk drives use a speech-coil actuator, older drives used a stepper motor.

**Areal density**    The amount of data that is stored on a disk per square inch. This is calculated by multiplying the number of bits per inch along the track, by the number of tracks per inch.

**Buffer**    Most drives have a small amount of RAM on-board, known as a buffer (between 128 kB and 1 MB) to temporarily store data to and from the drive. The buffer behaves in a similar way to the fuel tank in a car. The car engine consumes the fuel from the tank at a steady rate even though the tank is only filled periodically. In a similar way the computer can quickly fill the buffer up with data and then carry on with something else while the hard disk empties the buffer. The trouble is, this only works for small bursts of data. If continuous data transfer takes place, the buffer will soon fill to the brim and the computer will have to wait for it to empty before it can feed in more data, thus using up valuable processor time.

**Capacity**    The storage capacity of a hard disk is expressed in megabytes or gigabytes but the way it is defined causes some confusion. Instead of the normal decimal definition of kilo, mega and giga (i.e 1000, 1,000,000 and 1,000,000,000). In computer systems these quantities are defined using the nearest power to the base two (i.e. a kilobyte is defined as $2^{10} = 1024$ bytes, megabyte is $2^{20} = 1,048,576$ bytes and gigabyte is $2^{30}$ bytes $= 1,073,741,824$ bytes. This is also the way memory capacity is defined.

**CSEL** (Cable select)    An alternative method of setting the master and slave drive. Instead of using the master and slave jumpers, both drives jumpers are set to CSEL. The master and slave drive is then determined automatically by the drive's position on a special ribbon cable (i.e. when the CSEL signal is grounded the drive is configured as master and ungrounded it is configured as slave).

**Disk speed**    The rotational speed of the platters in revolutions per minute. This varies from 3600 rpm to over twice this speed.

**DMA** (Direct access memory)    An alternative protocol to PIO modes. ATA-2, ATA-3, ATA/ATAPI-4/5/6 defines the data rate for DMA transfer between the drive and the computer I/O interface. The latest UDMA modes are UDMA 66 and UDMA 100.

**Latency**    The average time in milliseconds that it takes for a particular sector to appear in front of the head after the head has reached the correct cylinder. It works out as the time the disk takes to do half a revolution.

**LBA** (Logical block addressong)    This is a method of uniformly addressing disk sectors, for example LBA0 is sector 1 of head 0, cylinder 0 and LBA1 is sector 2 of head 0, cylinder 0 and so on in a linear fashion. Many AT BIOS programs prior to 1996 address the hard disk using the CHS number.

**Mode 3 and 4 PIO**    The PIO modes are a range of protocols for data transfer between the drive and controller using programmed I/O rather then DMA, and are defined by the ATA-2 specification as 11.1 MB/s for mode 3 and 16.6 MB/s for mode 4.

**MIG head** (Metal-in-gap head)   The MIG Head is an improvement on the ferrite head, where metal alloys of aluminium, iron and silicon are deposited in the magnetic gap of the head. This substantially increases the magnetic field strength in the gap, thus allowing higher bit densities to be achieved.

**MR head** (Magneto-resistive head)   Pioneered by IBM in 1991. The MR head has a separate read and write head layered together. The write head is a conventional thin film inductive (TFI) head and the read head is based on an alloy film of nickel–iron that exhibits a change in resistance in the presence of a magnetic field. The major advantages of this type of head, compared to conventional heads are higher production yields, and greater sensitivity, allowing high-areal densities beyond 1 Gb/in.$^2$.

**Seek time**   The time taken in milliseconds, to move the heads to a required cylinder. Drive manufacturers often quote seek times as the average seek time (usually defined as the time it takes to move the head over one-third of the total number of cylinders). Track-to-track time (the time it takes to move from one track to an adjacent track) and full-stroke time (the time it takes to move the head from the innermost to the outermost track).

**Sustained transfer rate**   The read/write speed in bytes per second that the disk can maintain continuously.

**TFI head** (Thin film inductive head)   TFI read/write heads are made by depositing a tiny coil at the end of the head arm using photolithography–a process similar to that used to make integrated circuits.

**Zone recording**   A method of placing more sectors on the outer tracks where the tracks are longer and less on the inner tracks where the tracks are shorter. Older drives use the same no. of sectors for all tracks.

# Compact disc drives

The compact disc (CD) was launched in 1980 by the inventors, Philips and Sony, as a revolutionary new digital playback system for music. Named the 'red-book' standard, it soon became the most popular format for audio reproduction, to the demise of the vinyl record industry. Being digital, the CD system is capable of producing a superior, frequency response, signal-to-noise ratio and lower distortion, than vinyl records. and to date, over 3 billion CDs have been sold in the USA alone.

The data is stored as a series of microscopic bumps on the inner layer of the disc and these are read by a micro-fine laser beam, which is less than 2 μm diameter at the focus point. This tiny scale, allows up to 70 min of high-quality audio to be stored on one side of the disc – equivalent to both sides of an LP record, only much higher quality. As there is no physical contact during playback – unlike a vinyl record – CDs would have an almost infinite lifetime if we ignore bad handling practices.

As the CD market grew in size, the cost of the technology fell and this made it an attractive medium for computer storage. Realizing the high potential for this market, Philips and Sony announced a new specification in 1985, known as the 'yellow-book' standard. It sets out a method for the transfer of digital information from a CD to a computer. CDs conforming to this standard can be read, but new data cannot be stored on them, which coined the name CD-ROM.

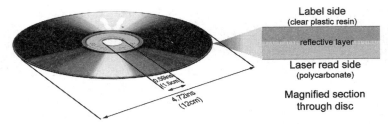

**Figure 2.79** *CD-ROM with edge view showing the layered structure*

Its lightweight, low-cost and high-storage capacity – over 650 MB – has made the CD-ROM the most widely used format for the distribution of computer software, from multimedia games to application programs. One CD has the equivalent storage capacity of over 400 high-density 3.5-in. floppy disks!

The CD-ROM has spawned a whole new range of standards for multimedia, including photo-storage, read and write storage and high-quality video, each being given a new 'book colour'. The most important for the computer world will be discussed later in this section.

## CD-ROM

A CD-ROM is a round, 12 cm diameter, plastic disc, with a 15 mm diameter centre hole. It is constructed in three layers and not from one piece of plastic, as one might have first thought. The base layer is injection moulded out of clear polycarbonate and a series of pits and lands are pressed onto one side. These hold the digital information. Next a thin layer of aluminium is vacuum deposited over the pits and lands, to reflect the laser beam. Finally, a coating of plastic lacquer is applied, which protects the delicate data layer from dust and scratches. This forms the label side of the disc (Figure 2.79).

Data is stored on the CD-ROM as a series of minute pits and lands on the inside surface of the polycarbonate substrate. The pits are 0.6 μm wide and less than 0.13 μm deep and vary in length from 0.9 to 3.3 μm (1 μm is one-millionth of a metre) The pits are recorded on the master disc in one continuous spiral, from the centre of the disc to the outer edge. Incidentally, if the spiral could be unwound, it would extend to over 3.5 km! The average distance between the adjacent tracks of the spiral is 1.6 μm. This minute spacing diffracts light into its component colours, which results in the colourful rainbow effect, when the disc surface is viewed in white light (Figure 2.80).

The data is read from the base layer by a pinpoint beam from an infrared laser diode and lens assembly. The beam is focused onto the top inside edge of the aluminium layer where it is reflected back into the lens. When a land moves in front of the beam it is scattered, and less light is reflected back into the lens (Figure 2.81).

As the disc rotates, the laser beam continually meets pits and land areas in the reflective layer. A land scatters the beam, and a pit reflects the beam, thus producing a binary signal. The reflected beam is picked up by a photo-diode in the lens assembly, and converted to logic '1's and logic '0's. It is interesting to note that a small scratch or fingerprint on the read surface of the disc is completely out of the focus range of the laser beam so it cannot corrupt the data. Deeper scratches will however effect the read operation.

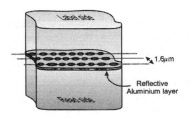

**Figure 2.80** *A highly magnified cross section through a CD-ROM disc*

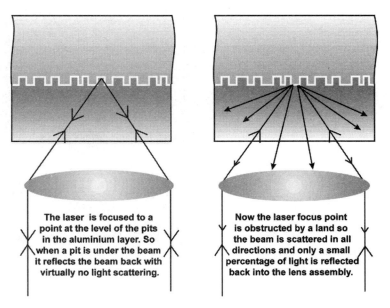

**Figure 2.81** *How a laser reads the absence or presence of a land in the reflective layer*

During the manufacturing process, a single speck of dust could destroy literally hundreds of bytes of information, so clean-room conditions are used throughout. Once a flaw free set of master, mothers and sons is produced, thousands of discs can be produced quickly at very low cost, typically less than 30 pence each in thousand-up quantities.

## CD-ROM drive

A very high-precision servo system is required to keep the laser beam focused to a spot on the land area, and to keep it centred on each track. The beam must be less than 1.7 μm at the tip of the focus point. In addition the track must be read at a constant linear velocity. This means that the disc must be made to spin faster when the head is at the centre of the disc and more slowly when the head is near the disc edge.

In spite of the high precision required, the actual construction of a CD-ROM drive is deceivingly simple. The elaborate laser optical system is contained in a single unit known as the optical block and this runs along a rack and pinion mechanism, in a similar manner to the head of a floppy drive.

The laser beam is automatically focused and positioned over the track, by moving the lens up and down perpendicularly and tangentially to the tracks on the disc, under the control of a servomechanism. The lens is suspended in the optical block and attached to two sets of coils. One pair of coils allows fine adjustment of the focus and the other pair adjusts the tracking. Each set of coils is centred in the field of a permanent magnet. By applying a control current to the coils, the lens can position and focus the laser beam to a high degree of accuracy over the track. A feedback signal from a photo-diode in the lens assembly ensures that the laser remains focused on the correct part of the track. See Figure 2.82.

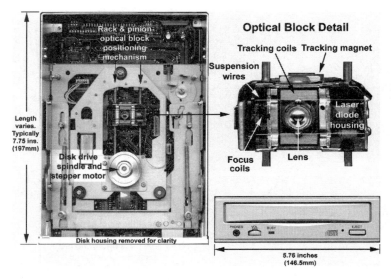

**Figure 2.82** *Anatomy of a CD-ROM drive*

The CD-ROM drive contains the drive mechanism and electronics control board in one neat unit, the same size as a 5.25-in. floppy drive. This allows it to fit into a spare drive slot on the front of the PC case.

The earlier CD-ROM drives spin the disc at a constant linear velocity between 200 and 530 rpm and achieve a data transfer rate of 150 kB/s. This is the normal 1X speed for red-book audio CD-ROMs. Modern drives can offer faster disc speeds at multiples of the original speed. Drives are available with disc speeds from 1 to 40X, corresponding to data transfer rates from 150 to 6000 kB/s.

The speed of rotation of the disc is one thing, but an '$n$'X CD-ROM does not mean that data will be read '$n$' times as quickly as a single-speed disc, it depends on the access time as well as the data transfer rate. The average random access time is measured by gauging the average time it takes to read data blocks across the whole area of the disc. This is performed several thousand times to arrive at a figure. For example, at 1X the average access time of a drive might be 320 ms, and at 12X it may drop to 110 ms – the access time is barely 3 times as fast, not 12 times as one might expect.

The faster disc speed makes a big difference when accessing large amounts of contiguous data such as video or large graphics files. Once the disc is set in motion, it can continue at high speed until all the data is read. There is only one spin up time interval (the focus settling time and motor start up time) and minimal seek time.

### Interfacing a CD-ROM

CD-ROM drives, like hard drives, can interface to either IDE or SCSI ports. The most popular on the PC is the IDE variety. The ATA-2 IDE standard and beyond, support a special CD-ROM specification known as *ATAPI*, which allows CD-ROM drives to connect to any PC fitted with an ATA interface. The ATAPI specification is based on a set of software drivers that essentially convert SCSI commands to ATA commands. The drivers are usually supplied with the CD-ROM drive. Modern motherboards have a generic CD-ROM driver built into the BIOS.

**Table 2.20** *Master/Slave Cabling arrangements*

| Primary cable | | Secondary cable | | Set up |
| --- | --- | --- | --- | --- |
| Master | Slave | Master | Slave | |
| CD-ROM 1 | | | | A CD-ROM drive on one cable |
| ATA hard-disk drive | CD-ROM 1 (slave) | | | A hard drive and a CD-ROM on one cable |
| CD-ROM 1 (master) | CD-ROM ) 2 (slave) | | | Two CD-ROM drives on one cable |
| ATA hard-disk drive | ATA hard-disk drive | | | Two hard drives on one cable |
| ATA hard-disk drive | | CD-ROM 1 (master) | | A hard drive on one cable and a CD-ROM drive on the second cable |
| ATA hard-disk drive | | ATA hard-disk drive | CD-ROM 1 (slave) | A hard drive on one cable and another hard drive plus a CD-ROM drive on a second cable |
| ATA hard-disk drive | ATA hard-disk drive | ATA hard-disk drive | CD-ROM 1 (slave) | Two hard drives on one cable and one hard drive and a CD-ROM drive on a second cable |

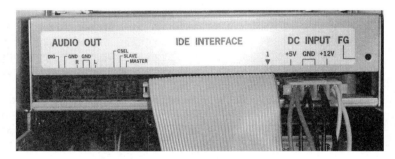

**Figure 2.83** *The rear view of an ATAPI CD-ROM*

Using an IDE CD-ROM drive on an IDE ATA port is straightforward, but it must be set as master or slave in the same way as an ATA hard disk. For example if the host computer has a single IDE cable or both a primary and secondary IDE cable, then the master/slave jumper must be set appropriately. See Table 2.20.

An ATAPI compatible CD-ROM drive has the same 40-way ribbon connector and power connector as a hard disk. Unfortunately obsolete single and double-speed CD-ROMs with non-standard interfaces also use 40-way cables. You can recognize an ATAPI CD-ROM drive from an older type by the master/slave and C/Select jumpers. The obsolete, manufacturer specific variety, uses a pseudo SCSI interface and has a set of jumpers labelled ID 0 1 2 3 (Figure 2.83).

Unit 3 contains a more detailed description on the installation of a CD-ROM drive.

## How data is recorded to CD

When data in the form of 8-bit binary numbers (bytes) is recorded it first has to be translated to a form understood by the CD-ROM system. Due to the way the pits and lands are formed, it is not possible with the existing technology, to have consecutive '1's recorded next to each other, as these examples have, 01100111, 11110001. To overcome this problem, the 8-bit data is recorded using 14 bits. The 256 binary numbers possible in 8 bits are converted to the same number of 14-bit numbers, using a ROM look-up table. With 14 bits, the numbers can be encoded to eliminate consecutive '1's. For example the binary number three (00000011) is converted to 10001000100000 and seven (00000111) is converted to 00100100000000. In fact, all 256 8-bit numbers from 0 to 255, are converted to a seemingly unrelated 14-bit number. Another problem exists when two 14-bit codes one ending in '1' and one starting with '1' are placed side by side. This results in two consecutive '1's. To overcome this, three so-called 'merging' bits are placed between the 14-bit numbers, so in total, 17 bits are used for each byte of input data.

### *Error detection and correction*
As if this was not enough, data errors can occur during playback due to dust and scratches and to imperfections in the CD-ROM itself . So the CD-ROM system must have some means of reducing errors. In the red-book audio standard, some minor errors are allowed for, as the most these can do is cause a faint click in the sound during playback. With the yellow and orange book CD-ROM standards however, a single error bit would most likely cause a program to crash or seriously corrupt data. To correct for errors, a much more elaborate error detection and correction scheme is called for.

The most simple error correcting scheme used on red-book discs is known as the 'Cross-Interleaved Reed Solomon Code' (CIRC). CIRC can reduce the error rate, on average, to one byte error for every gigabyte of data. This is okay for music but not good enough for computer data. For CD-ROMs another error scheme called 'EDC/ECC' is also used. This system is capable of reducing error rates to 1 bit in 1 TB, equivalent to 1-bit error in 2000 CDs! A terabyte is a 1024 GB!

The data on a CD-ROM is organized in a similar way to that of a floppy or hard disk, using tracks and sectors. The track on a CD-ROM is a continuous spiral, rather than the series of concentric magnetic tracks used on a hard disk. However the word 'track' is usually used to describe a portion of the spiral.

## CD-ROMS and the PC
When a CD-ROM is installed in a PC system, the OS must be able to recognize the data stored on the disc and treat it like another logical drive. For example the disc could be installed as the 'D:' drive. To achieve this, a special extension to the OS has to be loaded. In MS-DOS and Windows 95 OSs, the extension file name is MSCDEX.EXE. As mentioned previously, modern motherboards have a generic CD-ROM driver built into the BIOS, this is extremely useful as it allows a new OS to booted and installed directly from CD-ROM.

MSCDEX.EXE is a set of software extension routines, provided by Microsoft for the MS-DOS OS and Windows, which can read CD-ROMs conforming to the High Sierra ISO-9660 format. To set up a CD-ROM on a PC, DOS 3.1 or higher or Windows 95 is needed and a hardware-dependent device driver is required. The device driver is a piece of software normally provided by the OEM. In this example we will call the OEM driver 'MSCD0001'.

MSCDEX is installed from AUTOEXEC.BAT and MSCD0001 is installed from CONFIG.SYS. MSCD001 then uses the basic functions provided by MSCDEX to read data from the CD-ROM.

When the system first boots, MS-DOS loads the OEM device driver MSCD0001, from CONFIG.SYS. Then MSCDEX.EXE is invoked from AUTOEXEC.BAT. It immediately performs a system call to the OEM driver to locate the start addresses of the various driver routines. If it finds them, it reports a successful installation on the boot up screen and the CD-ROM is installed as a valid drive.

## CD-recordable

Recordable CD technology is specified in the 'orange book' standard. Part 1, specifies the format for *read and write* magneto/optical disks and Part 2 specifies a format for a CD *write once, read many* (CD-WO) system, commonly called CD-recordable (CD-R) .

CD-R is a natural extension of CD-ROM technology, allowing the user to record massive amounts of data onto a CD at low cost. CD-R can be used to make hard-disk backups or to create a pre-master disc for the production of high-volume CD runs, using conventional disc pressing methods. The CD-R system uses a special record/playback disc drive, and recordable discs known as 'gold' discs– they have a thin layer of gold, instead of aluminium, to reflect the laser beam. The dimensions of a gold disc are identical to a normal CD. The structure is however rather different, with a polycarbonate base layer, containing an unmodulated spiral track instead of the pits and lands. On top of the base layer is a thin photosensitive film of cyanine or phthalocyanine, a green, dye-like substance. This is then coated in a reflective layer of gold. Gold is used rather than aluminium, because it is relatively inert and does not combine chemically with the cyanine compounds. A protective layer of plastic lacquer then completes the structure.

A CD-R drive is almost identical to a normal CD-drive except for the laser diode which is much more powerful than a normal device. As the disc is recorded, the laser follows the preformed track and selectively melts 'pits' into the dye layer. Where the laser point is focused, it melts the dye and surrounding plastic slightly, causing it to become opaque rather than translucent. When the disc is read, the laser beam is diffracted by the opaque areas and reflected by the untouched parts of the track.

The data longevity of recorded CD-R discs is often called into question, and one often hears rumours of discs that loose their data when exposed to sunlight. A lot of these rumours are unfounded as the cyanine dyes are chemically adjusted to be sensitive to only narrow wavelengths of light, in the infrared region. Prolonged contact with hot surfaces could damage the data however. When not in use, it is wise to store them away from heat. CD-ROM manufacturers state a lifetime in excess of 70 years for correctly stored discs.

The price of a CD-R drive has fallen drastically over the last decade from several thousand pounds each, in the early 1990s, to a couple of hundred pounds today. The media has also fallen in price. 650 MB CD-R gold discs, are now available for less than £2.00 each. This represents a revolutionary storage medium for low cost, data backup at less than a penny a megabyte. To reduce manufacturing costs and offer an alternative to the gold coloured discs, silver coloured CD-Rs have now been developed. These cost about the same as a gold disc.

## CD-recordable/rewritable

CD-recordable/rewritable (CD-RW) is an advance on the CD-R system in that it enables a CD to be written, erased and rewritten many times, just like a floppy disk but with a much greater capacity.

This technology uses special CD-RW discs incorporating a heat sensitive phase-change layer in place the cyanine layer of a conventional CD-R disc. In its normal unrecorded state, the phase-change layer is polycrystalline. During the writing process the beam from the laser head is focused onto the unmodulated spiral track of the spinning disc and bursts of high-energy laser light, heat tiny points of the layer above its melting point, changing it from a crystalline state to an amorphous state. The resulting two states of the layer, representing binary 0s and 1s, have different light reflecting properties allowing the data to be read by the laser beam set to low power.

The nice feature of this system is that if parts of the track are now reheated by the laser to a temperature just above the crystalline state but below the melting point, the layer reverts back to its original polycrystalline condition, erasing the old data and allowing new data to be recorded in its place. A CD-RW drive costs about the same as a CD-R drive and the latest CD recorders support both technologies. This is a good feature as CD-RW discs are considerably more expensive than their CD-R counterparts. One unfortunate problem with CD-RW discs not found with CD-R discs is that they cannot be read by a conventional CD-ROM drive or audio CD player. This is because the strength of the reflected beam is too low to be read reliably. To get around this problem many new CD-ROM drives incorporate extra signal amplification and these are marketed as MultiRead drives.

## Digital versatile disc

In 1995, due to the huge success of CD optical storage over the first half of this decade, 10 of worlds major consumer electronics, computer and entertainment companies including Philips and Sony, formed a consortium to develop a single advanced CD standard for video playback and CD-ROM storage. The new specification known as 'digital versatile disc' (DVD), meets the specific requirements of the motion picture industry's Studio Advisory Committee for video playback. It also meets the specifications of the computer industry's Technical Working Group for a new DVD-ROM format.

The 10 companies known as the 'Consortium' are Toshiba Corp., Sony Corp., Matsushita Electric Industrial Co., Pioneer Electric Corp., Victor Co. of Japan, Mitsubishi Electric Co., Hitachi, Philips Electronics NV, Thomson SA and Time Warner Inc.

Compared to normal CD, DVD uses smaller pits and a finer track width and spacing. It also supports one or two data layers within a single disc the same size as a conventional CD-ROM. The result is a massive increase in storage density to 4.7 GB per layer compared to 650 MB for a CD-ROM. This allows over 8.5 GB of storage from one side of the disc. DVD discs can be reproduced using the same injection moulding technique used for normal CDs.

A conventional CD-ROM drive uses an infrared laser with a wavelength of 780 nm. These were cheap to produce and adequate for the required data density. For the finer pit spacing of DVD, 780 nm wavelength laser light is too long and a shorter wavelength, visible light laser is used at 635 and 650 nm. The beam focusing assembly has also been refined to produce a finer focus point. Despite the major differences, a DVD drive can reliably read conventional CDs but not CD-R discs.

The vastly increased storage capacity requires an even more elaborate error detection and correction system than that used on the yellow-book standard. A modulation scheme known as 'EFM plus' and a new error correction method 'RS-PC' (Reed Solomon Product Code), gives a 10-fold improvement in data integrity.

The specifications for DVD are laid out in books like the CD standards. *Book A* specifies DVD-ROM, *Book B*, DVD-Video, *Book C*, DVD-Audio, *Book D*, DVD-Write Once and *Book E*, DVD-Erasable and rewritable.

The data format used on DVD Books A to C is known as the universal disc format (UDF) bridge. UDF is a specification and trademark belonging to the Optical Storage Technology Association (OSTA). It defines how data is stored and retrieved on DVD and also defines the data structure, including sectors, blocks, volumes, partitions, allocation tables and files. The UDF bridge is a multipurpose, all singing all dancing, specification suitable for multi-platform, multi-application use, it is also backward compatible to the standard CD-ROM ISO-9660 standard.

A single-sided, dual-layer DVD is made by pressing two polycarbonate discs with the ultrafine data pits and then bonding them together to form two data layers. Before the bonding process, one half is coated with 100 per cent reflective aluminium and the other with a semi-transmissive material.

During playback the laser can focus onto the near semi-transmissive data layer or shine through this onto the deeper data layer. The laser first reads the deeper layer, and on reaching the end of the track, it automatically focuses onto the near layer and continues reading the data in a seemingly uninterrupted operation. To achieve this seamless operation, the DVD drive temporarily stores the data in an electronic buffer (Figure 2.84).

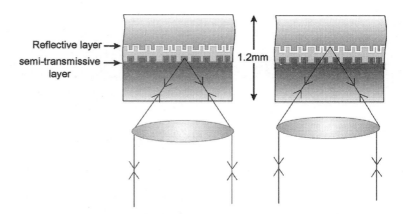

**Figure 2.84**  *Cross section through a DVD disc showing the dual focus laser system*

### DVD-ROM (Book A)

DVD-ROM is designed to provide a high-capacity storage medium for computer applications, multimedia and video games, to meet the ever-increasing demand for greater storage capacity. For example, some computer games are so graphic intensive that several CDs are required to store the full program. A single-layer DVD holds the equivalent of seven normal CDs, so elaborate games with high-quality graphics can be accommodated on a single disc.

### DVD-Video (Book B)

The Book B specification is destined to become the standard for domestic video playback, with picture quality close to studio production standard. Image sharpness and colour fidelity is better than video laserdisc and vastly superior to VHS. The other major advantage is the ability to instantly skip from one part of a recording to another without the need to rewind or fast-forward, as in a conventional tape system.

To set this standard rolling, the motion picture industry has released several hundred DVD discs of major cinema classics.

### DVD-Audio (Book C)

This specification is for very high quality multi-channel read only audio discs. Allowing the playback of audio at a higher quality than audio CDs. It also has provision for still images to be read during audio playback. DVD-Audio players will provide full downward compatibility with audio CDs.

### DVD-R (Book D)

DVD recordable (DVD-R) is similar to the CD-R orange book part II standard for write once, read many, only allowing a much greater storage capacity on a single DVD disc. A DVD drive however will not read existing CD-R discs as the laser wavelength is too short to be properly absorbed and reflected from the dye layer. To overcome this, special DVD compatible CD-R discs are now available. These can be recorded by a CD-R recorder and played back on a DVD drive.

The size of hard disks has increased enormously over the last half of this decade and an entry level drive is now around 2 GB. The ability to backup vast amounts of data from hard disk to DVD-R is an attractive one, as long as the price is right. Alternative removable hard disk technology and MO technology is capable of high-storage capacity and the price of this technology is falling steadily.

The most cost efficient system at the moment for backup purposes is CD-R at less than 0.08 pence per megabyte, about 50 pence per disc. DVD-R will only become popular if the media price can be brought down to a similar figure.

### DVD-RAM (Book E)

DVD-RAM is an exciting new product that stores 2.6 GB per side on a special re-recordable DVD that can be written to and erased many times.

This allows the user to record, playback and erase vast amounts of data in much the same way as a hard disk but at a much lower cost. Some companies have extended this format for use as a digital recording.

# Magneto-optical technology

MO technology combines the advantage of magnetic storage with the precision of laser optics to produce a reliable, high-density, fast-access, storage medium. A typical MO disk is capable of storing up to 1 GB on a 3.5-in. disk. These discs are made by coating a plastic or glass substrate with a highly reflective ferromagnetic compound of terbium, iron and cobalt.

To write data to the disk, both a laser beam and a magnetic recording head are used to change the light polarizing properties of the coating.

During the read operation, a low-intensity laser at 1.5 mW, is focused onto the surface of the spinning disk and the polarity of the reflected light is detected, and converted to a stream of '0's and '1's.

During the writing process, the intensity of the laser is increased to about 8 mW, which heats a tiny spot on the disk coating to its Curie temperature. This makes the spot susceptible to the magnetic field from the recording head. At this point the recording head applies the required magnetic polarity representing '0' or '1'. The laser is then switched off causing the temperature to drop below the Curie point and the magnetic polarity of the spot is fixed into the disks surface. The direction of magnetization determines the light polarization of the surface, which can be read with the low-intensity laser beam.

Once the data is 'frozen' into the disks surface it is immune from magnetic fields until the temperature is again raised to the Curie point, by the laser.

**Figure 2.85**   *The zip™ drive by IOMEGA, a portable floptical drive*

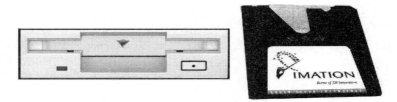

**Figure 2.86**   *LS120 floptical drive and media*

### Floptical disks

Unlike an MO drive, a floptical drive uses a laser solely as a means of guiding a conventional magnetic recording head along each track of the disk. The magnetic head is used to read and write data, in much the same way as a floppy drive.

Using the laser as a head guidance system, allows much finer tracks and more bits per track on a similar sized disk to a conventional 3.5-in. floppy disk. This results in a startling increase in storage capacity.

A floptical disk is very similar in appearance to a normal 3.5-in. floppy disk except it has a storage capacity of nearly eighty 1.44 MB disks. Some floptical drive variants actually read and write normal 1.44 MB disks as well as floptical disks. PC system manufacturers often fit a floptical drive in place of a standard floppy drive.

The price per megabyte of magneto-optical and floptical media is substantially higher than that of CD-R and CD-RW media and the read and write times are substantially lower. So floptical technology has not gained a wide acceptance in the PC industry. In an attempt to improve this situation some manufacturers have produced a high-capacity floptical system called LS-240 that records on special 240 MB super disks. It has the added bonus of allowing ordinary 3.5-in. floppy disks to store 30 MB instead of the lowly 1.44 MB. This will allow users to bring new life to stacks of old floppies that were previously discarded in favour of CDR and CDR/W discs (Figures 2.85 and 2.86).

## Self-assessment questions

**SAQ 2.1**   See page 34.

**SAQ 2.2**   What desktop size form factor uses motherboard mounted rear I/O panel?

**SAQ 2.3**   Why does the ATX 2.1 specification include an extra auxiliary power connector?

**SAQ 2.4**   What is the role of the motherboard chipset?

**SAQ 2.5**   What part of the chipset is interfaces the fast CPU-memory data highway?

**SAQ 2.6**   The maximum data transfer rate of the AGP 1.0 specification is approximately 246 MB/s. How is this derived?

**SAQ 2.7**   An external device is requesting the use of its I/O port to send data to the CPU. How can the CPU handle this as well as process the existing data?

**SAQ 2.8**   Why are separate maths co-processor chips no longer required?

**SAQ 2.9**   What is the data bus width of Socket 7 CPUs and the latest Socket A and Socket 370 CPUs?

**SAQ 2.10**  What major advance was made possible by the use of copper conductors instead of aluminium in CPU fabrication?

**SAQ 2.11**  Why will some CPUs will fail within seconds if an adequate heat–sink and fan is not used?

**SAQ 2.12**  What is the main purpose of the BIOS? *Hint:* Do not confuse this with the other routines stored in the system ROM?

**SAQ 2.13**  What system device shares the same battery-backed memory as CMOS setup?

**SAQ 2.14** What is the name of the special routine that executes after the POST and searches for a bootable disk?

**SAQ 2.15** To what two layers of the PC's command hierarchy does the OS interface?

**SAQ 2.16** What is the data bus width of a typical 184-pin DDR-DRAM DIMM?

**SAQ 2.17** What clock speed does PC3200 DDR-SDRAM operate at?

**SAQ 2.18** Why are lithium batteries used nowadays for the RTC? and how is it possible for the battery to power the RTC continuously for 7 years before a replacement is required?

**SAQ 2.19** A particular modem uses simple two tone FSK. If the bit rate is 300 what is the Baud rate?

**SAQ 2.20** If the paper suddenly runs out when the printer is printing, can the printer inform the computer of this condition to prompt the user to insert more paper?

**SAQ 2.21** In a modern PC/printer setup, what type of LPT1 port setting would probably be needed to allow the printer to inform the PC that a new colour ink cartridge has been installed and needs alignment?

**SAQ 2.22** State the two types of interrupt used on a PC.

**SAQ 2.23** What in essence is DMA?

**SAQ 2.24** A 1.44 MB 3.5-in. floppy disk has 80 tracks, 2 sides and 512 bytes per sector. Calculate the number of sectors on the disk.

**SAQ 2.25** What is the name given to the special read/write head design used on a floppy-disk drive that ensures that only narrow tracks are recorded on the disk?

**SAQ 2.26** How does the read/write head of a floppy-disk drive determine where a particular track is?

**SAQ 2.27** What major data conversion process does the FDC perform when the disk drive is reading and writing data?

**SAQ 2.28** Why are seven wires twisted on a floppy-disk drive ribbon cable?

**SAQ 2.29** What are the principle advantages of an MR head over a normal inductive head?

**SAQ 2.30** What is special about a UDMA 66/100/133 ribbon cable compared to an ordinary ATA 40-way cable?

**SAQ 2.31** What benefit does the CS facility offer?

**SAQ 2.32** What is LBA?

**SAQ 2.33** What is low-level formatting and how does it differ from high-level formatting?

**SAQ 2.34** What is an allocation unit?

**SAQ 2.35** What type of hard disc has a device ID jumper setting?

**SAQ 2.36** What name is given to the group of tracks under the read/write heads at a particular instant?

**SAQ 2.37** What promising features are offered by the SATA system?

**SAQ 2.38** State the capacity of a normal CD-ROM and compare this to the size of a single-layer DVD disc.

**SAQ 2.39** How is data read on a CD-ROM or DVD drive?

**SAQ 2.40** Where would you find focus and tracking coils.

**SAQ 2.41** State the data transfer rate of a normal 1X drive (same as the CD-audio standard).

**SAQ 2.42**   A CD-ROM is to be installed on a PC with a primary and secondary ATA interface. A single hard disk is present on the primary cable. How should the CD-ROM be set up?

(a) As a slave on the primary cable.
(b) As a master on the primary cable.
(c) As a master on the secondary cable.
(d) As a slave on the secondary cable.

**SAQ 2.43**   Why is the data to be recorded on a CD-ROM first converted from 8 to 14 bits?

**SAQ 2.44**   CD-ROM recording is prone to errors from minor blemishes in the disc manufacturing process and from surface scratches, how are these eliminated?

**SAQ 2.45**   What is the CD-ROM DOS extension file called?

**SAQ 2.46**   Where in the system is a CD-ROM set as a bootable device?

**SAQ 2.47**   What is the CD-ROM orange book standard?

**SAQ 2.48**   What is the major difference between a CD-ROM drive and a DVD drive?

**SAQ 2.49**   What combined technologies are utilized on a floptical disk drive?

# Unit 3 Peripherals

*Human/computer interfaces*

# Input peripherals

Any device connected and controlled by the computer to input data for processing is known as an input device or peripheral. The list below shows some of the bewildering variety of input devices used on a computer system:

- Bar code reader
- Digitizer or graphics tablet
- Document reader (optical character recognition (OCR))
- Joystick, paddle or joypad
- Keyboard
- Light pen
- Magnetic card reader
- Magnetic ink character recognition (MICR)
- Microphone for speech recognition
- Mouse
- Optical mark reader (OMR)
- Paper tape reader (obsolete)
- Point of sale (POS) terminal
- Punched card (obsolete)
- Scanner
- Terminal
- Touch-sensitive screen
- Digital camera

**Figure 3.1**   *Bar code reader*

### Bar code reader

Bar code readers are used for many applications were goods need to be identified and/or priced, quickly and reliably. The bar code is a series of black lines on a white background that can be read by a suitable scanning device. The code is often printed on a sticky backed label that can easily be attached to the goods to be identified.

They are used on items of food to motor car parts (Figure 3.1).

### Digitizer or graphics tablet

Digitizers or graphics tablets (Figure 3.2) are used in place of a mouse for drawing fine lines and curves and also for pointing and selecting from menu options. A must for a graphics designer or computer artist.

### Document reader

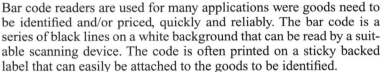

A document reader (Figure 3.3) is a fast scanning device that converts an image of printed text on a page to a form suitable to be entered and edited in a word processor program. A modern flat-bed scanner together with appropriate OCR software can carry out a similar process albeit more slowly.

**Figure 3.2**   *Digitizer or graphics tablet*

### Joystick, paddle or joypad

A Joystick, paddle or joypad (Figure 3.4) is used to play computer games. The joystick is ideal for flight simulator games where the position of the 'stick' sends positional information to the game to change the direction, attitude and altitude of the object being controlled (e.g. a plane or space craft). Buttons on the unit are used to input on–off information such as the firing of a gun or releasing a missile.

The joypad is a general purpose input device that also produces positional information via a finger pad and/or tracker ball. It also has a series of on–off control buttons.

**Figure 3.3**   *Document reader*

### Keyboard

A Keyboard (Figure 3.5) is used to enter alphanumeric characters. Each data key generates at least one unique character. The most common key layout is called QWERTY because the top left six alphabetical characters are the q, w, e, r, t and y keys.

### Kimball Tag

A Kimball Tag (Figure 3.6) is a paper card with a pattern of holes punched in it. Nowadays they are not so popular as bar codes but they are still used in some clothing stores. When an item is pu chased the card is torn in half and the free punched end is inserted in a special reader. This sends stock control information to a remote computer.

### Light pen

A light pen (Figure 3.7) is a hand-held pointing device used to make on-screen menu selections and draw graphic shapes, directly on the screen. The pen is used by pointing the tip of the pen close to the screen and pressing a button.

A light-sensitive photodetector and lens assembly inside the pen pick up the light emitted by the screen's phosphor dots immediately within the field of view of the lens and from the screen scanning signals, the position of the pen can be determined with great accuracy.

### Magnetic card reader

A magnetic card reader (Figure 3.8) is a simple input device used to read the data encoded on special cards containing a magnetic strip. Major uses are: inputting the data from credit, debit, and identity cards.

In use, the card is swiped in one direction through a slot in the reader device.

**Figure 3.4**  *Joystick, paddle or joypad*

**Figure 3.5**  *Keyboard*

**Figure 3.6**  *Kimball tag*

**Figure 3.7**  *Light pen*

**Figure 3.8** *Magnetic card reader*

**Figure 3.9** *MICR*

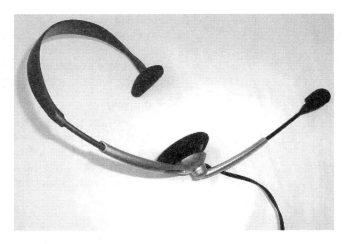

**Figure 3.10** *Microphone for speech recognition*

**Figure 3.11** *Mouse*

**Figure 3.12** *OMR*

### Magnetic Ink Character Reader
A MICR (Figure 3.9) is used to read data from bank cheques, etc.

### Microphone
A Microphone (Figure 3.10) is used to input audio to the audio-input port of a computer.

Nowadays you can control the computer directly using voice commands or speak words directly into a word processor. Special *speech recognition* software can interpret sounds via a microphone and convert speech phonemes into words recognized by a word processor.

### Mouse
A Mouse (Figure 3.11) is the main selecting device on desktop computers and terminals.

**Figure 3.13** *Paper tape reader (obsolete)*

### OMR
OMR (Figure 3.12) is used to automatically mark computer marked assignment papers in schools, universities and colleges.

### Electronic point of sale terminal (EPOS)
EPOS (Figure 3.13) is a common sight in shops and supermarket checkouts, etc. Used for entering and recording items of sale. Unlike a till this device can be automatically updated via a central computer.

**Figure 3.14** *Scanner*

### Scanner
Scanner (Figure 3.14) is a device used to make an image of a document or photograph. Can also make a 2D image of 3D objects. Most modern

scanners are capable of producing high-quality full colour photographic images.

### Terminal

A keyboard and visual display unit (VDU) combination is known as a terminal (Figure 3.15). It is used to enter data into a mini- or mainframe computer. It is called a terminal as it is remote from the main computer system. There are two types of terminal, dumb and intelligent. A dumb terminal simply allows keyboard characters to be echoed on the VDU as the characters are sent to the remote computer. An intelligent terminal contains its own central processing unit (CPU) and memory and allows a page of text, etc. to be completed and edited before being sent to the remote computer.

### Touch-sensitive screen

A Touch-sensitive screen (Figure 3.16) is used to make menu choices and selections easy for non-computer literate users. Ideal for inputting choices to control industrial equipment (e.g. a manufacturing plant). It is also a very useful input device for people with limited finger and hand movement.

One method used is to scan an $X$, $Y$ matrix of invisible infrared beams across the face of the monitor screen. A finger placed at a certain position on the screen breaks an $X$, $Y$ pair of beams and this generates a position signal.

### Digital camera

A digital camera (Figure 3.17) is a relatively new type of camera that exposes an image onto a light-sensitive chip rather than onto photographic film. The image is then read digitally from the chip into a non-volatile memory device. The stored image can then be downloaded to a computer. Once in the computer, it can be used immediately in a document. This is the major advantage of a digital camera over a conventional file camera. The time from image capture, to its use in a document, can often be measured in seconds, compared to the hours or days of conventional film photography.

**Figure 3.15**  *Terminal*

**Figure 3.16**  *Touch-sensitive screen*

**Figure 3.17**  *Digital camera*

# Output peripherals

Not surprisingly there is also a huge range of output devices used on a computer:

- Computer output on microfilm (COM)
- Liquid crystal display (LCD)
- Light emitting diodes (LEDs)
- Machine tools
- Plotters
- Printers
- Robots
- Terminal
- Turtle (or buggy)
- VDU (or monitor)
- Voice (sound)

### Computer Output on Microfilm (COM)

COM (Figure 3.18) was a popular method of archiving documents and images a few years ago before optical and other high-density

**Figure 3.18**  *COM*

**Figure 3.19** *LCD*

**Figure 3.20** *Machine tools*

**Figure 3.21** *Plotters*

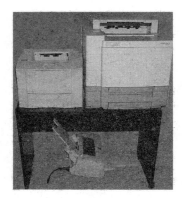

**Figure 3.22** *Printers*

storage systems became readily available. The output from the computer is connected to a special film-printing device. Hundreds of pages of information can be printed on a piece of microfilm (microfiche) approximately the same size as credit card, only a lot thinner.

Computers and CD/DVD storage have largely superseded this system today but due to the large amount of data still available on microfiche special readers are available that connect directly to a PC.

### LCD
LCDs (Figure 3.19) provide a high-quality flat alternative to the more obtrusive common type of cathode ray tube (CRT) monitor. The colour LCD shown here forms the display of a modern laptop personal computer (PC).

### Machine tools
Computers are extensively used in engineering to drive a whole variety of machine tools (Figure 3.20). The tool shown opposite is a drilling/milling machine.

### Plotters
Plotters (Figure 3.21) are sometimes used in place of a printer to create large-scale drawings and artwork on paper or film. Commonly used for technical drawings and schematics.

The output device is an ink pen on the end of an arm that traverses the work area in the $X$ and $Y$ directions.

### Printers
Apart from the monitor, the printer is the most widely used output device. It produces a high-quality printed image on paper and film. The printers shown in Figure 3.22 are mono-laser, colour laser and ink-jet. Figure 3.22 (a) mono-laser, (b) coloured laser and (c) ink-jet.

### Robots
Robots (Figure 3.23) are used extensively in industry to handle hazardous substances or to carry out mundane tasks such as pick and place.

### Terminal
For more details, see Input devices.

### Turtle/buggy
A turtle or buggy is a small computer controlled roving device programmed to carry out simple manoeuvres. Used to teach basic robotics and I/O programming (Figure 3.24).

**Figure 3.23** *Robots*

**Figure 3.24** *Turtle (or buggy)*

**Figure 3.25** *VDU (or monitor)*

**Figure 3.26** *Voice (sound card output)*

### *VDU – (or monitor)*

The monitor or VDU (Figure 3.25) provides an instant visual feedback of events taking place on the system. For example as you type in a word processor you can see characters appear on the screen. If you are editing a picture you can see the changes you make on the screen. It also allows you to select items from an on-screen menu using the mouse cursor or keyboard.

### *Voice (sound card output)*

The PC's sound system (Figure 3.26), provides an aural output for a wide variety of sources including: text–speech conversion, music, feedback sounds and a rich sound arena for games.

# Printer types and characteristics

The main output device of the PC system is the monitor. To see results, the computer and monitor must be powered up and the user must be sitting in front of the screen. This is fine for many applications but if a permanent and portable copy of data is required – one that can easily be assimilated by other human beings – then the monitor is clearly limited. Printers overcome this limitation by producing a permanent record on paper, a so-called *hard copy*. Although hard copy quickly accumulates and often takes up huge volumes of space, it is still the most versatile format for conveying data from person to person.

All printers produce *hard copy* in the form of printed characters on paper, the way this is achieved however, varies from printer to printer. There are several printing technologies available; the following types are used in desktop printing:

- Daisywheel
- Dot-matrix
- Ink-jet
- Laser
- Wax sublimation

These are categorized into two main types: *impact* and *non-impact* printing.

*Impact printers* press a solid metal type or pin-matrix against an inked ribbon and the paper. *Non-impact* printers, squirt, melt or electrostatically attract, 'ink' onto the paper, so there is virtually no physical contact between the head and the paper.

# Impact printers

The first impact printers were based on solid type, a direct follow on from the typewriter. The type set was moulded onto the surface of a solid metal ball or onto the petals of a daisy-shaped print wheel. These are known as *golf ball* and *daisywheel* print heads, respectively. The print speed of these printers was slow, but the print quality was superior to that of the subsequent generation of printers, the dot-matrix printers.

The golf ball printer was a cumbersome, noisy affair, which soon fell into obsolescence when the daisywheel appeared. Daisywheel printers were very popular for many years but their slow printing speed – up to 80 characters per second (CPS) – and the fact that they could only produce text output, left them vulnerable to other advancing printer technologies.

As computer technology advanced through the 1970s and 1980s, there was an increasing demand for printers capable of producing both text and graphics. Manufacturers responded by introducing a wide range of dot-matrix printers and printers based on other technologies including ink-jet and laser, and the demand for daisywheel printers slowly declined.

(a)

(b)

(c)

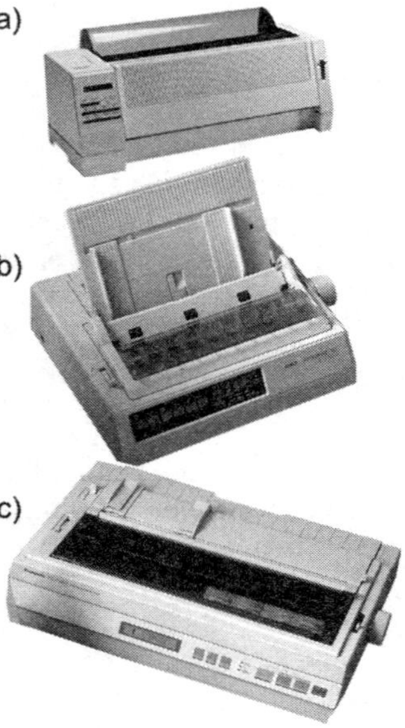

**Figure 3.27** *Typical dot-matrix printers: (a) Lexmark 4227, (b) OKI Microline 385 and (c) Panasonic KXP1654*

## Dot-matrix printers

The dot-matrix print head is formed from a series of solenoids (electromagnets). These fire an array of metal pins against an inked ribbon, which in turn transfers ink onto the paper. The greater the number of pins used on the head, the higher the print quality. However if too many pins are used, the head life is reduced considerably. As the number of pins increase they must be made correspondingly smaller in diameter, until after 40 or so pins they become too thin to achieve a reasonable head life. Some 48-pin dot-matrix printers were made but their cost put them in the same price league as laser printers, so they did not sell in great numbers.

Dot-matrix printers are capable of producing speeds beyond 800 CPS (i.e. about 10 times the print speed of a daisywheel).

A typical 24-pin dot-matrix printer (Figure 3.27) can generate draft quality at up to 800 CPS, near letter quality (NLQ) at 300 CPS and letter quality (LQ) at 100 CPS. Today dot-matrix printers are mainly used for printing till receipts at POS terminals and are seldom used for home and office use. Their main advantage today is the very low cost of consumables compared to laser and ink-jet technologies.

# Non-impact printers

Non-impact printing is a technique that transfers ink onto the paper without substantial physical pressure between the head, ribbon and paper. This type of printer is based on either thermal wax, ink-jet or laser technology.

## Thermal wax printing

Thermal wax transfer can produce excellent print quality but is rather slow compared to other non-impact methods. It is mainly

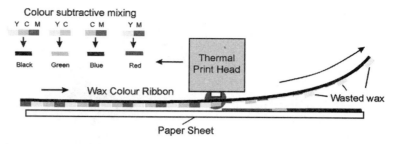

**Figure 3.28** *Diagram showing the thermal wax colour printing technique*

used for high-quality colour printing. In this process, a wax ribbon consisting of different coloured strips is heated by an array of resistive heating elements in the head. As the print head passes across the paper the heating elements are activated in sequence. This heats the ribbon sufficiently for the wax to turn to liquid which is then transferred onto the paper (Figure 3.28). The coloured pigments are bright, and the reflected light from tiny dots of wax, mixed subtractively to form other colours. The process can be expensive when large amounts of one colour are printed, as other colours on the ribbon are wasted.

In the simplest thermal wax printers a maximum of seven solid colours are possible without dithering.

A more advanced process known as dye-sublimation printing uses special inks that vaporize with heat. By varying the heat, many different colours can be produced without dithering. Such printers are very expensive and the running costs are high, however the print quality of colour photographic images is excellent.

### Ink-jet

Many ink-jet printers use variants on Canon's 'bubble jet' technology, where a tiny capillary tube (nozzle) containing ink is heated by a resistive heater. This causes the ink in the tube to 'boil' forming a microscopic bubble. The bubble forces ink out of the nozzle and onto the paper. To form a character, the print head contains a series of tiny nozzles arranged in a similar way to a dot-matrix head (Figure 3.29).

As the print head is moved across the paper, the tiny elements are heated by a pulse of electric current and ink bubbles are ejected onto the paper forming tiny dots. To print coloured images, cartridges containing the three subtractive pigments, magenta, yellow and cyan are used. The dots of coloured ink combine subtractively to form a range of colours. Often a fourth black cartridge is added to give a good-quality black for text and the shaded areas of an image. Photographic ink-jet printers usually have extra colours in addition to the subtractive pigments to help form more accurate colour reproduction.

Some ink-jet printers transport the coloured inks to the head via long lengths of flexible tubing using a pump arrangement to maintain a supply of ink. A more reliable system has evolved that places the ink cartridge on top of the print head thus eliminating the long ink carrying tubes. Going one step further Hewlett Packard have devised a system which combines the head and ink-cartridge

All pigmented waxes and inks produce additional colours by 'subtractive mixing'. The complementary pigments magenta, yellow and cyan combine to form hues of red, green and blue (RGB). For example closely spaced magenta and yellow dots produce a red hue. This is because when viewed in white light, magenta reflects red and blue light and absorbs green light, and yellow reflects red and green light and absorbs blue light. The eye therefore sees a red hue as the blue, green and a portion of red reflected light recombine into the white light.

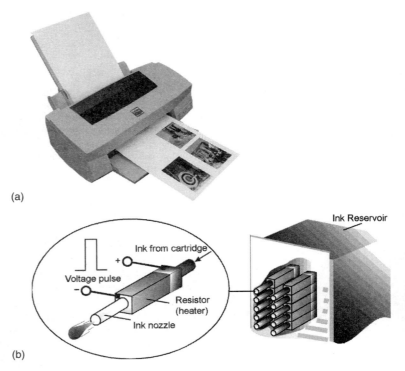

(a)

(b)

**Figure 3.29** *(a) The ink head of a typical ink-jet printer showing the tiny ink nozzles. The myriad of printed circuit tracks to each heater can just be seen. (b) Detail of the operation of a bubble jet print head*

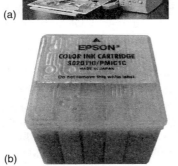

(a)

(b)

**Figure 3.30** *Epson Stylus photo printer and a typical five colour ink-cartridge*

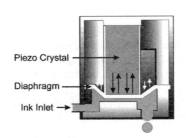

**Figure 3.31** *Detail showing one nozzle of a piezo print head*

into one replaceable unit. This has two advantages: (i) the reliability of the system is increased as no long ink tubes are required and (ii) each time a new ink-cartridge is fitted, the printer gets a new head. The high print quality is therefore maintained throughout the life of the printer. Surprisingly, the price of one of these combined ink and head cartridges is similar to an ink-only cartridge (Figure 3.30).

An alternative to the bubble jet technology uses a series of piezo devices to squirt microscopic jets of ink onto the paper. The head uses an array of tiny piezo devices and corresponding ink nozzles. The piezo devices change shape when a voltage is applied, thereby making them act as tiny pistons which repel the ink from the array of nozzles (Figure 3.31). Current piezo crystal ink-jet printers are capable of printing at resolutions beyond 1440 dots per inch (dpi).

The performance of today's ink-jet printers is impressive. Apart from being relatively inexpensive, they produce LQ output at print speeds beyond that obtainable with a dot-matrix printer and offer quiet operation and high graphics print quality as a bonus.

Photographic quality ink-jet printers (those with the extra ink colours) can produce excellent image quality. To achieve this, special coated paper must be used. Although the paper and ink is quite expensive the convenience of being able to print photographs directly from a digital camera via a PC is very attractive. The photographic quality obtainable from a digital camera and ink-jet printer now competes favourably with conventional 35 mm 'snap' photography.

### Installing ink-cartridges

If you are unfamiliar with a particular printer refer to the instruction manual. This sounds-like pure common sense and it is! However, it is surprising how often the uninitiated just dive in at the deep end and attempt to replace the cartridges the wrong way! These antics include, struggling to move the cartridge carrier into an accessible position by pulling on the carriage belts or by playing a special version of Russian Roulette by switching the printer on and off quickly in an attempt to catch the cartridges in the correct position! Even if access to the cartridge bay, is finally achieved, it is usually incorrect to replace an exhausted ink-cartridge simply by pulling the old one and plugging in a new one.

The important point to note here is that today's ink-jet printers have built-in intelligence and are able to communicate with the OS and the printer monitoring software. Unlike old first-generation ink-jets, to replace a cartridge the printer must be switched on but not printing. The printer control system must be made aware of what is going on.

A typical method used to install the ink-cartridges is by opening the printer cover and pressing the 'ink' button on the printer. On some printers the cartridges can only be replaced by running the printer monitoring software and choosing the 'replace cartridge option' or some such item.

When a cartridge is installed correctly, the printer is able to monitor the ink consumption and report on remaining ink levels. It will also inform you when the cartridge finally needs replacing. By not adhering to the recommended installation process the ink monitoring system will usually fail to report the true status of the cartridge and the user may end up replacing cartridges unnecessarily.

### *Laser printers*

Laser printers belong to a class of non-impact printer known as '*page printers*'. Page printers build up a dot-by-dot image of an entire page in read and write memory (RAM), before transferring it onto paper. The image is scanned onto a rotating photo-sensitive drum, by a fine laser beam. This system has several advantages over other printing systems – producing high-quality text and graphics, quickly, quietly and reliably. The printing rate and resolution is impressive. This ranges from four pages a minute, at 300 dpi, on a low-cost machine, to over 16 pages a minute at 2400 dpi, on a top-end machine.

To improve the resolution of printed text even further, most printers incorporate image enhancing techniques. The output of a 300 dpi printer is close to that of a 600 dpi printer without enhancement. Top-end machines often have *postscript* – Adobe's *page description language* (PDL) – built into the system. This special software (firmware), built into the printer, produces superior text and image quality, compared to a standard laser printer.

The print quality of a modern 600 dpi laser is similar to that of a typeset page, while 1200 dpi and higher machines actually match it. Today's ink-jet printers are also capable of producing high-resolution images beyond 600 dpi, but the absorbent property of normal paper produces a print quality which falls short of that found on laser

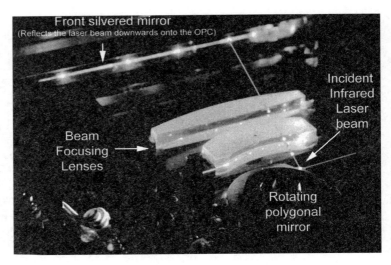

**Figure 3.32**   *Actual infrared photo inside a laser scanning unit*

printed pages. However for photo-quality colour images on special paper the ink-jet is superior to laser hardcopy.

A laser printer works by scanning a fine infrared beam from a *semiconductor laser diode*, onto the surface of a light-sensitive drum, known as the organic photo-conductor (OPC). The drum is slowly rotated a laser width row at a time, by a stepper motor. On a 300 dpi printer, one dot row corresponds to 1/300th of an inch rotation of the drum.

To provide the scanning action, the infrared beam from a fixed laser diode, is directed onto a rotating polygonal mirror. As it rotates, the laser beam is deflected horizontally across the face of a front silvered mirror. The reflected beam from the mirror is then directed onto the surface of the OPC, where it builds up an electrostatic replica of the original image. The laser beam effectively scans a row of dots, line by line, across the width of the drum. As it scans each row, the beam is turned on and off in sympathy with the dot by dot make up of the image (Figure 3.32).

The sealed scanner compartment of a laser printer (Figure 3.33) contains the optical equipment for modulating and directing the laser beam onto the drum. In the photograph of the inside of the laser scanning unit, the normal infrared laser diode has been replaced with a visible red laser diode to show the beam paths.

The laser printer uses electrostatic forces to create an image on the photo-sensitive drum. This is done in five stages:

1. Electrostatic image formation
2. Development
3. Transfer
4. Fixing
5. Cleaning

### 1. Electrostatic image formation

First the electrical resistance of the drum is made uniform by preconditioning it with a light source. At this stage the electrical potential of

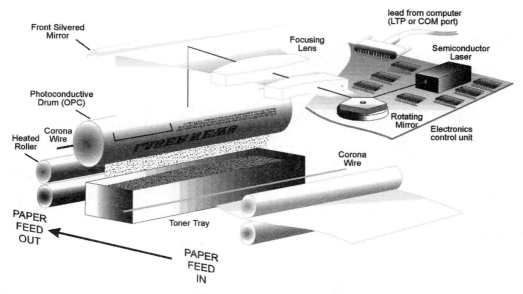

**Figure 3.33** *The basic laser printer system*

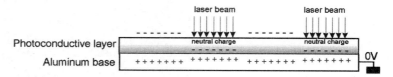

**Figure 3.34** *A section through the skin of the drum, showing how the electrostatic image is created on the outer surface*

the drum is about +100 VA. A primary corona wire placed close to and along the length of the drum, is connected to a voltage around −1000 V. This transfers a uniform negative charge across the drum's surface. As the laser beam draws a light image on the drum, the exposed areas neutralize the negative charge. This forms an electrostatic image on the drum's surface. The black areas to be are neutral and the blank areas are negative relative to the metal base (Figure 3.34).

### 2. Development
A developing tray containing finely divided particles of carbon and binder chemicals, is held at a negative charge equal to the charge on the unexposed areas of the drum. The exposed areas being at a higher potential, attract the toner particles and an image is formed on the drum.

### 3. Transfer
While the toner image is building up on the drum, the paper feed mechanism slowly moves the paper towards the drum. On its way, it brushes past another corona wire fixed at a high positive voltage. The toner is attracted off the drum and onto the paper, forming an image. At this stage the image is composed of loose toner powder, a light touch or gust of wind would easily destroy the image (Figure 3.35).

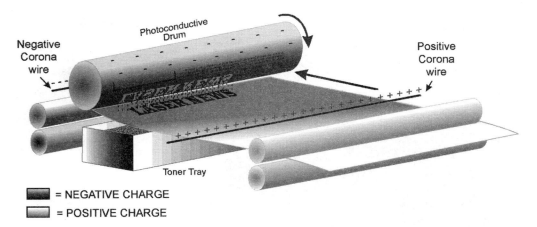

Negative Corona wire

Photoconductive Drum

Positive Corona wire

Toner Tray

▬ = NEGATIVE CHARGE

☐ = POSITIVE CHARGE

**Figure 3.35**

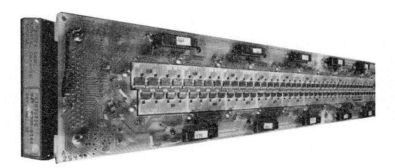

**Figure 3.36** *LED array*

### 4. Fixing

To make the image permanent the paper is fed between two fixing rollers one of which is electrically heated. The rollers are coated with a non-stick temperature-resistant material such as PTFE. Under the heat and pressure of the rollers, the binder compound in the toner fuses the particles and fixes the image permanently onto the paper.

### 5. Cleaning

Any toner remaining on the photo-sensitive drum after the paper has fed out of the printer, is removed electrostatically or wiped off smoothly with a rubber cleaning blade, in readiness for the next page.

## LED printers

A technology using LEDs has been adopted as an alternative to the laser, by some manufacturers. An LED printer is very similar in principle to the laser printer but it uses an array of tiny LEDs in place of the laser scanning mechanism. The LEDs are fabricated closely together to match the resolution available on most modern laser printers (300–1200 dpi) (Figure 3.36).

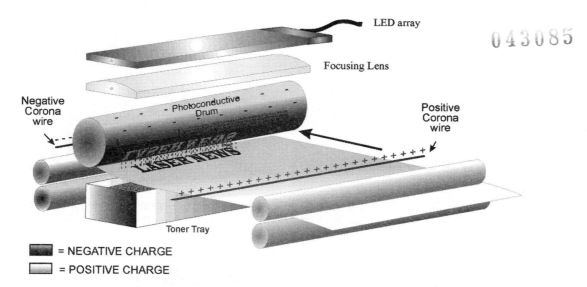

043085

**Figure 3.37** *The basic mechanism of an LED printer*

The advantage of this system is that the imaging system is solid state with no moving parts. A laser printer on the other hand relies on an elaborate optical system based on a laser beam and rotating mirror assembly.

The LED array extends the width of the page and works by flashing specific LEDs on and off in sympathy with the image. The LED image is focused onto a photoconductive drum is much the same way as a laser printer (Figure 3.37). To maintain consistent brightness along its length, the array has a built-in electronic circuit that monitors the brightness of each LED. It uses this information to set the duration of the *turn-on* pulse for each LED. A bright LED only needs a short duration pulse, whereas a dimmer specimen, requires a longer pulse. This technique is known as pulse width modulation.

### Colour laser printers

The price of colour laser printers have at last fallen to a level that small offices and shops can afford. This is largely due to new toner technologies and the falling price of electronic components, including CPUs and memory – two key items in a colour laser printer.

A colour laser is, in essence, four laser printers combined in one housing. To render a near photographic image, the laser uses three primary-subtractive colour toners, cyan, magenta and yellow, in addition to black. The image is built up in stages by splitting the original graphic file from the computer, into four separate overlay files. To process the four images in an acceptable amount of time, the printer incorporates a very fast RISC processor and a large amount of RAM. This is usually upgradeable to 128 megabytes (MB) or more, using SIMMs or DIMMs (Figure 3.38).

Until recently, the three colour toners were based on conventional pigmented toner, similar to that used in a monochrome laser. Now a more advanced toner is used. This has revolutionized the cost and quality of the printed image.

**Figure 3.38**  *The black, yellow, magenta and cyan toner cartridges from a Brother HL2400 colour laser printer (one of the first printers to use this new technology)*

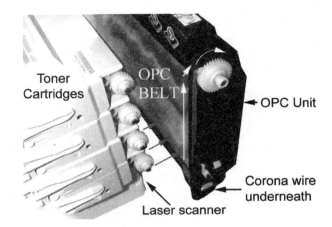

**Figure 3.39**  *Exploded view of the toner cartridge and OPC unit*

The new toner is made from microscopic beads of plastic wax. The cyan, magenta, yellow and black toner is applied, in four separate passes, onto a rotating drum. The resulting toner image on the drum, is a dithered version of the original graphic. This is then transferred electrostatically onto the paper sheet. The toner is then blended and fixed onto the paper by passing the page through a set of oiled, heated rollers. The heat combined with the oil, melts and merges the individual particles of coloured toner to form a much more acceptable image than the powdery images produced on early models.

### How a modern colour laser printer works
The process is similar to that of a conventional laser except for the additional coloured toners and fixing oil.

1.  The image for each overlay is scanned via infrared laser onto a large OPC belt where an even electrostatic charge has been previously applied via a corona wire.
2.  The respective coloured toner cartridge is then moved, via a cam arrangement, up close to the OPC where the toner is attracted to the charged image (Figure 3.39).

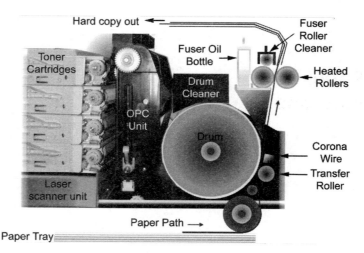

**Figure 3.40** *Cut-away view of the colour print engine minus the electronic control unit*

3.  Each overlay image is scanned in turn onto the OPC, effectively a repeat of stage 1 and 2, until all four overlays have been applied to the OPC belt (Figure 3.40).
4.  Now the four colour toner image built up on the OPC is transferred electrostatically onto a wide silicon rubber coated drum with a circumference large enough to accommodate the maximum length of page supported usually A4 or A3. This is called the imaging drum or drum for short.
5.  At this point a sheet of paper is introduced from the paper tray. This first passes over a corona wire which applies an even electrical charge across the paper. The paper is then pressed up against the rotating drum by a spring loaded roller, known as the transfer roller. The charge on the paper pulls the toner particles away from the drum where they adhere to the paper. The image on the drum is therefore transferred onto the paper.
6.  The paper, complete with its toner image, is then passed through a pair of heated fixing rollers coated in silicon oil – a high temperature tolerant liquid. The heat melts the toner particles and the oil helps to blend the different coloured waxes to form the final image.

The consumables required for a colour laser printer are as follows:

- Laser compatible cut sheet paper (plain copier paper is often satisfactory).
- Black, cyan, magenta and yellow toner cartridge (replaced approximately every 3000–8000 pages).
- Fuser oil (replace after approximately 15,000 pages).

### Serviceability

Most parts of a laser printer nowadays are field replaceable units (FRUs). For example the OPC, fuser unit, drum cleaner unit, transfer rollers, paper transport mechanism, etc. can be replaced by a service technician in the field. Whereas the consumables, including the fuser oil and cleaners, can easily be replaced by the user.

As with all items of equipment it is always good practice to refer to the user manual whenever you come across an unfamiliar model.

# Paper feed methods

There are many different sizes and types of paper used with printers. For example, letters, reports and brochures need high-quality single sheet paper whereas memos, lists, invoices and delivery notes are quite acceptable on low-cost 'listing' paper. Special applications may require long continuous rolls of paper such as banners and signs. To accommodate these different paper types and sizes, printer manufacturers offer a choice of paper feed mechanisms. The different types are illustrated below.

## Tractor feed

Tractor feed systems use continuous paper, which is inexpensive, low weight and punched with holes and perforations (Figure 3.41). The holes line up with the tractor wheel spokes on the printer. When the wheels revolve, the paper is either pushed or pulled through the printer. Some dot-matrix printers incorporate a lever to select the paper feed direction (i.e. they can be set to push or pull).

Continuous paper is used mainly for non-presentation purposes such as packing notes, invoicing and computer program listing (Figure 3.42). It is ideal for the latter purpose, as individual program listings often run into hundreds of pages. The fact that the pages fan fold together, allows the program listing to be stored neatly and in one piece.

## Friction feed

Friction feed is used to transport single sheets of paper in and out of the printer. The sheet is gripped between a series of rubber rollers which pull it through the paper path.

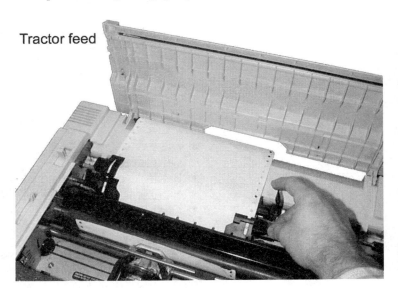

Tractor feed

**Figure 3.41** *Continuous tractor feed paper loaded into a dot-matrix printer*

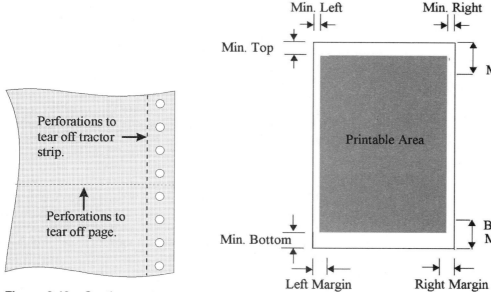

**Figure 3.42** *Continuous trac-tor feed paper*

**Figure 3.43**

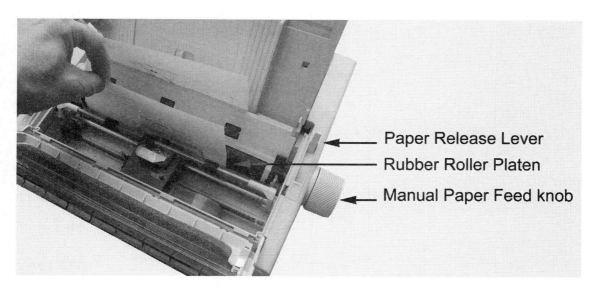

**Figure 3.44** *Loading single sheet paper into a dot-matrix printer. Friction feed detail showing how the paper wraps around the platen*

The margins around the printed page may be set by the application software. However there is a minimum margin size, beyond which printing is not possible. On dot-matrix printers this is just a few milli-metres from the page edges. Other printer types such as ink-jet and laser printers usually have a much larger minimum margin range (i.e. less printable area) (Figures 3.43 and 3.44).

# Printer input output and control

The printer is essentially an output device with print data flowing from the computer to the printer. However to achieve reliable data transfer, the printer must inform the computer when it is busy and when it is ready to receive more data. The computer must also be told when the printer is out of paper and when it is *off-line*. The communication between computer and printer is therefore a two-way process, even though the main print data only goes one way.

The most fundamental method of achieving two-way communication between computer and printer is to use separate control wires, in a process called *handshaking*. The PC's traditional parallel port (LPT1) and serial ports (COM1 and COM2) utilize handshake lines.

Legacy parallel port printers simply relied on the handshake lines to inform the PC when it was out of paper, etc. Current printers use 'intelligent' two-way communication and therefore need a bi-directional parallel port – most Pentium-style motherboards incorporate Intel's enhanced parallel port (EPP) specification on the LPT1 port, which is bi-directional. In these printers the eight data lines of the LPT1 port are used to both send and receive data.

A typical printer uses an elaborate microprocessor-based control system which can be simplified into five main sections as follows (Figure 3.45):

- CPU, read only memory (ROM) and RAM containing the main control program and character sets.
- RAM-based data storage area known as the *printer buffer*.

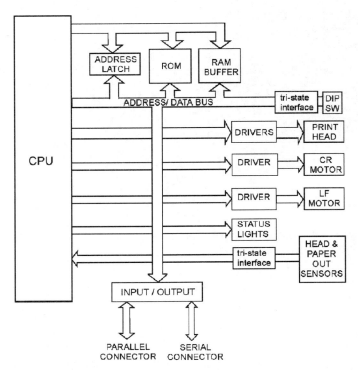

**Figure 3.45**   *The microprocessor-based printer control system of a typical printer*

- Drive electronics for the various motor-driven subsystems.
- Input circuits for the various sensors and selector switches.
- An input/output interface circuit for connection to a computer.

# Character sets

All printers use American Symbolic Code for Information Interchange (ASCII) and its variants, to print characters. To accommodate the special characters used in different countries, a specific *character set* can be loaded into the printer's memory. This is done in one of two ways: either by downloading it via an application program and OS or by selecting it at the printer itself using dip switches or software control. The later method was used in matrix printers by altering dip switches at the back of the printer. Today's modern printers are automatically configured by the application software and do not contain any user adjustable dip switches.

In the Windows environment, drivers are included for most popular printers. These allow the full range of Windows *TrueType* fonts to be used by the printer. TrueType fonts are specially designed to appear on screen the same as they appear on the printed page. They can also be scaled to a range of point sizes. Special country specific characters like the *pound* sign are included when Windows is first installed – these settings can also be altered from control panel via the *keyboard* and *regional settings* icons.

In addition to scalable text characters, TrueType ᴛᴛ fonts allow special characters to be inserted into documents and scaled to a range of sizes. A few examples are shown in Figure 3.46.

Prior to sending characters, special commands known as *control codes* are sent to the printer. These set different printer modes and options and directly control certain printer functions such as carriage returns, line feeds and page orientation, margin and font settings.

The ASCII-8 code represents each control code and printable character as a unique 8-bit binary number (byte). Using 8 bits, a maximum of 256 ($2^8$) different numbers are available for control and character codes. In the ASCII system, 32 codes are used for control purposes. These occupy the numbers (0–31). The rest of the code numbers are used for characters and symbols. The 8-bit ASCII code is listed in Table 3.1.

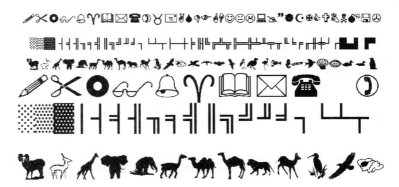

**Figure 3.46** *Sample characters from special ᴛᴛ font sets*

**Table 3.1** *ASCII code*

| Dec | Hex | Chr | Dec | Hex | Chr | Dec | Hex | Chr | Dec | Hex | Chr |
|-----|-----|-----|-----|-----|-------|-----|-----|-----|-----|-----|-----|
| 0 | 00H | NUL | 32 | 20H | Space | 64 | 40H | @ | 96 | 60H | ' |
| 1 | 01H | SOH | 33 | 21H | ! | 65 | 41H | A B | 97 | 61H | a |
| 2 | 02H | STX | 34 | 22H | " | 66 | 42H | C D | 98 | 62H | b |
| 3 | 03H | ETX | 35 | 23H | # | 67 | 43H | E | 99 | 63H | c |
| 4 | 04H | EOT | 36 | 24H | $ | 68 | 44H | F G | 100 | 64H | d |
| 5 | 05H | ENQ | 37 | 25H | % | 69 | 45H | H | 101 | 65H | e |
| 6 | 06H | ACK | 38 | 26H | & | 70 | 46H | I | 102 | 66H | f |
| 7 | 07H | BEL | 39 | 27H | ' | 71 | 47H | J | 103 | 67H | g |
| 8 | 08H | BS HT | 40 | 28H | ( | 72 | 48H | K L | 104 | 68H | h |
| 9 | 09H | LF VT | 41 | 29H | ) | 73 | 49H | M | 105 | 69H | i |
| 10 | 0AH | FF CR | 42 | 2AH | * | 74 | 4AH | N O | 106 | 6AH | j |
| 11 | 0BH | SO S1 | 43 | 2BH | + | 75 | 4BH | P Q | 107 | 6BH | k |
| 12 | 0CH | DLE | 44 | 2CH | , | 76 | 4CH | R S | 108 | 6CH | l |
| 13 | 0DH | DC1 | 45 | 2DH | - | 77 | 4DH | T U | 109 | 6DH | m |
| 14 | 0EH | DC2 | 46 | 2EH | . | 78 | 4EH | V | 110 | 6EH | n |
| 15 | 0FH | DC3 | 47 | 2FH | / | 79 | 4FH | W | 111 | 6FH | o |
| 16 | 10H | DC4 | 48 | 30H | 0 | 80 | 50H | X Y | 112 | 70H | p |
| 17 | 11H | NAK | 49 | 31H | 1 | 81 | 51H | Z | 113 | 71H | q |
| 18 | 12H | SYN | 50 | 32H | 2 | 82 | 52H | [ | 114 | 72H | r |
| 19 | 13H | ETB | 51 | 33H | 3 | 83 | 53H | \ | 115 | 73H | s |
| 20 | 14H | CAN | 52 | 34H | 4 | 84 | 54H | ] | 116 | 74H | t |
| 21 | 15H | EM | 53 | 35H | 5 | 85 | 55H | ^ | 117 | 75H | u |
| 22 | 16H | SUB | 54 | 36H | 6 | 86 | 56H | _ | 118 | 76H | v |
| 23 | 17H | ESC | 55 | 37H | 7 | 87 | 57H |  | 119 | 77H | w |
| 24 | 18H | FS GS | 56 | 38H | 8 | 88 | 58H |  | 120 | 78H | x |
| 25 | 19H | RS US | 57 | 39H | 9 | 89 | 59H |  | 121 | 79H | y |
| 26 | 1AH |  | 58 | 3AH | : | 90 | 5AH |  | 122 | 7AH | z |
| 27 | 1BH |  | 59 | 3BH | ; | 91 | 5BH |  | 123 | 7BH | { |
| 28 | 1CH |  | 60 | 3CH | < | 92 | 5CH |  | 124 | 7CH | \| |
| 29 | 1DH |  | 61 | 3DH | = | 93 | 5DH |  | 125 | 7DH | } |
| 30 | 1EH |  | 62 | 3EH | > | 94 | 5EH |  | 126 | 7EH | ~ |
| 31 | 1FH |  | 63 | 3FH | ? | 95 | 5FH |  | 127 | 7FH | DEL |

Most printers naturally need more than 32 control codes to produce all the facilities available. So to get around this limitation, printers use group of control codes.

# Installing and testing a printer in windows

Windows application programs have several major advantages over the old disk-operating system (DOS)-based applications. Not least, is the fact that one set of drivers is common to all application programs. In DOS, each application program must provide its own set of drivers.

When a printer is first installed on a Windows-based PC system, a set of driver routines must also be added to the operating system (OS). These are usually provided on a disk supplied with the printer. Each new edition of Windows also has a built-in set of drivers for most of the popular printers that were available prior to its launch. It also accommodates new drivers supplied by the printer manufacturers. Once a printer driver is installed, it appears in the *printers* folder in *control panel*. If several printers are available, drivers can be installed for each. Whichever printer is currently connected to the PC

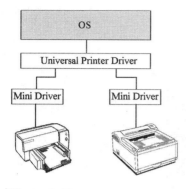

**Figure 3.47**

can be selected as the default printer, simply by clicking on it in the *printers* folder.

In Windows 95/98/ME/NT/2000/XP, the printer device driver routines are divided into two parts: a *universal driver* and a *mini-driver*. Windows provides the universal driver – known as an application program interface (API). This includes information relevant to all printers and also carries out the essential communication between other parts of the OS (Figure 3.47). The printer manufacturer provides the mini-driver – which communicates between the universal driver and the printer itself. This arrangement allows the printer manufacturer to concentrate on making the printer specific software as user friendly as possible, rather than spending the programmer's time writing interface routines.

## Practical work SAP 3.1: Installing a new printer in Windows 95/98/ME/XP

### Items required:

- Pentium 200 MHz equivalent or above, with 32 MB minimum memory, and Windows 95/98 or ME installed.
- A packed printer with all the accessories including: driver disk, manual, printer leads and consumables (paper and ink or toner cartridge).

The following steps are necessary to install a new printer in the Windows environment:

(i) Remove the printer from its box and remove the packing material and restraining items and place the packing back in the box. Refer to the printer manual if necessary to ensure that all the restraining material has been removed from inside the printer.

(ii) Make sure the computer and printer are switched off. Read the printer manual and ensure that the installation instructions and driver, match your OS.

(iii) After ensuring that you have the correct drives and installation instructions for your OS and printer, follow the installation instructions therein.

*Note*: If the drivers and installation instructions do not match, log onto the manufacturer's web site and download the latest drivers and installation instructions applicable to your printer and OS. Do not attempt to install the wrong drivers.

### *Using the 'Add New Printer' Wizard in 95/98/ME*

You should always install a printer by following the manufacturer's instructions. However, sometimes the specific installation instructions are not available. This can happen for a number of reasons. If the printer is quite old the original instructions may have been mislaid or it may no longer be supported on the manufacturer's web site.

Generally, you can install the printer using the *Add Printer Wizard* in Windows. The procedure is shown below:

(i) In the Windows 95/98 desktop choose **My Computer | Printers | Add Printer**. Alternatively or if you have Windows ME choose **Start | Settings | Printers | Add Printer**.

(ii) The Add Printer Wizard panel appears as shown in Figure 3.48.

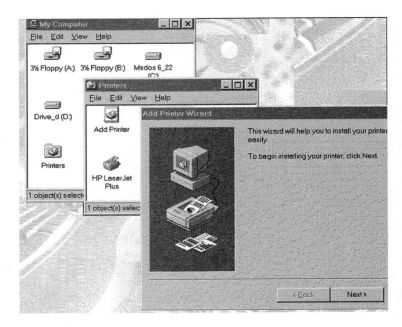

**Figure 3.48**

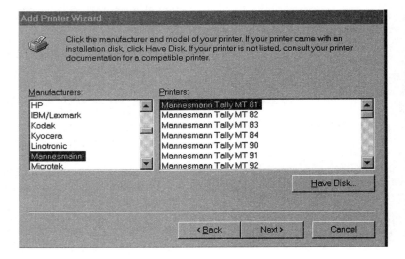

**Figure 3.49**    *Select your printer manufacturer and model*

If your printer is not listed, then click on '***Have Disk***' and follow the instructions directing the wizard to where the correct drivers are located. If you have downloaded drivers to a folder on the hard disk or to a floppy disk, then use the browse facility to locate the relevant files.

(iii) Click **Next**> and a list of printer manufacturers (Figure 3.49) and printer types appears. Choose your printer from the list.
(iv) Click **Next**> and select the appropriate printer port. The default setting is LPT1, the parallel 'Centronics' port. If this is correct, click **Next**> (Figure 3.50).

If you are installing an RS232C serial printer, select COM2, as COM1 is normally used for the modem. Press the 'Configure port' button and enter the required RS232 parameters. It is best to use the default settings of 9600 baud, 8 data bits, no parity, and 1 stop bit, as shown below. Then click **OK** (Figure 3.51).

**Figure 3.50**

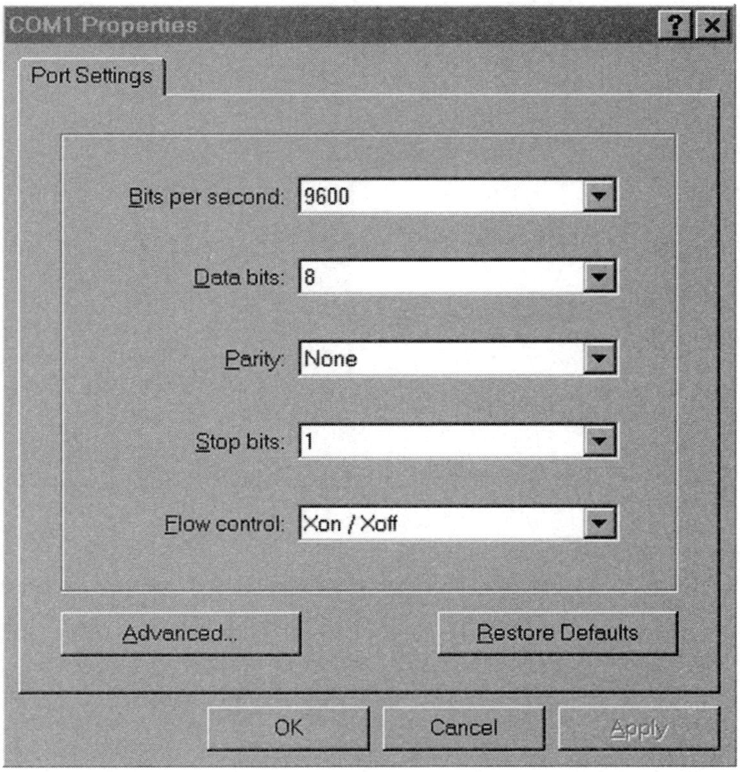

**Figure 3.51**

(v) From the port selection window click **Next**>.
(vi) You can change the printer name and set it as the default printer. Click **Finish** to complete the information gathering stage (Figure 3.52).
(vii) If you wish, choose the option to do a test print to check the print quality. Finally click **Finish** to complete the installation (Figure 3.53).

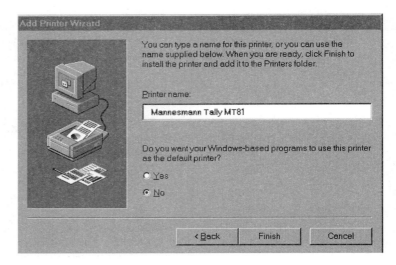

**Figure 3.52**

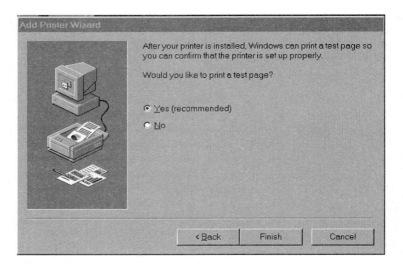

**Figure 3.53**

### Installing hot-wired plug-and-play printers

Current plug-and-play (PnP) printers based on the so-called hot-wired serial bus systems, USB and Firewire™, are usually installed using a different procedure to the one used on traditional Centronics and RS232 printers. For example, the manufacturers drivers and utility software are often installed before the printer is connected to the bus therefore bypassing the Add Printer Wizard. Again you must refer to the manufacturers installation instructions for the correct procedure. Do not just bash on regardless ignoring the instructions, unless of course you have installed exactly the same printer model before. Here is a typical hot-plug installation procedure to show you just how different and user friendly the installation procedure can be:

- Make sure the printer and PC is turned off connect the printer power lead and signal lead. Now turn on the computer but leave the printer switched off.

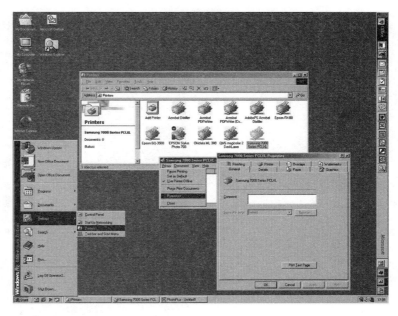

**Figure 3.54**

- Let Windows boot and install the printer driver and utility software.
- A message will then prompt the user to switch on the printer.
- The print head then moves to the ink-cartridge installation to allow you to fit the cartridges.
- Just follow the on-screen instructions to complete the installation.

### Specifying printer properties

You can change the print options for the installed printer at any time. If you are working within an application and want to change a printer parameter, for example you might want to change the page mode from portrait to landscape. You do this in most applications by choosing **Files** | **Print** in the pull down menu or by pressing **Ctrl** | **P**.

To set the printer parameters outside an application do the following:

(i) Choose **My Computer** | **Printers**, or **Start** | **Settings** | **Printers**.
(ii) Double click the icon of the printer you want edit.
(iii) Choose Printer | Properties. The Properties window appears. In the example above (Figure 3.54) a Samsung laser printer is selected.
(iv) As you can see from the example menu above, in the properties menu you can click on the various tabs to view and edit a wide range of printer options.

### Using the 'ADD New Printer' Wizard in Windows XP

Windows XP provides a more user friendly version of the Add Printer Wizard however activating it is slightly different to the Windows 95/98/ME version. Here we will go through the installation of a Centronics ink-jet printer.

**Figure 3.55**

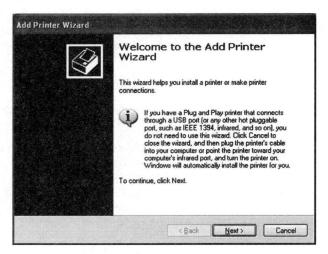

**Figure 3.56**

(i) Choose **Start | Control Panel | Printers and Faxes** (Figures 3.55 and 3.56).

(ii) The Add Printer Wizard window is displayed. Note the information message advising you not to use the wizard to install hot-pluggable PnP printers but rather let windows detect and install the printer for you. Click **Next**> (Figure 3.57).

(iii) Choose whether the printer is attached to this PC (local) or otherwise. Then click **Next**> (Figure 3.58).

(iv) Choose your printer manufacturer and model. If your printer is in the list of printer all well and good. However if you need to install it from disk it should ideally be certified by Microsoft as a compatible driver. Click **Next**> (Figure 3.59).

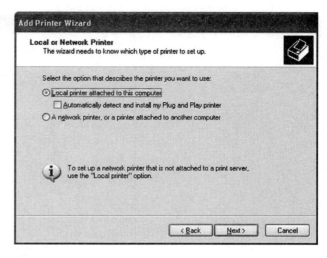

**Figure 3.57**

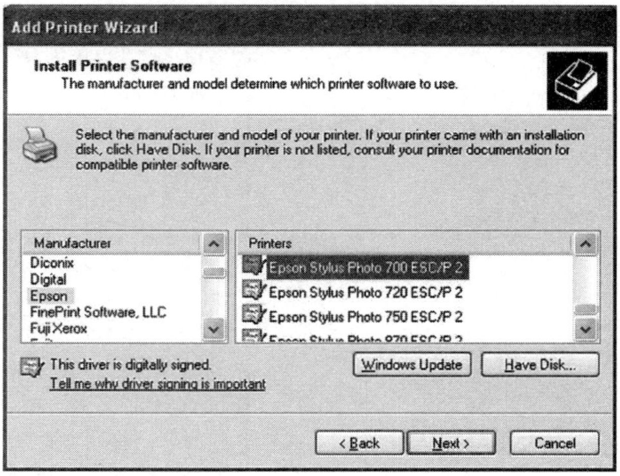

**Figure 3.58**

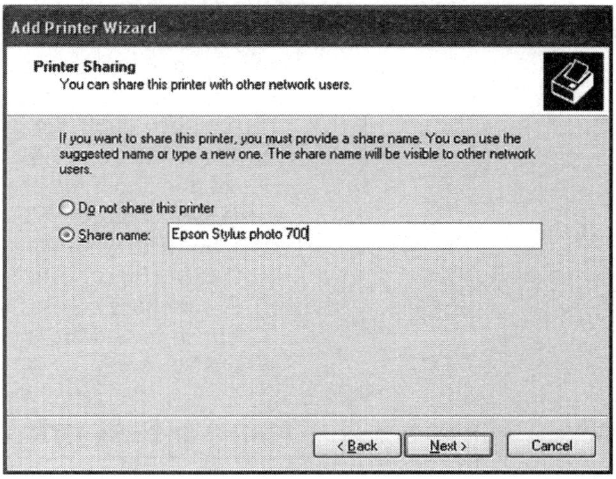

**Figure 3.59**

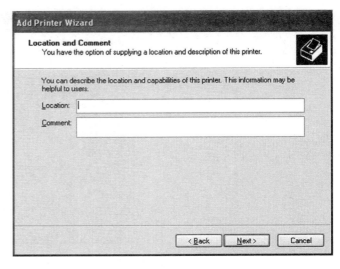

**Figure 3.60**

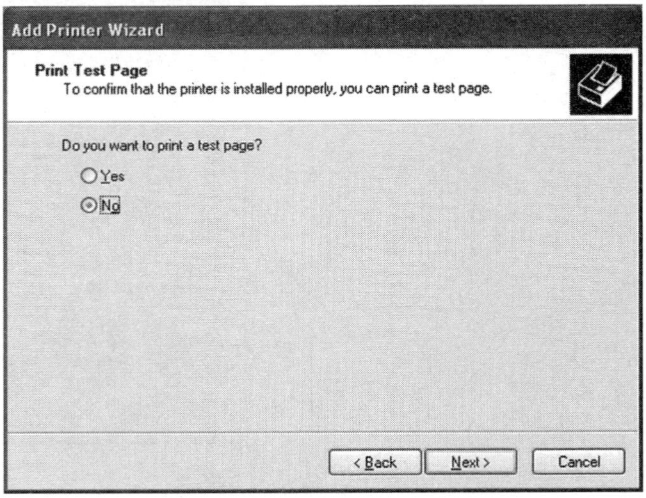

**Figure 3.61**

(v) If you intend to share this printer with other network PCs then click the share name button and enter a name of your choice and then click **Next**> (Figure 3.60).

(vi) For a shared network printer it is useful to enter some information about the printer and its location on the network. Click **Next**> (Figure 3.61).

(vii) Choose if you want to perform a printer test and then click **Next**> (Figure 3.62).

(viii) A summary of your selections is displayed before you choose Finish to end the installation or back track to change a previous selection.

## Doing a test print using a word processor

Windows is provided with two basic word processors called '**Notepad**' and '**Wordpad**'. These applications can be used to print a wide range

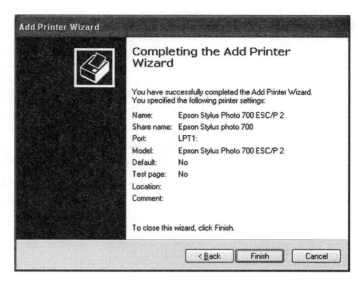

**Figure 3.62**

**Figure 3.63**   *An example test print using TrueType fonts*

of 'TrueType' fonts in different point sizes and colours. Thus testing the printers capabilities. The example below (Figure 3.63) shows a simple test print from a colour printer. (*Note*: The print on this page may not be in colour.)

# Scanners

A scanner is a peripheral device that captures an image and converts it to digital form for the computer to store, edit and display. The image can be text, a photograph, a line drawing or in the case of flat-bed scanners, even a small 3D object.

A scanner can be used to do the following: Scan a graphics image; scan printed text without retyping; scan documents for faxing, scan an image or document and print it, just like a photocopier.

## Types of scanners

There are three main types of scanner:

- Hand-held
- Sheet-fed
- Flat-bed

Several years ago, flat-bed scanners were expensive – approximately the price of a complete PC system – so the cheaper hand-scanners and sheet-fed scanners became popular even though they offered inferior performance. The flat-bed types were mainly used for professional text and image capturing.

Hand-held and sheet-fed scanners use fairly basic mechanisms and do not offer the flexibility of the flat-bed types. A sheet-fed scanner, cannot scan solid objects or scan a page from a book, but the scan quality from a single page is often as good as that obtained from a flat-bed (Figure 3.64).

With a hand-scanner it is difficult to scan anything accurately, apart from small areas of text and graphics, as the linearity of the scan is determined by the speed and accuracy of your hand movement.

Flat-bed scanners have recently dropped in price significantly, pushing the other types of scanner out of the market. An entry level flat-bed now costs less than a hand-scanner cost only a few years ago.

## How a scanner works

In this description, the operation of the flat-bed type of scanner will be investigated. The other types of scanner work on a similar principle.

Scanners work by shining white light from a fluorescent lamp onto the object being scanned (Figure 3.65). The reflected light from the object is then passed through a coloured filter and lens assembly onto the surface of a charge coupled device (CCD), to form a focused image of the part of the object being scanned. The filter, lens, CCD and light source are mounted on a moving carriage that travels along the bed of the scanner.

A CCD is an array of metal oxide semiconductor (MOS), capacitors formed on the surface of a silicon chip. When light falls on a single MOS capacitor (cell), it stores a charge depending on the light intensity. An array of such devices is therefore able to store a pattern of charge representing the incident image. The capacitors are electrically coupled together in a serial array, so that the charge from each cell can be shifted along from cell to cell until it reaches the output wire. The CCD is in fact an *analogue shift register*. The charge is shifted along from cell to cell on the application of a clock pulse.

(a)

(b)

(c)

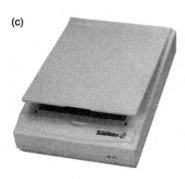

**Figure 3.64** *The three main types of scanner: (a) hand-scanners, (b) sheet-fed type and (c) flat-bed type*

Most CCD chips incorporate their own timing circuits so the only electrical connections required on the CCD are an external direct current (DC) supply and a clock pulse. *See the pages on Digital cameras.*

The analogue signal from the CCD is then converted to a digital signal by an analogue-to-digital converter (ADC) chip. The digital signal is then processed and passed on to a small computer systems interface (SCSI) circuit or Centronics interface, to allow easy connection to the computer.

The image is scanned in lines. After each line scan, the whole CCD assembly including the light source is moved a tiny fraction further along the scanner bed, by a stepper motor and toothed rubber belt. Then the next line is scanned. This action is repeated until the scan head reaches the far end of the set scan area or the end-stop of the carriage (Figure 3.66). The step distance can be set by the user from a low resolution of say 50 lines per inch up to 600 lines per inch or more, on most scanners.

### *Types of flat-bed scanner*

There are two types of colour flat-bed scanners available, namely *three-pass* and *single-pass*.

### Three-pass scanner

As the name suggests, in a three-pass scanner the object is scanned three times, once for each of the primary colours, RGB. Before each scan, one of the three primary filters is moved in front of the lens, thereby only allowing light of that colour to reach the CCD. After three consecutive scans, the RGB images are combined to form a full colour image.

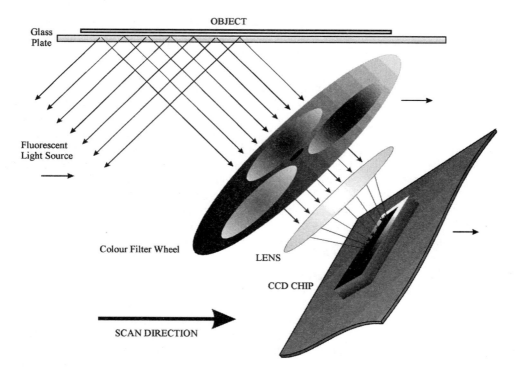

**Figure 3.65**

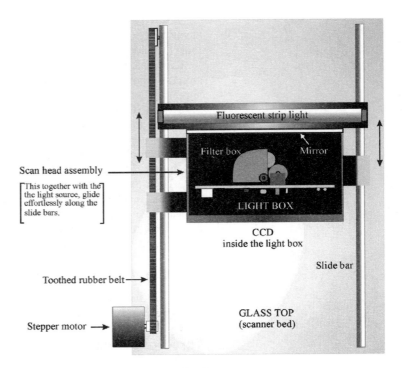

**Figure 3.66**   *The scanner mechanism*

There are two disadvantages with this system: (1) it takes time to perform three separate scans and (2) it is susceptible to registration errors. The latter will occur if the object moves between each scan. This is however unlikely when scanning from a document but it can happen when scanning 3D objects, due to the inherent vibration from the stepper system.

## Single-pass scanner

A single-pass scanner scans the object in one single traverse of the CCD assembly. The colour wheel therefore must be changed three times for each line. An alternative and much more expensive method, is to use three separate CCDs, one for each primary colour. In the latter system, the light from the image is passed through a dichroic mirror where it is split simultaneously into RGB components. At the factory, the three CCDs are registered accurately onto the image formed by the lens/filter assembly.

## *Resolution*

The resolution of a scanned image is a measure of the maximum possible amount of detail recorded by the scanner. This is based on two factors: The number of individual lines per inch, stepped by the scanner head, and the actual resolution of the CCD device.

The first factor depends on the maximum number of individual steps per inch that the scan head mechanism can achieve. More expensive scanners incorporate more sturdy mechanisms and a better-quality stepper motor.

The second factor is determined by the number of light-sensitive elements provided by the CCD. Each element determines the smallest unit of detail in the image, so the more there are, the better the resolution. High-resolution CCD arrays cost considerably more than their lower-resolution counterparts and they usually require more elaborate support chips and a better-quality lens unit.

The majority of entry level flat-bed scanners offer optical resolutions of $300 \times 600$ dpi and this can be artificially enhanced by software algorithms up to the equivalent of $2400 \times 2400$ dpi. The enhanced or *interpolated* resolution is not really a true guide to the scanner's capabilities. When comparing scanners it is the optical resolution not the interpolated resolution, that distinguishes a good scanner from a run of the mill device.

## Storage requirements for image scanning

One thing often overlooked when a scanner is purchased is the system memory and disk storage space required when an image is scanned. If we take a typical 7 in. $\times$ 5 in. colour photograph and scan it at say 300 dpi using 24-bit colour, the required storage space is found from: $7 \times 300 \times 5 \times 300 \times 3$ bytes (24 bits) $= 9,450,000$ bytes (i.e. over 9 MB). At 600 dpi the same image would take a staggering 36 MB. The resolution has doubled and the file size has quadrupled. In other words, the file size varies as the square of the resolution increase.

When a graphic image is scanned it often needs to be edited in a photo-editing program before the processed image is saved to disk. Unfortunately these programs also need a lot of memory. Most photo-editing programs running in Windows need approximately three times the image size to run satisfactorily. So to scan and edit our previous 5 in. $\times$ 7 in. photograph, we need at least 36 MB of system memory for a 300 dpi scan and 108 MB' for a 600 dpi scan.

When Windows runs out of conventional system memory (RAM) it resorts to using the hard disk as *virtual memory*. This is fine, but unfortunately it runs hundreds of times more slowly than RAM. So if a large, high-resolution image is scanned, the system will slow down to a snails pace and the image will seem to take an age, just to appear on the screen. If large high-resolution colour images are being scanned it is essential to fit plenty of RAM.

For most purposes high resolutions are not really necessary, Table 3.2 shows the typical resolutions required for different output requirements.

From Table 3.2 you can see that high scan resolutions are only necessary when the printer or printing process is capable of producing high-resolution hard copy.

## Practical work SAP 3.2: installing scanners

To install a scanner follow the manufacturer's instructions. Hot-pluggable PnP scanners based on the USB and Firewire™ are quite

**Table 3.2**

| Printer type | The original image to be scanned | | |
| --- | --- | --- | --- |
| | Colour | 256 grey scale | Line art |
| 300 dpi Mono laser or dot matrix | | 75–100 dpi | Same resolution as the printer |
| Colour ink-jet or thermal wax printer | 1/3 to 1/2 the printer resolution | 1/3 to 1/2 the printer resolution | Same resolution as the printer |
| Dye-sublimation printer | Same resolution as the printer | Same resolution as the printer | Same resolution as the printer |
| Litho printing, typesetting | Same resolution as the printing process | Same resolution as the printing process | Same resolution as the printing process |

straightforward to install. You usually install the driver and utility software first with the scanner turned off or disconnected and let the automatic installation routine guide you through the installation.

The following guide takes you through a typical process of installing a top-end SCSI flat-bed scanner. This scanner is supplied completely with a PnP peripheral component interconnect (PCI) SCSI controller card and leads.

### Installing an SCSI scanner

There are several things you need to do to install a SCSI scanner and this is a short list of the important steps to carry out:

(i) Unpack the scanner and read the packing list to ensure that everything you need has been supplied and is in good mechanical condition. If items are missing or broken first contact the distributor and state what is missing or broken. Keep the packaging as this will come in handy if the unit needs to be returned for any reason.

(ii) Unlock the shipping device (transit screw) – most scanners incorporate some form of locking device to stop the head carriage from moving whist in transit. If the scanner needs to be returned, make sure the transit device is reinstalled.

(iii) If the scanner is mains powered, check the voltage setting on the back of the scanner and confirm that it is set correctly for your mains supply.

(iv) Ensure that the scanner and PC are switched off and remove the system-unit cover. Wearing an antistatic wrist band unpack the SCSI controller card from its protective wrapper (Figure 3.67).

(v) Now insert the SCSI card into a spare PCI slot, making sure that the edge connector is pushed fully into the slot. Secure the card with a fixing screw. If no internal SCSI devices are to be fitted, refit the case cover.

(vi) Connect the power lead to the scanner and plug it into the mains but do not switch it on. Connect the SCSI cable between the computer and scanner.

**Figure 3.67**   *An Ultra 160 SCSI controller card*

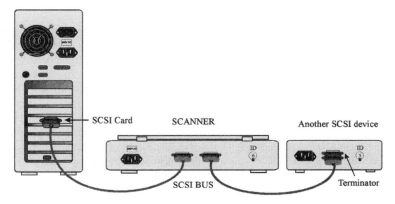

**Figure 3.68**

(vii) The SCSI card is usually preset with ID 7 and bootable SCSI hard disks are preset to ID 0. Look on the back of the scanner and set the ID to an unused number (e.g. ID 5 or ID 6) should be ideal. If the scanner is the last device on the bus, insert the recommended terminator into the scanner's upstream SCSI connector. If there is only one connector you must fit an in-lead terminator (Figure 3.68).

(viii) Switch on the PC and scanner after reading the installation instructions and install the SCSI driver software as instructed. Windows will automatically detect the presence of the adapter card and the installation wizard will prompt you for the driver disk if the correct driver is not present in the Windows built-in driver list.

(ix) The next step is to install the scanner application software. Again read the manufacturer's installation instructions for the particular version of Windows in use.

(x) After the software has been installed correctly, you will probably be asked to reboot the computer. Do this and watch the scanner ready light, if one is fitted, it should come on indicating that the scanner is correctly set up. On some scanners the ready light will blink if there is a hardware problem.

If the scanner is not the only peripheral connected to the SCSI bus (i.e. there are two or more peripherals daisy chained together), then it is important to ensure that each device has its own unique ID number and that the device at the end of the chain is terminated correctly. If this is not done, the SCSI bus will not function correctly. The SCSI controller utility software usually displays a list of the devices active on the BUS and their respective ID numbers and this helps greatly if you are unsure of the device IDs. System administrators take great care to ensure that the details of every PC system and changes to that system, are logged and kept up to date. SCSI IDs are no exception.

### Installing a parallel port scanner

Parallel port scanners reached a peak in popularity in the late 1990s as they were easier to install and less expensive than their SCSI cousins. Today however their popularity has dwindled, due mainly to the introduction of even less expensive and even easier to install USB scanners.

Parallel port scanners are designed to use the same parallel port as the printer. You might think it odd trying to connect an input device like a scanner to an output device but this is entirely feasible as today's parallel ports support the advanced bi-directional standards known as the 'EPP' and Enhanced Capabilities Port (ECP). The original IBM Parallel Port uninspiringly named 'LPT1' (from 'Line Printer One'), was indeed an output only port. However since the introduction of the IBM AT computer and its equivalents, the advanced bi-directional specifications are now standard on all AT and ATX PCs. The port is still however known as 'LPT1'. Some PCs have a second parallel port labelled 'LPT2' again supporting EPP and ECP.

To enable the scanner to run on the same port as the printer, it uses a special 'pass-through' port. The scanner usually connects to the PC via the LPT1 port using a 25-pin D female to 25-pin D male, lead. The printer then connects to the scanner using a standard parallel printer lead. The pass-through port usually requires power to work correctly so the scanner must be powered up continuously to allow the printer to function normally.

### Troubleshooting parallel printer/scanner problems

Problems relating to combined printer scanner installations where the printer fails to work correctly can often be traced to the scanner being turned off by the operator. The port is shared, so both the scanner and printer must be turned on. Other problems relate to the port settings in the BIOS. As always, follow the scanner makers instructions but in general, parallel scanners need LPT1 set to EPP and/or ECP mode. To set these, go into CMOS setup, enter the 'Integrated Peripherals' menu and select EPP, ECP or ECP + EPP.

Some printers that use bi-directional communication can cause problems with a parallel scanner attached. Such conflictions may be overcome by disabling bi-directional support in the printer driver.

To do this in Windows 95/98/ME/2000: Open the Printer window | **Start** | **Settings** | **Printers** |, right click on the relevant printer icon. Now choose | **Properties** | **Details** | **Spool Settings**. Click the '**Disable Bi-Directional Support for this Printer**' radio button.

To do this in Windows XP: Choose | **Start** | **Control Panel** | **Printers and Faxes** |, right click on the relevant printer icon. Now choose | **Properties** | **Ports** |. Uncheck the '**Enable Bi-Directional Support**' radio button.

Try this to see if this cures the problem, if it does not, return the setting to its previous state. The ultimate solution may be to install a second parallel printer port using an add on card, thus giving the scanner its own unique port.

### Installing a USB scanner

The majority of scanners available today are supplied as USB versions. The USB provides a simple hot-pluggable system that is easy to install and inexpensive enough to satisfy the mass market. In a PnP OS-like Windows, generally it is a case of simply connecting the power source to scanner and connecting the scanner to the USB port – you do not even have to switch off the computer system before you do so. Thanks to PnP, the OS automatically detects the scanner and prompts the user for the driver disk. Most USB scanners are generally easy enough for the novice to install without assistance. Whoever installs the scanner, it is important to refer to the manufacturers installation instructions for the version of the OS in use and follow the recommended procedure.

### OCR

Apart from their usefulness in scanning photographs and small flat objects, scanners are ideal for scanning printed text for insertion into a text file. Rather than inserting the scanned text into a document as an image (which it would be if it was output straight from the scanner into the document), it can be input to an OCR, utility. The OCR program cleverly peruses the image of the scanned text looking for recognizable words and characters and converts it to a proprietary text file. Compared to just scanning the text as an image it drastically reduces the size of the resulting file and allows the text to be inserted and edited in a word processor.

Current OCR algorithms are powerful enough to reproduce images, tables, and text straight off the scanned page and into a word processor, spreadsheet or database.

# Modems

A modem is used to transfer data serially over long distances using the Public Switched Telephone Network (PSTN). Unfortunately telephone lines were originally designed for analogue voice communication and therefore have a rather limited bandwidth of around 3000 Hz. Most telephone exchanges in the UK have now been updated to digital systems, with fibre-optic cables and microwave links, to other exchanges. However, many local lines still use the old analogue cables and these are unsuitable for digital data transfer.

To enable a computer system to use the extensive PSTN system, the modem must therefore convert digital signals at the computer, to an analogue signal for transmission over the telephone line.

Modems suitable for use on a PC are usually supplied either on an adapter card to fit the ISA or PCI bus or as an external unit that plugs into COM1 or COM2. These are known as internal and external modems, respectively. External modems were generally regarded as the most convenient choice as they did not use up valuable port addresses and IRQ lines and were also easier to install. Since PnP, this is no longer an issue and the internal types are generally less expensive.

An external modem will be considered in this example: During transmission, the modem converts RS232 signals from the serial port (A in the diagram) to a modulated tone signal at the telephone line (C). The amplitude, frequency and phase of the tone can be altered to represent the equivalent of several bits of data. The basics of tone modulation were covered in Unit 2.

During reception the converse applies, the modem converts the modulated tone from the PSTN (C) into an RS232 signal at the COM port (B). With good line conditions and a good error detection/correction rate, speeds up to 56,000 bits per second (bps) can be achieved with this system (Figure 3.69).

### Handshaking

Apart from the modulation/demodulation process, a means of controlling the data flow along the link is required. This process is known as '*handshaking*'. Handshaking is necessary to allow an orderly flow of data from the sending device to the receiving device. For example, it allows the receiving device to inform the sending device that it is busy or waiting to receive more data, and the sending device can inform the receiver that it has new data to send  (Flow Control).

Handshaking can be achieved by using extra wires between the two ends of the link. This is known as *hardware handshaking* and is used in both the RS232 serial system and the parallel Centronics system.

Handshaking is particularly necessary when a fast device is linked to a slow device, like a computer/printer link, for example. The slow device must have a means of signalling to the faster device when to send and stop sending data.

Hardware handshaking is not possible on the PSTN with only two wires, so *software handshaking* is used. This is based on sending special commands along with the data to control the data flow along

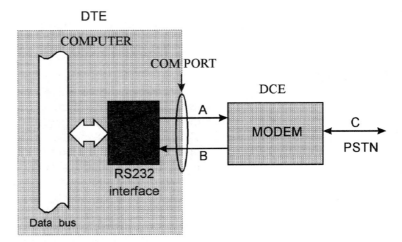

**Figure 3.69**   *A modem connected to a computer and telephone line*

the PSTN. The simplest commands are basic flow control commands such as XON/XOFF (transmission on/transmission off). The particular commands used depend on the particular data transfer protocol in use.

### Protocols

When a computer is linked to the PSTN via a modem, special communication software is used to manage the transfer of data across the link. This software uses sets of universally recognized rules and commands, known as *protocols*. The protocols are usually drawn up by major computer/telecommunication companies, which govern the organization and transmission of data. There are many different protocols in use, each with their own advantages and disadvantages. Their purpose is to allow intelligent control of data flow across a communication link with the following two main aims:

1. *Handling breakdowns in transmission*: for example the cable could be accidentally disconnected, in which case an intelligent system would allow transmission to continue where it left off.
2. *Handling errors*: for example if an error is received, the system should allow for detection of the error and re-transmission of the corrupted character/s.

Modern communication software contains programs to oversee the complete communication system installation, from the setting of basic parameters such as data rate and character frame parameters, to actually establishing a communication link between the terminal and its host and allocating a particular command protocol.

The modem application program usually allows the user to choose a particular file transfer protocol, the four main ones used on simple PC to PSTN links are XMODEM, YMODEM, ZMODEM and KERMIT. These are described briefly below.

### XMODEM

This is a simple protocol developed by Ward Christianson in 1978, to give computer users an easy to use and reliable system. The XMODEM protocol places data into blocks of 128 bytes and adds a checksum. The checksum is a sum of the 128 unique bytes making up each block. The receiving device calculates its own checksum by adding each of the 128 received bytes and compares it with the received checksum. If the checksums do not match, then the software initiates a repeat of the last block. If they do match, then the next block is requested. This protocol relies on the transmitter pausing and waiting for an ACK message (data ACKnowledged) or NAK (data Not AcKnowledged) after each block of data is sent, this inevitably results in a slow transmission rate.

### YMODEM

YMODEM is based on XMODEM but transfers data in blocks of 1024 bytes.

### ZMODEM

This is an elaboration on XMODEM and allows faster file transfer and automatic mode setting at the receiver. This protocol is commonly used on PC to Bulletin-board, links.

### KERMIT

This is totally different from the other protocols mentioned as it converts all data into codes to improve error detection and correction. It also has its own built-in communications software. This gives it the ability to access other software files as well as handle raw data. KERMIT is supported in the UK by Lancaster University.

### TCP/IP

In order to transfer data reliably, across international telephone networks, a set of internationally agreed guidelines were drawn up by the International Standards Organization (ISO) known as the Open Systems Interconnection reference model (open system interconnection, OSI). The guidelines are based on seven layers of logical structures for defining protocols for a variety of network applications, from the *physical layer* (layer 1) – the actual electrical and electronic considerations – to the so-called *application layer* (layer 7) – concerned with the user–software interface.

There are many protocols for local-area networks (LAN) and wide-area networks (WAN) in use, that broadly follow the OSI Guidelines. The most notable and widely used network protocol is the *Transport Control Protocol/Internet Protocol*, commonly known as TCP/IP. This is a vendor independent and hardware independent protocol that can be used by any WAN, LAN or simple PC to PC link without regard to the actual hardware configurations in use. It is the protocol used on the *Internet* – the vast, extremely popular, international network of computer users.

## Communication interface standards

Devices can only communicate across a modem link when a standard communication method is adopted. The bit rate, modulation method, compression technique and error correction/detection system must conform to a common standard.

**Table 3.3** *List of the ITU standard modem speeds*

| ITU(CCITT) version | Maximum data rate (bps) | Duplex |
| --- | --- | --- |
| V.17 | 14,400 | Full |
| V.21 | 300 | Full |
| V.22 | 1,200 | Half |
| V.22bis | 2,400 | Full |
| V.27 | 4800 | Half |
| V.29 | 9,600 | Half |
| V.32 | 9,600 | Full |
| V.32bis | 14,400 | Full |
| V.32ter | 19,200 | Full |
| V.34 | 28,800 | Full |
| V.34bis | 33,600 | Full |
| V.90 | 56,000 | Full |

*bis* and *ter* mean second and third version respectively.

Several international bodies have been set up purely to maintain and develop common standards. The important ones for communication interface standards are: the International Telecommunications Union (ITU), set up under the United Nations, the ISO and the International Telegraph and Telephone Consultative Committee (CCITT), now under the wing of ITU.

Some of the communication interface standards set by the ITU are shown in Table 3.3.

---

**For the technically minded:**

In 1948, Claude Shannon, showed that the maximum possible bit rate $T_{r(max)}$ in bps, of a practical transmission line of bandwidth $B_w$, and signal to noise ratio $S_n$, is given by the formula:

$$T_{r(max)} = B_w \times \log(1 + S_n)/\log 2$$

An analogue telephone line, for example, has a maximum bandwidth of 3000 Hz and a maximum signal-to-noise ratio of about 1000:1. So from Shannon's formula, the maximum bit rate achievable is about 30,000, that is:

$$T_{r(max)} = 3000 \times \log(1001)/\log 2$$

i.e. approximately 30,000 bps

---

Most modems manufactured prior to 1998 support the V34 and V34bis speeds. They also support the V42 and V42bis error correcting and data compression standards. Many modems nowadays are sold as fax/modems as they support the Group 3 FAX modes, V17,V21,V27 and V29.

### Pushing the speed even higher

Two major American modem manufacturers US Robotics and the Rockwell have independently developed 56,000 bps modems that purport to run on standard PSTN lines. The two systems are known as X2 and K56Flex, respectively. In February 1998 the ITU agreed these new speeds in the ITU-T V.90 standard. V.90 provides compatibility across the two systems and allows either type of 56kB modem to connect to any other modem type. Several Internet service providers (ISPs) have agreed to support the new transfer rates and have achieved download rates (ISP to USER) of 56 kbps and upload rates (USER to ISP), on an average of 33.6 kbps on standard PSTN lines.

### British Approvals Board for Telecommunications/ European Union approval

The telecommunications system in the UK is no longer a monopoly, and line equipment can be obtained from a variety of suppliers, instead of just one company. To ensure that only approved equipment is connected to the PSTN the British Approvals Board for Telecommunications (BABT) was set up. One of its roles was to set rules and standards upholdable in a court of law that make it illegal to place incorrect currents, voltages and loads on the PSTN. Such

**Figure 3.70** *Logos to indicate that equipment conforms to an accepted standard for use on European Public Telephone Networks*

malpractice could if blatantly disregarded, jeopardize the safety of PSTN line workers or result in line malfunction.

Up to April 2000 all telecommunications equipment including telephones and modems sold in the European Union (EU), had to be independently certified as being compatible with the PSTN as determined by standards bodies such as BABT. However under an EU Directive in April 2000, the ruling no longer holds. Manufacturers are now independently responsible for compatibility. However most reputable manufacturers still conform to the standards originally laid down by BABT and the ITU (Figure 3.70).

## Practical work SAP 3.3: Installing and configuring a modem

There are both internal and external modems available for the PC. An internal modem plugs into a spare PCI slot inside the PC using the internal power supply and data bus, whereas an external modem sits outside the PC and therefore requires a separate mains adapter and a vacant serial port, i.e. COM1 or COM2.

An external modem package usually contains the following items:

(i)  An approved external modem unit
(ii)  Mains powered DC supply (adapter)
(iii)  RS232 cable
(iv)  Telephone cable
(v)  User manual
(vi)  Driver software

*Note:* If you are planning to use an internal modem do not be tempted into buying a cheap one. Some of these inexpensive brands use a host signal processor (HSP). HSP modems use the OS to carry out the digital signal processing instead of using a dedicated digital processor chip on the modem card itself. At the very least they hog processor time when transferring data via the PSTN. Also the signals from these modems are weak compared to their dearer counterparts. The result is that the user may experience connection problems. Some remote modems may fail to connect at all.

### Connecting the hardware

Many modems have two linked BT sockets to allow both a telephone and modem to share one BT wall socket. The telephone plug is inserted into the modem and a fly lead connects the modem to the wall socket.

To connect the modem to the computer, the RS232 cable is connected between the 25-way 'D' socket on the back of the modem and a spare serial port on the computer (i.e. COM1 or COM2). Usually COM1 has a serial mouse attached, so COM2 is used.

Finally the mains adapter is pushed into a convenient mains outlet and the low-voltage output lead is connected to the DC power socket on the modem.

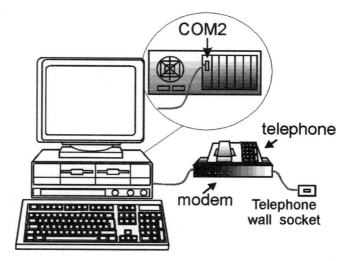

**Figure 3.71**

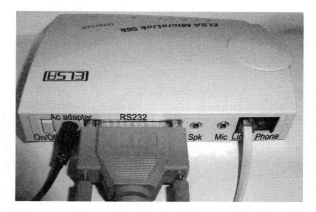

**Figure 3.72** *The rear panel of a modern modem showing the various connections*

**Figure 3.73**

The signal connections between the modem and the computer are shown in Figures 3.71 and 3.72.

## Installing a modem in Windows 95/98/ME

Installing a modem in Windows is very straightforward. Follow the simple steps below:

(i) Choose **Start** | **Settings** | **Control Panel** | **Modems** (Figure 3.73).

(ii) Click **Add**. An option screen appears asking if you want Windows to auto-detect your modem or choose one from a list. Select the latter, then click Next (Figure 3.74).

(iii) Select the modem manufacturer and model type from the list. If your modem is not listed, then use the Windows 95/98 driver software supplied by the manufacturer and click **Have Disk** ... (Figure 3.75).

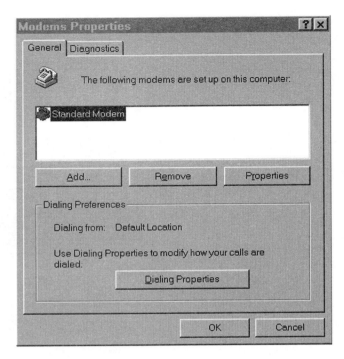

**Figure 3.74**

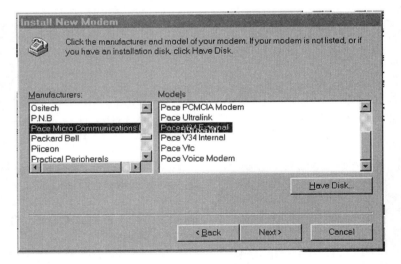

**Figure 3.75**

(iv) Now select the serial port you want to use. The default COM2 is usually highlighted. Select this and then click **Next**> (Figure 3.76).

(v) You should then be informed on a successful installation and told how to make subsequent changes to the modem settings (Figure 3.77).

(vi) To set the modem/telephone dialling properties Choose **Start** | **Settings** | **Control Panel** | **Telephony**. Enter your country/region and area code and any other relevant dialling properties. Click **OK** (Figure 3.78).

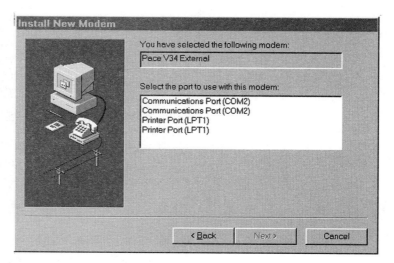

**Figure 3.76**

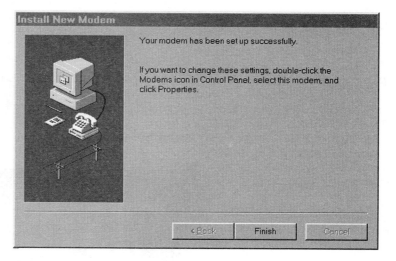

**Figure 3.77**

# Installing a modem in Windows XP

The modem installation procedure in Windows XP is slightly different to that of Windows 95/98/ME:

(i) In Windows XP, choose **Start | Control Panel | Phone & Modem Options**. Enter your phone/modem area code etc, then select **OK** (Figure 3.79).
(ii) Select **Modems** (Figure 3.80).
(iii) Click **Add** … (Figure 3.81).
(iv) Choose to let Windows XP auto-detect your modem or select one from a list. **Click Next**> (Figure 3.82).
(v) Choose a standard modem from the list or click **Have Disk** … (Figure 3.83).

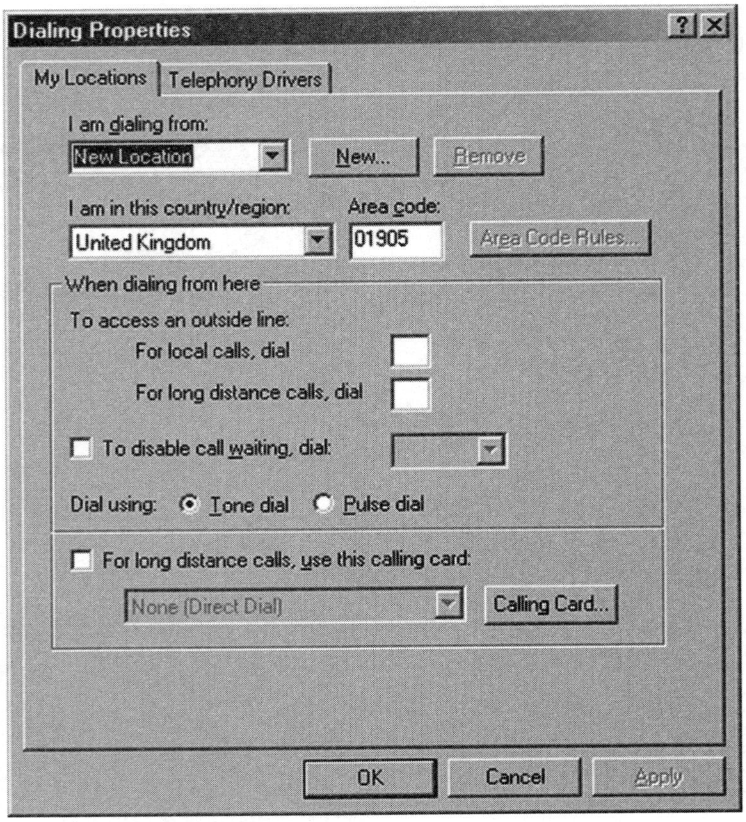

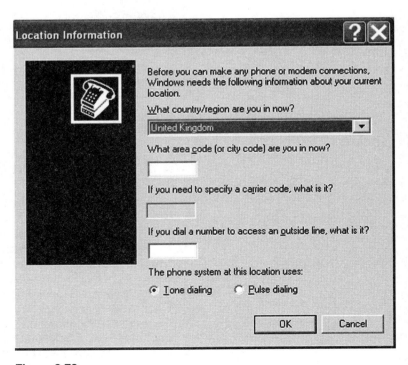

**Figure 3.78**

**Figure 3.79**

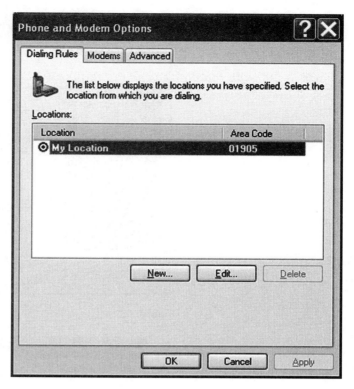

**Figure 3.80**

**Figure 3.81**

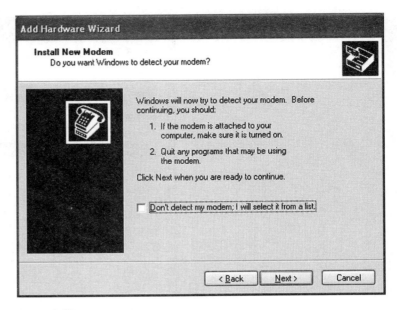

**Figure 3.82**

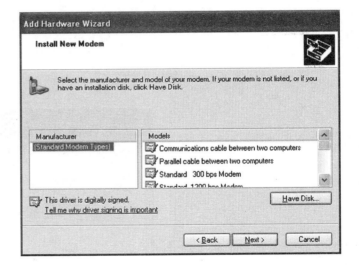

**Figure 3.83**

(vi) Choose an I/O port for the modem. Select **Next**> (Figure 3.84).
(vii) The modem installation is now complete. Click **Finish** (Figure 3.85).

## Practical work SAP 3.4: Setting up an Internet connection

After installing a modem, the next step is to set up an Internet dialup connection to a remote service provider. This will allow the user to access the Internet and send and receive e-mail.

Service providers or ISPs, as they are now popularly called, are companies with a gateway to the Internet backbone. The most

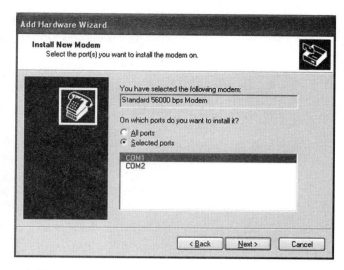

**Figure 3.84**

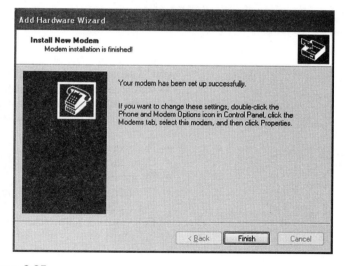

**Figure 3.85**

common type of connection offered to clients is a dialup account. The name derives from the fact that the client connects through a modem and telephone line. The ISP shares the gateway to a large number of clients for a monthly fee. Often the fee includes the telephone call charge as well as the use of their service.

If you are setting up the dialup account for a client it is important that they are fully informed of the terms and conditions drawn up by the ISP and have consented to the contract.

### Internet installation

Large ISP companies usually ship out thousands of Internet access compact disk read only memory (Internet access CD-ROMs) in magazines and through the post with junk mail. Each one contains a unique password and an automatic setup program to help the user

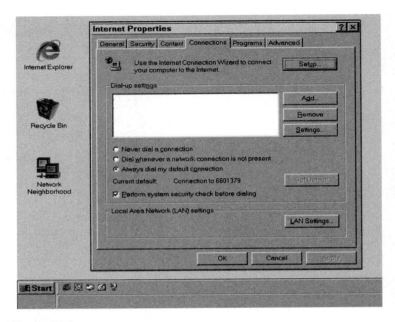

**Figure 3.86**

make a connection quickly and easily. Alternatively some ISPs allow you to set up an account over the phone or via the post in the form of a written contract.

In the following scenario, we have already set up an account with an ISP and obtained the following connection information:

- Password
- User name
- Phone number
- IP addresses (usually these are not required as the ISP will set these automatically)
- POP3 server name (the ISPs e-mail server name for your incoming mail)
- SMPT server name (the ISPs e-mail server name for your outgoing e-mail)

### Windows 95/98/ME and 2000 Internet Connection Wizard

- Right Click the Internet Explorer icon on your desktop. In the dialogue box, select the 'Connections' tab. **Select Always dial my default connection** then click on **Setup** (Figure 3.86).
- This invokes the Internet Connection Wizard menu as shown in Figure 3.87.

You now have a choice of options including a short tutorial on the Internet.

If you have an ISP lined up and you know their connection phone number then choose the third manual setup option – this is the recommended option. However, if you do not know what ISP you want to sign up to then you could choose option 1. The wizard will automatically dial a local sever and give you a list of possible ISPs to choose

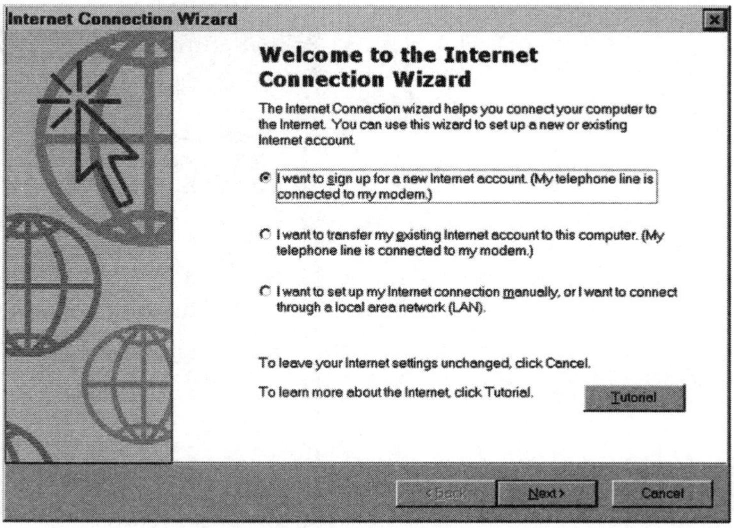

**Figure 3.87**

from. Click **Next>** and follow the directions in each subsequent window.

The full step-by-step guide that follows if for Windows XP:

## Windows XP Internet Connection Wizard

(i) Choose **Start | Control Panel | Internet Options** or right click the Internet Explorer icon on the desktop or in the Start menu and choose **Properties**. Choose **Programs** and enter details of the programs you want Windows to use for e-mail, etc. The html editor and other items can be left blank if they are not required. Click **OK** (Figure 3.88).

(ii) Choose **Connections**, click on **Always dial my default connection** then Click **Setup** ... (Figure 3.89).

(iii) The **New connection Wizard** is displayed. Click **Next>** (Figure 3.90).

(iv) Choose **Connect to the Internet**. Click **Next>** (Figure 3.91).

(v) Choose how you want to set up the Internet connection. We have a password and ISP phone number so we chose **Set up my connection manually**. Click **Next>** (Figure 3.92).

(vi) Choose **Connect Using a dialup modem**. Click **Next>** (Figure 3.93).

(vii) Choose the modem driver you want to use to use for the Internet connection. Click **Next>** (Figure 3.94).

(viii) Choose a name to identify your ISP. Click **Next>** (Figure 3.95).

(ix) Enter the area code and phone number of your ISP. Click **Next>** (Figure 3.96).

(x) Enter the **User name**, **Password** and check the user options as required. Click **Next>** (Figure 3.97).

(xi) This completes the information gathering stage. Click **Finish** to create the connection (Figure 3.98).

(xii) Now the Internet connection is set up you can change the default Internet home page that appears when you connect to

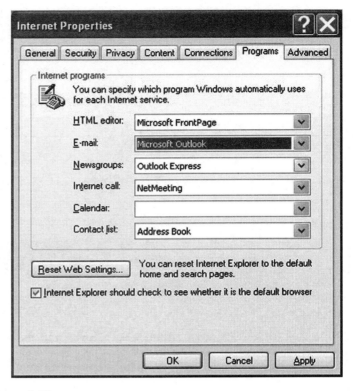

**Figure 3.88**

**Figure 3.89**

**Figure 3.90**

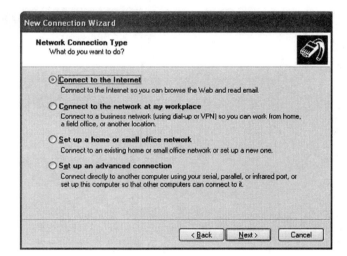

**Figure 3.91**

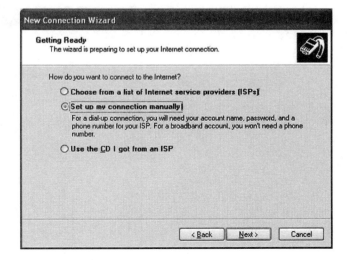

**Figure 3.92**

**Figure 3.93**

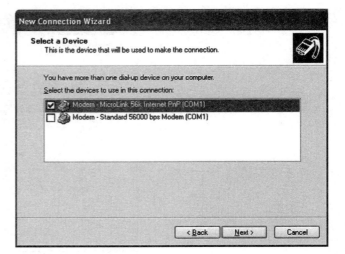

**Figure 3.94**

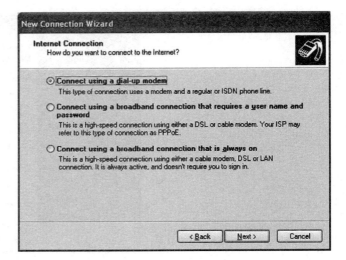

**Figure 3.95**

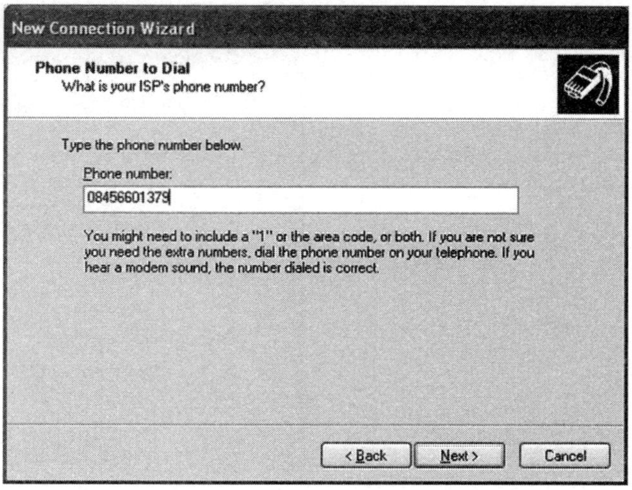

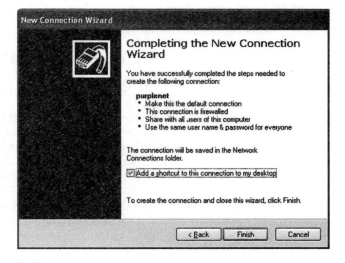

**Figure 3.96**

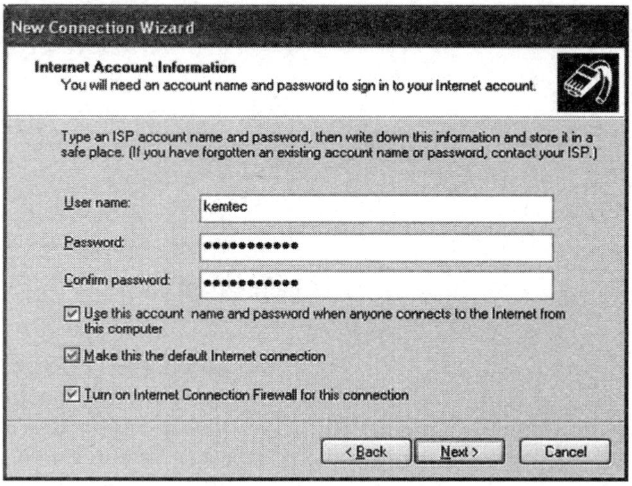

**Figure 3.97**

**Figure 3.98**

**Figure 3.99**

the Internet. Choose **Start** | **Control Panel** | **Internet Options** or right click the Internet Explorer icon on the desktop and choose **Properties**. Choose General and enter the URL of the user's favourite web site (Figure 3.99).

(xiii) Finally if you want to use Microsoft's Internet Explorer as your browser click on the Internet Explorer icon on your desktop or in the case of XP, in Start. Select the option to make it your default Internet Browser.

## Practical work SAP 3.5: Setting up e-mail

Setting up e-mail is just as straightforward as setting up the Internet. Here we show you how to set up Microsoft Outlook. Outlook is provided with Microsoft's Office applications software. If you do not have Outlook use Outlook Express provided with Windows. The installation process is similar.

(i) Double click the Outlook icon on the desktop to run the Outlook Startup Wizard. Click **Next>** (Figure 3.100).

(ii) Choose an e-mail service option for stand-alone users the first option **Internet Only** is usually best (Figure 3.101).

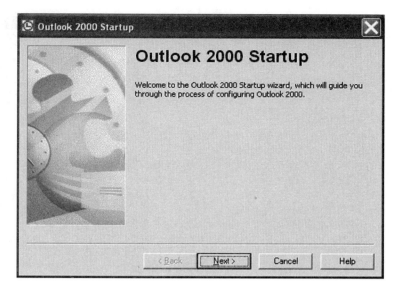

**Figure 3.100**

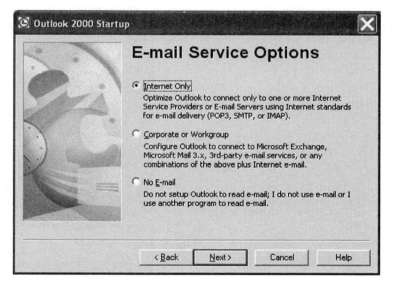

**Figure 3.101**

(iii) Enter a valid e-mail address as provided by your ISP. Click **Next>** (Figure 3.102).

(iv) Most ISPs are POP3 servers so select this option. Then enter the POP3 server names provided by your ISP. You should have one for POP3 incoming e-mail and one for outgoing SMTP e-mail. Click **Next>** (Figure 3.103)

(v) Enter your account name and password supplied by your ISP. Click **Next>** (Figure 3.104).

(vi) Choose the method you use to connect to the Internet. Click **Next>** (Figure 3.105).

(vii) Select your modem. Click **Next>** (Figure 3.106).

(viii) Make sure your ISP's phone number is correct. Click **Next>** (Figure 3.107).

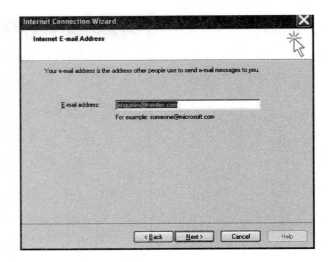

**Figure 3.102**

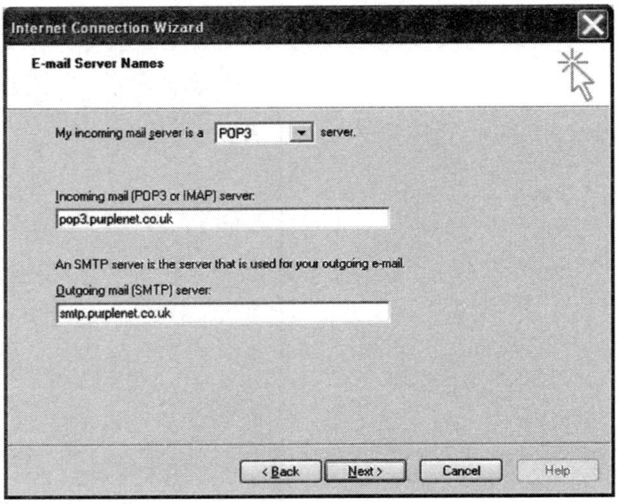

**Figure 3.103**

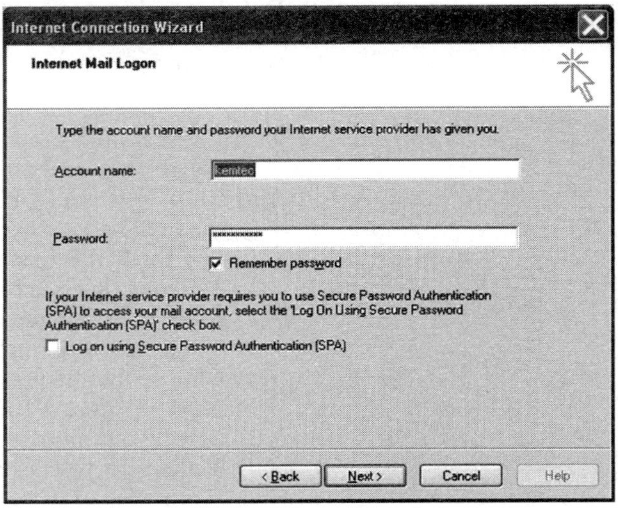

**Figure 3.104**

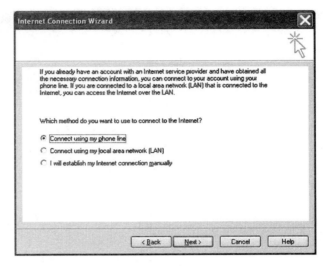

**Figure 3.105**

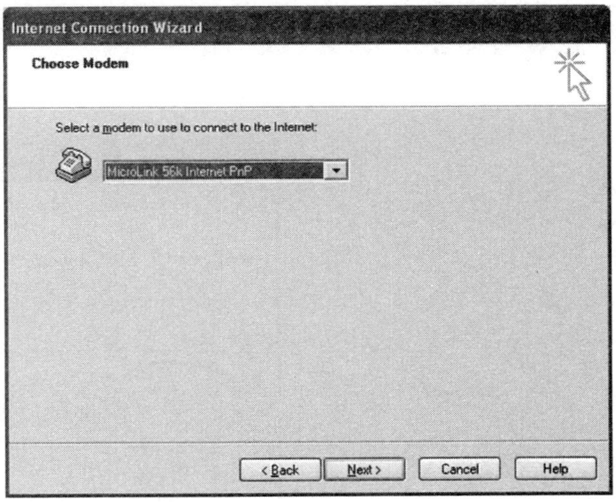

**Figure 3.106**

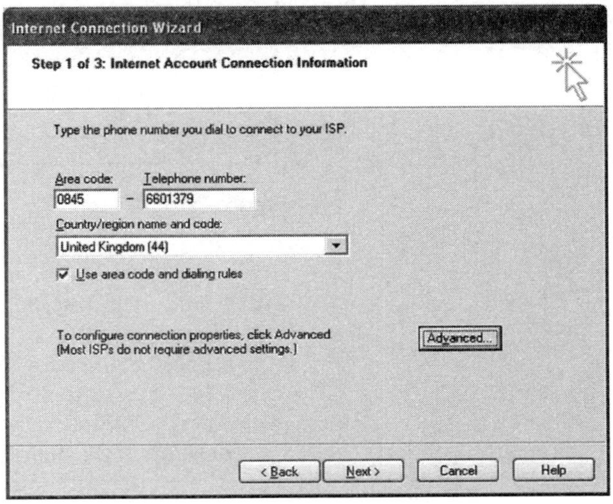

**Figure 3.107**

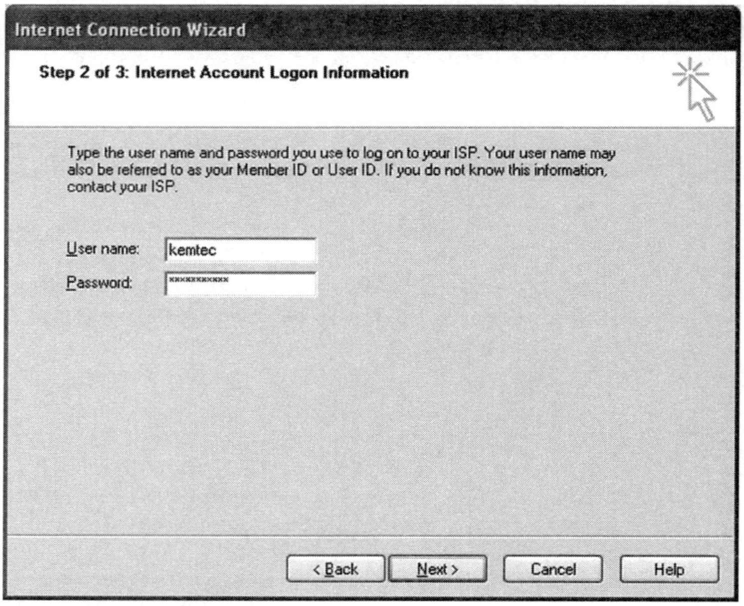

**Figure 3.108**

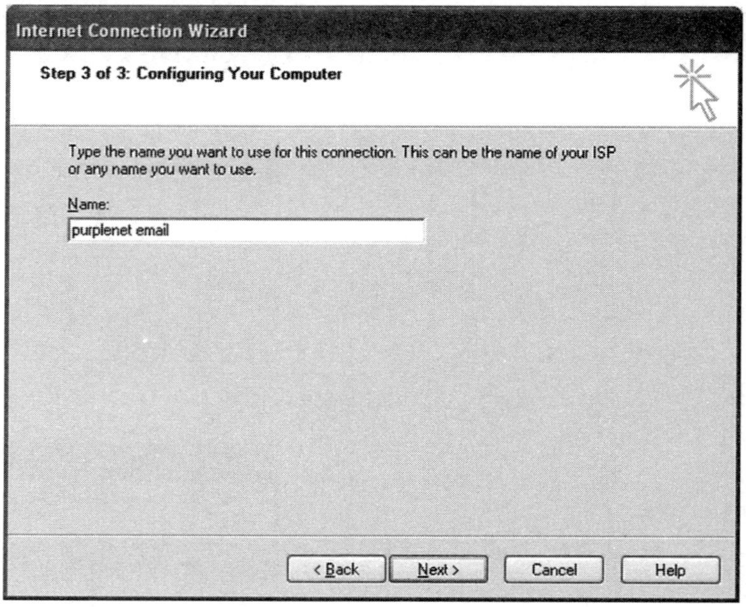

**Figure 3.109**

(ix) Enter your Internet User name and Password. Click **Next>** (Figure 3.108).

(x) Choose a name for the connection. Click **Next>** (Figure 3.109).

(xi) Click **Finish** to complete the Outlook setup (Figure 3.110).

(xii) To use Outlook for your e-mail. Click on the Outlook icon in the desktop. If the dialog box below appears click **Yes** to set Outlook as your default e-mail manager (Figure 3.111).

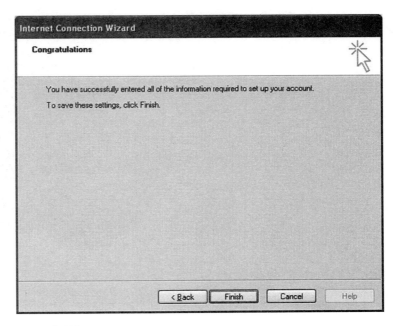

**Figure 3.110**

**Figure 3.111**

# Digital cameras

## What is a digital camera?

A digital camera is a relatively new type of camera that exposes an image onto a light-sensitive chip rather than onto photographic film. The process of forming the image onto the light-sensitive plane is the same for both types of camera. After that the two technologies differ significantly. Instead of the chemical process of transferring the film's latent image onto light-sensitive paper, as in conventional photography. The image is read digitally from the light-sensitive chip into a non-volatile memory device. The stored image can then be downloaded to a computer. Once in the computer, it can be used immediately in a document. This is the major advantage of a digital camera over a conventional file camera. The time from image capture, to its use in a document, can often be measured in seconds, compared to the hours or days of conventional film photography (Figure 3.112).

Digital cameras have been about for nearly a decade but their popularity has only just started to match that of the conventional film camera. This is partly due to the increase in Internet use and partly due to falling prices and increasing quality of the imaging system.

When the first affordable digital cameras appeared on the market the image quality left a lot to be desired. The average camera was only capable of producing fuzzy looking images, only suitable for non-demanding applications such as web page pictures. In short the image

**Figure 3.112** *The Cybershot DSC-S70 3.3 megapixel digital camera by Sony*

**Figure 3.113**   *A lens assembly from a digital camera showing the tiny 'focus' and 'zoom' motors*

quality was well below that of even the most rudimentary 35 mm film camera.

Today the story has changed significantly. Although inherently still not matching the fine resolution offered by film, the latest 2–3 million pixel (megapixel) cameras are producing images that virtually match those obtained from middle of the range 35 mm film cameras.

### Film versus CCD

For the technically minded, a typical film camera can resolve 50 line pairs per mm at a contrast ratio of 1000:1. On a 35 mm film frame of 24 mm × 36 mm, this equates to 8,640,000 pixels (24 × 36 × 100 × 100). Today's high end digital cameras are equipped with high-resolution CCDs with resolutions topping this. So in terms of theoretically achievable resolution, top-end digital cameras are producing results on par with 35 mm film.

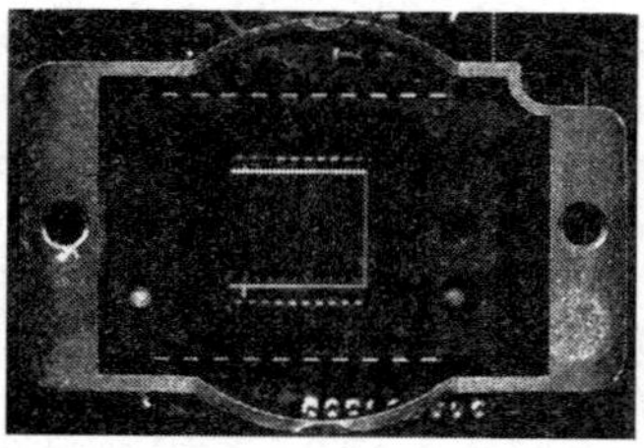

When other factors are taken into account, such as the films light sensitivity and the quality of the film to paper developing and printing process, the results obtained from a middle range digital camera are also very impressive. For desktop publishing (DTP) and website design, the digital camera is unchallenged. Images can be quickly and effortlessly captured in a standard image format (tif, gif, jpg, pcx, bmp, etc.) and inserted into the document being composed (Figures 3.113 and 3.114).

**Figure 3.114**   *The CCD, the heart of the digital camera*

## How a digital camera works?

The image of the object being photographed is focused onto the optically flat surface of a CCD (Figure 3.115). This is a superfine array of tiny light-sensitive MOS field effect transistors (MOSFETs). The gate electrode and underlying silicon substrate of each MOSFET form a tiny capacitor. Each capacitor forms an individual light

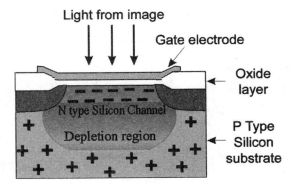

Light from image

Gate electrode

Oxide layer

N type Silicon Channel

Depletion region

P Type Silicon substrate

**Figure 3.115** *Diagram of a CCD cell*

sensor cell. When light falls on each cell, electrons are released from the semiconductor material forming a tiny electrical charge on the capacitor. This charge is directly related to the intensity of light falling on the cell. An array of such cells is therefore capable of storing an image. Once an image has been exposed, a light shutter covers up the cells while the charges are being read out of the array, to prevent re-exposure. After processing the image is stored in a read/write memory device such as battery backed RAM or magnetic disk.

To extract the charge relating to each element of the image, the cells are arranged in rows. Each cell in a row is *coupled* to its neighbour – hence the name CCD. The way it works is analogous to a line of sand buckets in a bucket brigade. The quantity of sand in each bucket represents the electrical charge on each cell which in turn relates to the light intensity. The charge stored in a single row is extracted by passing the charge from one cell to another in a sequential manner. The whole process is controlled by a quartz-crystal-controlled clock pulse.

The sequence of events for extracting a single row of information (also, see Figure 3.116) is as follows:

- Clock pulse 1: the first bucket empties its charge (*the charge on the first cell is output*).
- Clock pulse 2: the second bucket empties into the first bucket.
- Clock pulse 3: the third bucket empties into the second bucket while the first bucket empties (*the original charge on the second cell is output*).
- Clock pulse 4: the fourth bucket empties into the third bucket, and the second bucket empties into first.
- Clock pulse 5: the fifth bucket empties into the fourth, the third empties into the second and the first bucket empties (*the original charge on the third cell is output*).

This process continues until all the charges have shifted along the row to the output end where they are read. The CCD's output circuit converts the charge from each cell into a signal voltage. This is achieved using a capacitive circuit. For a fixed value capacitor, the voltage output is directly proportional to the charge from each cell.

Piles of sand representing the charge in each bucket

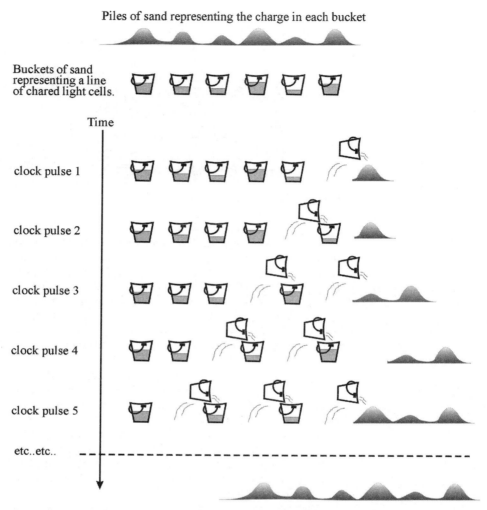

**Figure 3.116**   *Representation of the action of a bucket brigade CCD device. After a finite number of clock pulses, the output from the line of cells is a serial stream of analogue pulses whose amplitude is directly related to the charge stored in each cell. The number of pulses is equal to the number of cells in the row*

There are naturally thousands of cells to a row and thousands of rows in a megapixel CCD. All the rows making up the CCD array are clocked simultaneously by a pixel scanning circuit. The signal voltage from each row is combined into a single digital signal via an ADC chip before being fed to a digital signal processing circuit. Here the signal is reconstituted as a standard full colour bitmap image and then saved to memory.

### How colour information is sensed by the CCD?

There are two main techniques used in commercial digital camera to obtain a full colour image from a CCD they are:

- RGB 3-CCD cameras
- Single CCD cameras

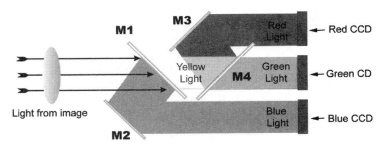

**Figure 3.117** *The RGB light splitter arrangement of a three colour CCD camera*

## RGB 3-CCD cameras

The best technique used for high-quality imaging, is to use one CCD for each of the primary additive colours. To achieve this, light from the image is split three ways by a prism or a set of dichroic mirrors. A dichroic mirror allows the light from two primary colours to pass through it, while reflecting a third primary colour. These mirrors are manufactured by vacuum depositing various minerals onto the surface of optically flat glass plates. In Figure 3.117, light from the image passes through dichroic mirror M1 where the blue component of the image is reflected and the red and green component passes through. Dichroic mirror M4 then reflects the red component and the green component passes straight through onto the 'green' CCD. The reflected blue and red light component is further reflected by the plain mirrors M2 and M3 onto their respective CCDs.

There are additional correcting lenses and filters not shown in the diagram to remove any astigmatism and to adjust the respective light levels.

Using this method, the full CCD resolution is available for all three colours. The only major down fall with this method is the cost of the prism and the two extra CCDs. There is also a slight trade off in resolution due to the precise registration required to superimpose the three images accurately, but the results from this system are still superior to that of the single CCD camera using the same type of CCD.

## Single CCD cameras

By far the most common type of digital camera is the single CCD colour camera. To retrieve colour information from the image, a special filter is placed immediately over the CCD. This consists of an array of ultra miniature transparent filters arranged in groups of four (quads). Each of the tiny filters fits over just one light cell. As the green component of the image contains the most brightness detail there are two green filters and only one of red and one of blue (Figure 3.118).

Using this arrangement each cell only receives light of a certain colour. The red and blue components of the image respectively use only 25 per cent, of the overall CCD resolution and the green image uses 50 per cent. Despite this, when a high-resolution megapixel CCD is used, surprisingly good results are obtained (Figure 3.119).

## The camera to computer interface

After a photographic session, the next step is to download the images from the camera's memory to the computer. The first generation of digital cameras interfaced to the computer via the serial RS232 port.

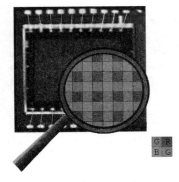

**Figure 3.118** *A close-up of the single CCD RGB filter*

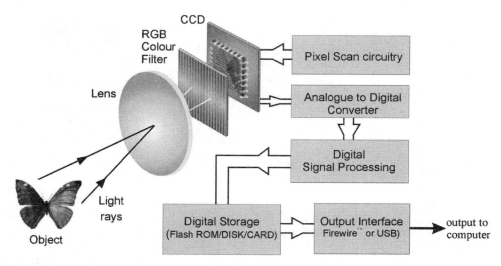

**Figure 3.119** *Block diagram of a single CCD camera system*

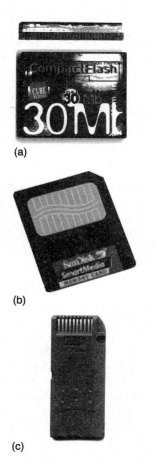

**Figure 3.120** *Flash Memory Cards: (a) Compact Flash, (b) Smart Media and (c) Memory Stick*

Transferring images from the camera was a time-consuming process, as the standard data transfer rate on COM1 and COM2 is at best around 115 kilobits per second (kbs). This means that a camera with 30 MB of memory, will take about 36 minutes to download, and this is not taking into account the image processing time. On one old 1.2 megapixel camera tested, each image in high-resolution mode, took 2 minutes to download via the RS232 port. The camera stores 38 images in 30 MB of memory, so the total download time is well over an hour! This shows that the processing time can also be quite significant. Compare this with conventional film developing and times and even this does not seem too long!

Today's digital cameras interface to the USB or Firewire™ port which really makes a huge difference to the image transfer rate. USB 1 provides a data transfer rate of 12 Mbps, some hundred times faster than the PC's standard serial port. USB 2 and Firewire™ cameras improve on this even further. So instant digital photography is now a reality!

People who own first-generation digital cameras with removable memory devices can greatly improve the download time by using a USB Flash Card Reader. After taking pictures the memory card is simply removed from the camera and inserted into the Flash Card Reader. The computer then reads the images from the Flash Card just like a hard disk drive only it is a lot faster.

## Memory devices for digital cameras

The most common image storage device used on digital cameras is Flash Memory. This is a special type of RAM device that retains data even when the device is unplugged from the camera. It is usually packaged as a flat plastic card. Unfortunately there is little standardization in the packaging. Some of the types available are 'Smart Media™', 'Compact Flash™' and 'Memory Stick™'. They all use different connections and packaging.

'Smart Media' is simply Flash Memory embedded in a thin card with the connections exposed on top.

Compact Flash Cards (Figure 3.120) have a controller chip built-in as well as the Flash Memory. The connections are also well protected.

The 'Memory Stick' by Sony.

## Practical work SAP 3.6: Installing and configuring a digital camera

Installing a USB digital camera is probably the most straightforward peripheral of all to install but as always, refer to the manufacturer's installation instructions for your particular OS.

A typical installation procedure goes something like this:

(i) Follow the instructions and install the camera battery and memory card. If the camera is new you may have to charge the battery first.

(ii) Turn on the computer and let Windows boot.

(iii) Connect the USB cable between the camera and PC and turn on the camera.

(iv) Insert the manufacturer's CD-ROM.

(v) The Windows Add Hardware Wizard starts and prompts you for the manufacturer's driver disk. Click Have Disk … select Browse and choose the appropriate drive containing the manufacturer's CD-ROM.

(vi) The Add Hardware Wizard installs the driver/s.

Do not be surprised if the Add Hardware Wizard starts up twice. This is quite normal on modern cameras supporting both still and moving images. This is because two USB drivers are utilized.

Once the drivers are installed the camera is treated just like another drive. To save pictures onto your hard disk, connect the camera to the USB cable and wait a few seconds and Windows will automatically recognize the camera as a particular drive letter. This is usually Drive E: or some other drive letter depending on how many hard disks and compact disk (CD) drives are installed.

You can use **Windows Explorer** or **My Computer** to access the 'new drive' and copy and paste your pictures at will.

Most digital camera packages provide image viewing and editing software which can make the task of managing your pictures even more straightforward.

# VIDEO AND SOUND SYSTEMS

## VDU types and characteristics

A computer is only useful if it can provide intelligible output. When you press a key or click on a mouse button, you expect to see a visual cue in response to your actions. The most satisfactory way of doing this is with a VDU. Your actions and the computer's response to your actions is echoed on the VDU screen in a form that is intelligible to you. Imagine trying to write a string of commands from the keyboard with nothing echoed on the screen, i.e. no visual feedback!

In some applications the VDU screen is the main output. For example, when playing a computer game, the computer's response to input data from the keyboard, mouse or joystick, is a change in the image displayed on the screen. The screen effectively provides immediate visual feedback to the players' actions.

Word processing, DTP and other similar applications rely on the VDU screen to provide the user with a high-quality display of the text and images that go to make up the printed page. The page under construction is often displayed on the screen exactly how the final printed page will appear. This is known as 'WYSIWYG' (What You See Is What You Get, pronounced 'wizzywig') – a name coined when this facility first became available. Today, most text and image editing applications provide a WYSIWYG display.

The VDU is therefore essential to give the operator a direct visual cue to commands or data typed at the keyboard or entered via a mouse. It enables you to monitor your actions by providing visual feedback. Nowadays the word 'Monitor' is preferred to 'VDU' but the terms are often used interchangeably.

There are several monitor display technologies available today but the two most common types used on PC systems are CRT monitors and LCD, monitors. The latter type will be discussed later in this unit.

## The CRT

The main component inside the traditional desktop monitor is the CRT, often referred to as the 'tube'. This is a very bulky item that makes up most of the volume and weight of the monitor. It consists of a large evacuated glass bulb, narrowed at one end to accept an electron gun and flattened at the other end to form a rectangular viewing screen. The inner surface of the screen is coated in phosphorescent material that fluoresces (gives off visible light) when electrons strike it at high velocity. The image is formed by focusing, directing and modulating a beam of electrons onto the back of the screen to form a moving spot of light.

To form a display, the spot of light must be scanned across the screen in a series of fine horizontal lines, one below the other to form a 'raster'. The various shades of the original image are formed by varying the brightness of the spot as the raster is scanned (Figure 3.121).

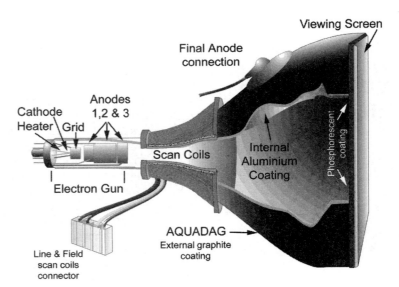

**Figure 3.121**  *Cut-away view of a monochrome CRT complete with scan coils*

The CRT is composed of four main parts (not including the evacuated glass envelope):

- Electron gun
- Final anode
- Phosphorescent screen
- Scan coils

### Electron gun

At the heart of the CRT is the electron gun assembly. Its purpose is to shoot a finely focused beam of electrons at the screen to create a small spot of light. The electrons are generated at the surface of the cathode electrode. This is a small metal tube coated in strontium and barium oxide with a heater element through its centre.

When power is applied to the heater it becomes red hot, just like the filament of a light bulb, and the cathode heats up. The heat causes the cathode coating to emit masses of electrons which form an invisible negatively charged cloud around the cathode. A series of high-voltage electrodes called anodes, at the opposite end of the gun, attract electrons from the cathode at a sufficiently high velocity to hit the screen and create light.

A small distance from the cathode is another electrode called the GRID. This is hat shaped, with a hole in the top to allow electrons to pass through. The grid is normally held at few volts more negative than the cathode which repels some of the electrons away from the anode electrodes. If the grid-cathode voltage is set to a high-negative value, nearly all the electrons are repelled away from the anodes and the electron beam is cut off.

As the negative potential at the grid is reduced, more and more electrons pass through the gun and hit the screen. So the intensity of the electron beam and therefore the brightness of the final image can be finely controlled simply by varying the DC voltage on the grid.

In line with the grid but spaced several millimetres from it, is the anode assembly. This is a set of three cylindrical electrodes. Each anode is connected to a successively higher voltage. These high potentials accelerate and focus the electron beam to form a small spot on the screen. In fact one anode has a connection brought out to a pin on the tube base to allow an external focus potential to be applied. This is appropriately named the 'focus anode'.

### Final anode

The final anode, is a thin bright mirror-like coating of aluminium deposited on the inside of the tube at the screen end. The aluminium layer is connected externally to a metal stud on the top of the tube. This electrode operates at a high-positive DC voltage between 18,000 and 25,000 V, depending on the tube type and size. The high-positive voltage creates a strong electrostatic field which attracts the negatively charged electron beam from the gun and accelerates it, at a high velocity, towards the screen.

### Phosphorescent screen

Coated on the inside surface of the glass viewing screen is a phosphorescent material made from various compounds of phosphorus.

The actual composition of the coating determines the colour, and the length of time (persistence) that the image remains on the screen, after the beam has passed. Long persistence phosphors reduce screen flicker but unfortunately create 'after glow' which creates a ghost image when the display changes quickly from bright to dark.

On top of the phosphor layer is a very thin layer of aluminium and this is connected internally to the final anode. When the CRT is operating, the electron beam from the gun is attracted at high velocity towards the high voltage field surrounding the final anode. The beam hits the screen coating with sufficient force to penetrate the thin aluminium layer and strike the phosphorescent coating. The energy from the impact causes the coating to fluoresce (give off visible light). The stronger the electron beam, the brighter the emitted light. The aluminium coating helps to reflect the light towards the viewer and also removes the build up of negative charge from the electron beam.

### Scan coils

The scan coil assembly contains both horizontal and vertical coils, known as the line-scan and field-scan coils, respectively. These coils produce transverse magnetic fields that deflect the electron beam both horizontally and vertically. With the correct waveforms applied to the coils, the electron beam and therefore the light spot, can be deflected into a series of vertically separated lines known as a '*raster*'.

The raster is produced by applying a high-frequency saw-tooth current waveform to the line-scan coils and a similar shaped, but low frequency current waveform to the field-scan coils. The resulting raster forms a regular repeating framework of lines that the picture can be built upon. An image is formed by varying the brightness of the light spot as it traverses each line of the raster.

In IBM's monochrome display adapter (MDA) graphics standard, used on the original XT system, the horizontal (line) scan rate is 18,432 Hz and the vertical (field) scan rate is 50 Hz. So 50 complete fields are drawn on the screen each second. Other standards have different horizontal and vertical scan frequencies.

One complete set of horizontal lines used to form a single screen image is called a frame, much like the individual frames of a 35 mm photographic film.

The more lines used per frame, the better the vertical resolution of the image. In the UK 625 line TV system, one frame is composed of 575 lines, the rest of the 625 lines are used for frame blanking , similar to the dark blank regions between the frames of a 35 mm film.

The lines can be scanned one after the other in a sequential manner or scanned in two separate scans by interleaving one inside the other. The later is known as *interlaced scanning*. In this method, all the odd lines are scanned first and then all the even lines are scanned in between the odd ones. The odd and even sets of lines are known as fields.

At the end of a frame or field, the beam must traverse back to the top of the screen to start the next one. This is called the frame and field fly-back period, respectively.

In both the non-interlaced (sequential) and interlaced scanning systems, the vertical fly-back period is equivalent in time to several line-scan periods. So the number of lines actually displayed on the

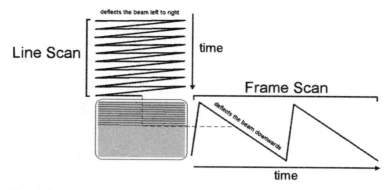

**Figure 3.122**

screen (the so-called active lines), is always less than the total number of lines available (Figure 3.122).

# Scanning methods

There are two methods of generating a raster, using non-interlaced (Figure 3.123) and interlaced (Figure 3.124) scanning. Interlaced scanning is used when the monitor cannot cope with high-frame rates. To work at high-resolution, high-frame rates, such monitors use two interlaced fields to produce one frame (a complete picture). One field scans the odd lines and the other the even lines.

## Non-interlaced scan

In the sequential or non-interlaced system, each line is drawn one below the other in sequence.

From Figure 4.123 line 1 is drawn as the beam traverses from A to B. The beam then flies back quickly from B to C and then draws line 2. This process continues until a whole frame of lines is drawn. The beam then flies back from D to A to start a new frame.

The process of redrawing each new frame is known as 'vertical refresh'. The time the beam takes to move from B to C, etc. is known as the *line fly-back period*, and the time to go from D to A, is known as the *frame fly-back period*.

During the fly-back portions of the scan, the beam must be turned off to make the trace invisible. The beam is cut off by feeding large negative going line and frame, blanking pulses to the grid during the fly-back intervals. This process is known as *line* and *frame blanking*.

## Interlaced scan

The interlaced scanning system is used on some inexpensive monitors to reduce the frame frequency and therefore the complexity of the scanning circuits when running in high-resolution modes that require high-frame rates. More expensive monitors use non-interlaced scanning for all screen modes.

In the interlaced scanning method, first all the odd numbered lines are scanned and then the even numbered lines are scanned in between the odd lines.

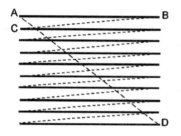

**Figure 3.123**   *Non-interlaced scan*

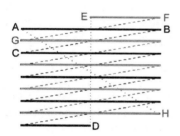

**Figure 3.124**   *Interlaced scan*

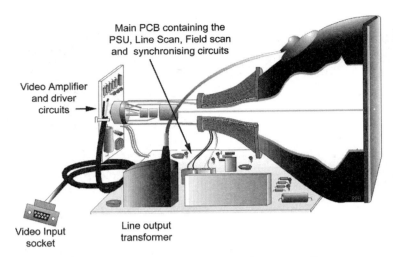

Main PCB containing the
PSU, Line Scan, Field scan
and synchronising circuits

Video Amplifier
and driver
circuits

Video Input
socket

Line output
transformer

**Figure 3.125**   *The inside of a typical monochrome monitor*

Line 1 is drawn from A to B, then the beam flies back to C to scan line 3 and so on until all the odd numbered lines are scanned, then the beam flies from D to E to start the even field scan.

At low-frame rates, interlacing two fields to form one frame reduces the effect of screen flicker. The UK TV system uses two inter-laced fields at 50 Hz instead of a non-interlaced frame rate of 25 Hz, which would otherwise produce an intolerable flicker.

# Monochrome CRT monitor system

To show an image on the screen, the brightness of the spot must be varied to match the digital image stored in the memory of the graphics card. In a monochrome monitor (Figure 3.125) the video signal from the graphics card contains the brightness information for each line of the image. (In a colour TV system the video signal contains both brightness and colour information known as the *luminance* and *chrominance* signals, respectively.)

To ensure that the start of each line and frame of the raster occurs at precisely the right instant in time to match the video signal, *sync pulses* as well as the video signal itself, must be sent to the monitor. At any instant in time, these signals contain information about the brightness and precise position of the dot at any particular instant in time.

The brightness of the image is determined by the voltage level of the video signal. For example, a standard TV video signal has a maximum voltage of 1 and uses 0.3 V to represent total darkness and 1 V to represent maximum brightness. These levels are known as *black level* and *peak-white*, respectively. Voltages between these two limits represent continuous shades of grey from black to white.

In addition to the video information, line and frame synchroniz-ing signals are sent from the graphics card. These force the line and frame scan generators in the monitor to fly-back at the correct instant in time.

A simplified block diagram of a complete monochrome monitor system is shown in Figure 3.126. The video signal, and synchroniz-ing pulses from the graphics adapter are applied to the monitor inputs, shown on the left of the diagram.

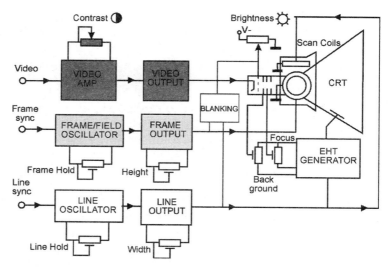

**Figure 3.126** *Block diagram of a typical monochrome monitor system*

# Inside a monochrome monitor

The signal from the graphics adapter is applied to a video amplifier which increases the signal to a level required by the video output stage. At the output of the video amplifier the signal level is greater than actually needed, so a contrast control can be added. This allows the user to adjust the level of the signal voltage and therefore alter the contrast of the displayed image.

From the video amplifier, the signal is passed to the *video output* stage, where the signal is further amplified, before being fed to the cathode electrode of the CRT.

Under normal conditions with an image on the screen, the cathode is several tens of volts more positive than the grid, and varying in amplitude with the signal. The grid is therefore always negative with respect to the cathode.

The more negative the grid, the dimmer the displayed image and vice versa. The grid electrode is therefore connected to a potentiometer which enables the DC voltage between the grid and cathode, and therefore the image brightness, to be manually adjusted by the user.

During the line and frame fly-back intervals, the spot on the screen beam must be extinguished, to prevent fly-back lines appearing on the screen. One easy way of achieving this is to apply negative going fly-back pulses, to the grid of the tube. This is called '*line and frame blanking*'.

## Frame oscillator and output circuit

To ensure that each field or frame of the raster is drawn exactly in step with that of the image, sync pulses from the video card are applied to a field oscillator circuit. This locks the frequency of the oscillator to the exact field frequency set by the graphics card.

In the monochrome MDA system, the field oscillator frequency is 50 Hz, but it can be well over 100 Hz in a modern high-resolution colour system. The number of frames per second set by the field oscillator is known as the *vertical refresh rate*.

The synchronized pulses from the field oscillator are then applied to the field output stage which generates a saw-tooth-shaped current waveform in the vertical scan coils. The amplitude and shape of the waveform can be adjusted to vary the height and vertical linearity of the displayed raster.

### Line oscillator, line output and extra high tension circuit

In a similar way to the frame scan stage, line sync pulses from the graphics card, lock the free running line oscillator to a precise horizontal scan frequency. This varies from 18.432 kHz on the MDA system, to well over 50 kHz on a modern high-resolution system.

The synchronized horizontal scan waveform is then applied to the line output stage, which incorporates a line output transistor and transformer. The purpose of the line output stage is two-fold; it generates the required saw-tooth current waveform for the horizontal scan coils, and provides a set of high-voltage DC outputs for the CRT.

The amplitude and shape of the line waveform can be manually adjusted to vary the width and horizontal linearity of the displayed raster.

At the high line-scan frequency, the fly-back time is just a few microseconds. This fast collapse in current in the line output transformer (LOPT) during fly-back, causes a large voltage of several thousand volts to be induced in the primary winding of the transformer. This is stepped up in the secondary winding of the LOPT and rectified, to provide the EHT required by the final anode of the tube – usually around 24,000 V for a colour CRT.

The high voltages present on the LOPT are also used for the electron gun anodes. By varying the DC voltage on Anode 1, the displayed level of black, known as the *background level*, can be finely adjusted. This is set to give a totally black screen when no information is being displayed. Black in this context is the colour of the CRT screen when the gun is turned off. In a well-lit room the background level appears less black than the same monitor in a darkened room.

The voltage on 'Anode 2' can also be adjusted to set the overall focus of the displayed image. Both the background and focus adjustments are usually mounted on one side of the LOPT and are pre set and sealed by the manufacturer.

## Colour CRT monitor system

To achieve a full colour display, the video signal is composed of three separate colour components containing the RGB information of the image. The monitor therefore requires three separate signal channels, we will name them R, G and B. The R, G and B channels extend from the graphics card right through to the CRT itself.

As you would expect colour CRT is far more elaborate than its monochrome counterpart (Figures 3.127 and 3.128). It requires three electron guns, an aperture grill and a tri-colour screen made up from millions of tiny phosphor dots arranged in triads. The red electron gun only impinges on the red phosphor dots, the green electron gun only impinges on the green phosphor dots and so on for the blue electron gun. Each triad contains an RGB phosphor dot. When each dot of a triad is bombarded by its respective electron beam the light from the three dots merge to form a single hue. By varying the relative intensity of each electron beam, each triad can emit millions of colours.

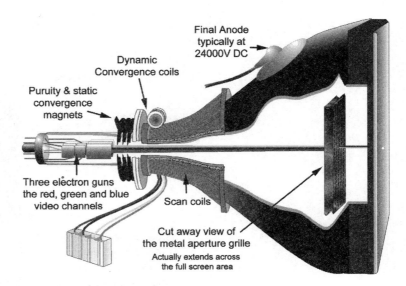

**Figure 3.127**  *A typical colour CRT*

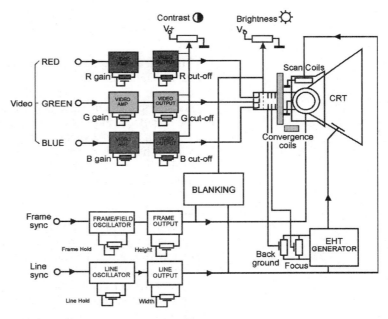

**Figure 3.128**  *Block diagram of an RGB colour monitor*

A colour tube requires more complicated circuitry than a monochrome monitor. Apart from two extra video channels, additional circuitry is needed to ensure accurate colour fidelity and registration.

The three electron beams emanate from slightly different positions due to the displacement of each electron gun, so three coloured rasters of RGB are offset from one another on the screen. The beams must therefore be adjusted to allow them to converge at all points on the screen plane. The convergence system uses permanent magnets and convergence coils to apply correction fields to each electron beam.

A block diagram of a typical colour monitor is shown above. The video signal is made up of three separate colour signals, RGB. Each signal is amplified separately and applied to its respective electron gun.

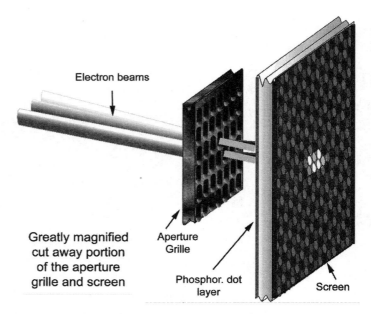

Electron beams

Greatly magnified
cut away portion
of the aperture
grille and screen

Aperture
Grille

Phosphor. dot
layer

Screen

**Figure 3.129** *The aperture grille enables each electron beam to fall on the correct coloured phosphor*

The gain and beam cut-off point for each gun must be adjusted so that each channel has the same relative amplitude across the whole brightness range. Separate 'gain' and 'cut-off' controls are included in each channel to accommodate this. Normally, these controls are only set at the manufacturing stage or when the monitor is being serviced.

When a grey-scale test pattern is sent from the graphics card, the screen should show an even graduation of grey shades from black to peak-white, free of colour tinges. The practical procedure of finely adjusting the gain and cut-off controls to achieve a perfect grey scale is known as 'grey-scale tracking'. You will see more of this procedure later in this unit.

Today's high-resolution colour CRTs have a triad pitch of 0.26 mm or smaller, which means for example, that a 17-in. screen has approximately four million phosphor dots.

Even with all these dots, each electron beam, must only fall on dots of the correct colour. For example, the red electron beam must only illuminate the red phosphor dots. This is achieved by placing an *aperture grille* or *shadow mask* as it is sometimes called, behind the screen. During the manufacturing stage, the position of each gun is precisely aligned with the grille and corresponding phosphor dots. As the beams are scanned, they either fall on their respective phosphor dots or hit the grille (Figure 3.129).

## Picture elements

A picture element (pixel) is the smallest possible definable spot of light on the screen at a particular screen resolution set by the OS. For example with the screen resolution set at $1024 \times 768$ there are 786,432 pixels making up the image on the screen.

Each pixel of a colour monitor is composed of groups of RGB phosphor dots. The absolute smallest pixel possible on a CRT monitor would be that from a single triad of phosphor dots but usually the screen resolution set by the OS is much lower than this, as other physical effects become more noticeable at high resolutions, such as screen flicker.

# Additive colour mixing

To the human observer, different intensities of RGB light from the screen phosphors add together to form a wide range of different hues. The three colours are known as the additive *primaries*.

Paints and dyes mix subtractively to produce different colours. For example when viewed in white light, cyan and yellow pigments mixed in relative proportions, create a green pigment. This happens because the cyan pigment is absorbing red light, and the yellow pigment is absorbing blue light. When the two are mixed, both the red and blue primaries are absorbed and green light is reflected.

For convenience, the relative intensity of each of the primary colours required to produce white light, will be treated as 1 unit. So to make white light we need to add 1 unit of red light, 1 unit of green light, and 1 unit of blue light. Each unit of RGB light actually contributes, 30, 59 and 11 per cent, respectively, of the total intensity of white light. Colours other than white can be produced by adding different intensities of each primary.

*Note:* Full colour images of this section can be found on the Elsevier website at http://books.elsevier.com/companions/0750660740

By additively mixing whole units of light from the primaries, a total of seven colours including white are produced. Black is the absence of light. This is shown in Figure 3.130.

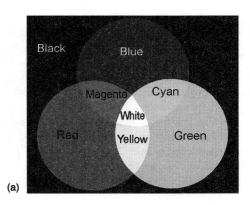

(a)

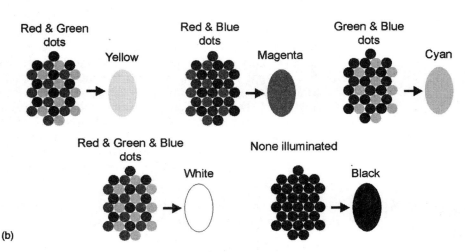

(b)

**Figure 3.130** *(a) Overlapping of RGB lights. (b) Additive colour mixing of fixed intensity, RGB lights*

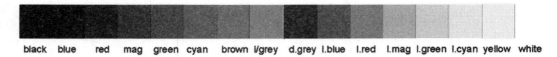

| black | blue | red | mag | green | cyan | brown | l/grey | d.grey | l.blue | l.red | l.mag | l.green | l.cyan | yellow | white |

**Figure 3.131** *The 16 colours available on the old CGA text display*

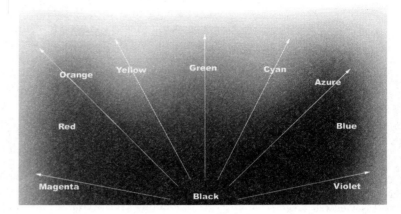

**Figure 3.132** *The 16 million colours available in the SVGA 24-bit 'true colour' mode*

Figure 3.130(a) shows RGB light beams overlapping to produce seven coloured light beams + black. Black is simply the absence of all three colours. In a similar fashion, the triads of fixed intensity, RGB phosphors of a colour CRT, shown in 3.130 (b) also add together to produce these colours. This is achieved by turning on and off the various electron beams.

By varying the intensity of each light source, millions of different colours can be produced. For example, one of millions of possible pastel green hues is obtained with 100 per cent green, 60 per cent, red and 60 per cent blue.

The various video standards used on the PC system have different colour ranges. The first colour display system used on the PC was known as the colour graphics adapter (CGA). It used a very basic digital system which consists of turning the RGB guns on and off to produce eight basic colours including black and white. To increase the colour range of this basic 3-bit system, a further intensity bit was included. This gave eight dull colours with the intensity bit off (logic 0) and eight bright colours with the intensity bit on (logic 1). Unfortunately these 16 colours were only available in text mode (Figure 3.131). If you wanted colour graphics you had to make do with just four colours at a very low resolution of 320 × 200 pixels.

Today's super-video graphics adapter (SVGA) display standards, use 8, 16, 24 or 32 bits per pixel to show a maximum range of 256, 65,536, 16 million or over 4 billion colours, respectively (Figure 3.132).

# Types of video signal

Monitors are classified by the types of signal used to present the information on the screen. The main types are:

- Composite video
- S-video

- Transistor/transistor logic (TTL) Mono (obsolete)
- TTL RGB (obsolete)
- Analogue RGB.

## Composite video

A composite video signal, as the name suggests, contains all the information needed to display an image, in one single signal. The luminance, colour and synchronizing signals are all combined in one composite signal. A composite signal only requires a single coaxial cable between the monitor and computer.

Composite video is seldom used on desktop PC systems as most adapter cards use a standard SVGA 15-pin DB socket with separate RGB and sync signals. This eliminates the inevitable signal distortion that occurs when signals are mixed into a composite signal at the video card and then subsequently separated again at the monitor. Some laptop computers however have a composite video output socket in addition to the normal SVGA output to enable it to connect to a domestic television.

## S-Video

Rather than use a composite signal with its inherent distortions, some graphics cards and laptop computers have an auxiliary *S-Video* output socket. The S-Video (Figure 3.133) standard separates the colour information from the luminance and synchronizing signals and produces an overall higher-quality image than that provided by a composite signal.

The luminance or 'Y' signal is a line by line monochrome version of the image with synchronizing signals incorporated. Colour information is sent separately and this is referred to as the 'C' signal. S-Video leads contain two separate coaxial screened cables one for the 'Y' signal and ground and one for the 'C' signal and ground.

Laptop users often have problems when they try to connect their laptop to a TV that does not have an S-Video input socket. This can be overcome by fitting an S-Video to Composite video converter plug. The cheapest converters simply recombine the 'C' signal with the 'Y' signal using passive components. The result is usually adequate but

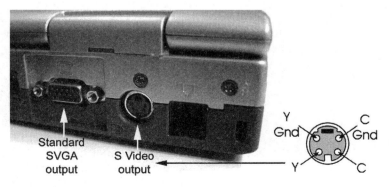

**Figure 3.133**

the image quality is not as good as that obtained with an active S-Video to RGB converter.

# Analogue video

Around the time of the PC's launch in 1983 and for several years to follow, computer display systems used a digital connection system between the graphics card and monitor. The connection was based on the TTL system. In this system, Logic 1 is defined as any voltage between $+2$ and $+5\,V$ and Logic 0 as any voltage between 0 and $+0.8\,V$).

Monitors and graphics cards based on TTL were inexpensive to manufacture as TTL chips were widely used in electronics manufacturing and the system offered consistent performance from one system to another. However as the demand for graphics intensive applications grew, the market was ripe for a better more advanced display system. IBM's then very popular extended graphics adapter (EGA) system was the last PC graphics standard to use TTL. To obtain a much larger number of colours, it is necessary to vary the level of each colour signal over a wide range. The monitor tube is inherently an analogue device so is capable of displaying a virtually unlimited range of colours.

## *Analogue signals*

Our human response system is based on analogue stimuli. We perceive a seemingly infinite range of colours and brightness levels. For example when we look at a pastoral scene, we see continuous shades of greens, browns and blues and other colours that merge smoothly together through shadows and shafts of light. It is virtually impossible to count the number of different colours and shades. Our brightness range is also pretty impressive; we can distinguish hundreds of different light intensity levels from total darkness to full light.

These two extremes in light intensity from black to peak-white represent the lower and upper limits of our visual response system. Between these limits there is a infinite range of possible levels. However it has been shown in practice that 256 shades of grey can adequately represent our perception of light intensity.

If we examine the light intensity from a typical moving scene using an oscilloscope and an analogue light-sensitive device (e.g. a photo cell), we see a signal waveform similar to that shown in Figure 3.134. The photo cell converts the incident light intensity into a varying voltage analogue. The brighter the incident light the higher the output voltage.

Notice how the waveform varies smoothly from one level to the next.

So an analogue signal is one that varies in sympathy with the original signal and has an infinite number of possible values between an upper and a lower limit.

To display a large number of colours and brightness levels on a CRT-type monitor – as required for photographic quality images – the graphics card must interface the computer's inherently digital circuits to the monitor's inherently analogue system. No prizes for guessing that this part of the graphics card circuit is called the

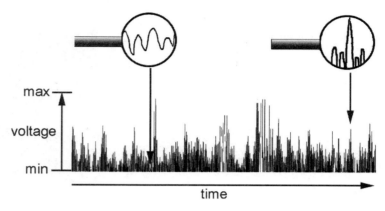

**Figure 3.134** *A typical analogue waveform*

*digital-to-analogue converter* (DAC). Basically the DAC converts binary values representing the brightness of each colour signal to a corresponding voltage level.

For example, if 8 bits are used for each colour signal (24 bits per pixel), the DAC can produce 256 different output voltage levels for each of the three colour channels. This corresponds to 256 shades respectively of RGB. That is $256 \times 256 \times 256$ or 16.7 million colours per pixel, including black and peak-white.

This is pretty good compared to the four graphics colours of the early CGA system! At the start of this discussion we mentioned that the eye can perceive a seemingly infinite range of colours but 16.7 million is not bad and some top-end graphics cards can produce four billion colours per pixel using 32 bits.

Due to its inherent analogue design, a CRT-type monitor can display a virtually infinite range of colours between upper and lower limits but the graphics card cannot possibly achieve this and it does not have to! When the graphics card uses 24 bits per pixel (8 per colour) or more, the colour fidelity is virtually indistinguishable from that of a true analogue signal. This is known as 'True Colour'.

> The graphics display mode used on a modern colour monitor is known as an 'all points addressable' (APA) system, because each individual pixel making up the display can be set to a particular colour and intensity.

# Video cards

Current graphics cards are a far cry from the traditional SVGA cards used on the first Pentium PCs. Today's top-end cards offer incredible 2D and 3D processing features including real-time 3D rendering effects, and brilliant image quality.

The graphics card featured in Figure 3.135 is an excellent example of the top-end performance now available from an advanced graphics port (AGP) graphics card. Some of the features of this card are listed below:

- Maximum palette: 4 billion colours (32 bits per pixel)
- Video memory: 256 MB data direct synchronous dynamic random access memory (DDR-SDRAM)
- Bus: AGP 8X
- Maximum resolution: 2048 × 1536 pixels
- RAM DAC clock speed: 400 MHz
- Video processor clock speed: 325 MHz
- Supported APIs: Open GL, Direct 3D, Direct X 9.0, Active X

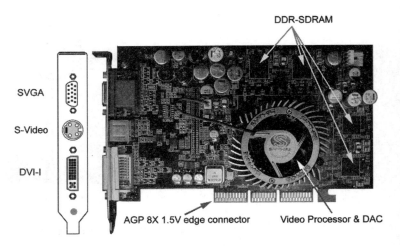

Figure 3.135 *The Radeon 9700 graphics card*

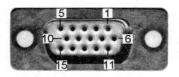

**Figure 3.136** *VGA/SVGA 15-pin high-density 'D' connector pin-outs. Pin nos. and functions are: (1) red analogue signal, (2) green analogue signal, (3) blue analogue signal, (4) monitor ID bit 2, (5) ground, (6) red ground return, (7) green ground return, (8) blue ground return, (9, 15) not connected, (10) sync ground return, (11) monitor ID bit 0, (12) monitor ID bit 1, (13) H sync and (14) V sync*

**Figure 3.137** *(a) DVI-D, 24-pin connector and (b) DVI-I, 29-pin connector*

## Video output connectors

The traditional SVGA connector provided on current graphics cards is the 15-pin high-density 'D' connector (Figure 3.136). This connects the graphics card to CRT-type monitors and other types including LCD screens and projectors.

Due to the rise in popularity of LCD monitors which are inherently digital, many graphics cards also include a special digital visual interface (DVI) connector. As the name suggests, the DVI connector provides direct digital signals rather than the normal analogue signals required by traditional CRT SVGA monitors. Using a DVI connection for an LCD screen makes sense as it keeps the signal in its native digital format and eliminates the inevitable loss in image quality in converting the data from digital to analogue via the DAC on the graphics card, and then from analogue back to digital at the LCD monitor.

There are two types of DVI connector in use (Figure 3.137):

- DVI-D: This type of DVI connector uses 24 pins and only carries digital video.
- DVI-I: The 29-pin DVI connector is becoming more widely used. This will connect digital or analogue video signals. So it can be used to connect an LCD monitor or a traditional SVGA CRT monitor.

The architecture of a typical SVGA graphics (Figure 3.138) card can be simplified into the three following functional blocks:

- Graphics engine (video chipset)
- RAM DAC
- Video memory

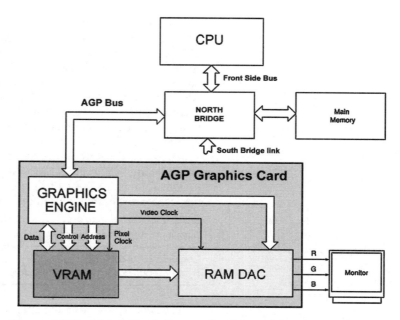

**Figure 3.138** *Simple block diagram of a high-performance graphics card*

## Graphics engine

The graphics engine contains an elaborate set of hardware and firmware devices to provide a comprehensive 2D/3D graphics processor, PCI or AGP bus interface, and controller/interface for the video memory and RAM DAC. It also contains a *CRT controller* (CRTC), to generate the horizontal and vertical synchronizing signals, a *character generator* to provide on-screen fonts and a *graphics ROM*. The latter acts like an extension to the system BIOS ROM.

### *RAM DAC*

The RAM DAC incorporates a number of functions in one chip allowing colour ranges up to 24-bit true colour. It integrates a colour palette RAM with three (RGB) high-speed DACs. Each DAC essentially converts the digital data stored in the video memory into analogue voltage levels for the monitor. For example a certain binary number stored in the video memory represents the brightness and hue information for a single screen pixel. The DAC converts the binary value into three unique voltages for the RGB monitor-drive outputs. The RAM DAC usually features a screen cursor generator and a three colour overlay palette RAM.

## Video memory

A few years back there were two main categories of memory chips used on video cards known as, *single-ported RAM* and *dual-ported RAM*. Single-ported RAM, as the name implies, uses one port to input and output data. So while the RAM is receiving data, it cannot physically output data. A dual-ported device on the other hand is able to receive and output data at the same time.

On a graphics card, data is stored in the video memory in frames, where one frame represents one complete screen picture. The portion of video memory used to store a frame is known as the *frame buffer*. At high vertical refresh rates and high picture resolutions, single-ported video memory struggles to keep the frame buffer updated, while outputting data to the RAM DAC. It can only do one of the tasks at a time. The video circuitry therefore has to wait for the screen to be refreshed before it can send new data to the frame buffer, this inevitably results in a bottleneck.

Dual-ported RAM allows the frame buffer to be refreshed while data is being sent to the RAM DAC, thus avoiding the bottleneck.

The following types of single and dual-ported video memory devices were commonly used on graphics cards:

| Single-ported video memory | Dual-ported video memory |
| --- | --- |
| Dynamic RAM (DRAM) | Video RAM (VRAM) |
| Synchronous graphic RAM (SGRAM) | Windows RAM (WRAM) |

The data representing the image travels along the PCI or AGP bus and then enters the video chipset where it is processed via commands from the application/driver software. The data then passes to the frame buffer – which is part of the video memory – where a complete screen image is built up. With single-ported RAM, the full image is then passed to the RAM DAC where the digitally stored image sent straight out to the DVI connector or converted to an analogue RGB signal suitable for a CRT monitor. With dual-ported RAM, as a new image is being built up in the frame buffer, the existing image is sent to the RAM DAC.

### Synchronous dynamic random access memory

In the late 1990s, synchronous dynamic random access memory (SDRAM) became the most widely used DRAM device for the main memory of the PC. It has also gained popularity for use in a wide range of applications requiring fast read/write memory. Its ability to synchronize to the PC's system clock has maintained its popularity particularly as prices have fallen substantially over the years.

The speed and low cost of SDRAM made it an ideal choice for video memory. Many cards still use SDRAM in preference to other forms of video memory.

SDRAM contains two independent array banks that can be accessed individually or interleaved. This allows enhanced performance compared to conventional DRAM. The fact that it synchronizes to a common clock also adds to the performance. All data is written or read in burst fashion. Given a single starting address, the SDRAM internally accesses a sequence of locations based on that address. The length of the burst sequence can be programmed to 1, 2, 4, 8 or 256 accesses. This makes it ideal for reading and writing image data as this is usually in consecutive blocks of data.

### DDR-DRAM

Today DDR-SDRAM is the popular choice of memory for graphics card manufacturers. It is inexpensive, due to its use as the main RAM in modern systems, and it offers excellent performance. DDR-SDRAM features all the benefits of SDRAM while transferring data twice in each clock cycle on the rising and falling edges of the clock.

## The AGP

For 2D graphics applications the PCI bus is fine as it easily copes with the data transfer rate between the graphics accelerator chip, CPU and memory. However when fast 3D operation is required for games, DTP and 3D modelling, the extra processing and data throughput needed for texturing, etc., places a heavy demand on the PCI system. The nominal 33 MHz clock frequency and 32-bit wide bus of the PCI system is just too low for such demanding applications. Intel and other manufacturers realized that a new high-bandwidth bus was needed and set up a committee to look into it. The AGP specification was the result.

The original AGP design is based on the PCI 2.1 specification. It uses the same 32-bit wide bus architecture but it is clocked at 660 MHz – double the rate of the PCI bus. It also features enhanced memory read and write hardware known as graphics addressing remapping table (GART). The GART effectively keeps items of image data together as contiguous memory addresses for the graphics accelerator even though in reality they are spread over several different memory locations in RAM. This allows the graphics engine to process the data more quickly.

The AGP can therefore transfer data at least twice as fast as the 127 MB/s of the PCI bus (i.e. 254.3 MB/s). In addition by using special hardware techniques, data can be transferred on the rising and falling edge of each clock pulse thus doubling the data transfer rate to 508.6 MB/s. This is known as AGP 2X.

Today, AGP 4X and 8X graphics cards are available enabling transfer rates of over 1 and 2 GB/s, respectively. These high data transfer rates can only be realized with a compatible graphics card and motherboard.

The 'games market' provides the main impetus for the development of enhanced graphics hardware. This is due to the public's insatiable demand for better-quality graphics at realistic frame rates. The latest version of games like 'Quake' and 'Tomb Raider' with their high-resolution 3D textured worlds, push graphics cards to their limit. These programs are often used as benchmarks to 3D performance.

## Direct X

Direct X supplied by Microsoft and built into Windows 98/2000 and ME is a suite of intermediate software drivers for multimedia applications called application programming interfaces (APIs). Programmers can now concentrate on writing their programs to use the APIs rather than trying to write directly to hardware drivers which vary considerably between systems. This allows any Windows 9*/2000/ME PC with Direct X, to play games using the full range of installed multimedia devices with virtually no compatibility problems. This was never achieved with DOS-based multimedia applications. The Direct X suite contains many APIs including the following.

- Direct Draw     Provides fast access to the video accelerator cards
  Direct Video    special features.
  Direct 3D

- Direct Sound     Provides access to the sound card's features and
  synchronizes sound and video sequences.

- Direct Music     Handles the musical instrument digital interface
  format and allows programmers to apply sur-
  round sound techniques and handle user musical
  input devices.

- Direct Input     Handles all manner of user input devices joysticks, game pads, and USB items.

- Direct Play     Provides standard tools for Multiplayer gaming
  over serial cables and networks, including the
  Internet.

- Direct Show     This API provides an interface for video and sound
  playback using audio video interleave (AVI) and the
  Motion Pictures Experts Group (MPEG) formats.

These APIs are updated quite regularly to keep pace with hard-ware advancements. You can download the latest Direct X from Microsoft (see below). Make sure you choose the correct version for your OS. The Windows 95/98/ME, APIs are different to the Windows 2000 and XP versions.

If you have access to the Internet, Microsoft's Direct X can be downloaded from: http://www.microsoft.com/directx/homeuser/downloads/default.asp

You can also find out more about Direct X on: http://www.microsoft.com/directx/homeuser/aboutdx.asp

# Installation and adjustment

## Choosing a monitor

It is surprising how often, the monitor is given the least consideration when a new PC system is being purchased. There is plenty of emphasis on finding out the system speed, RAM size and hard disk capacity but often very little attention is given to the performance of the monitor. The monitor is however, one of the most important items in the system. A good monitor not only makes work on the computer more pleasant, it can also make it more productive. The on-screen display (OSD) is, after all, the main feedback between you and the computer. The incidence of eye strain and operator fatigue reduces considerably when a high-quality, flicker-free monitor is used. The old adage, 'you only get what you pay for' definitely holds true for monitors. Paying a few more pounds for a high-quality flat screen monitor is money well spent, as sooner or later an inferior monitor ends up dumped in a corner.

European legislation on electromagnetic radiation (EMR) and screen flicker ensures that all new monitors at least meet a minimum acceptable standard. Most monitor manufacturers now produce so-called '*green monitors*', to the German TCO and Swedish MPR-II standards. These feature low-radiation and high refresh rates and usually incorporate display power management signalling (DPMS). A DPMS monitor will automatically switch to 'standby mode' when the monitor screen has been inactive for a predetermined time, thus conserving energy.

Your final choice of monitor should be arrived at by carefully considering the use and characteristics of monitors generally, some pertinent questions are:

## Monochrome or colour?

The only applications these days that do not need a colour display are those that use alphanumeric characters only, such as simple DOS-based spread sheet, database and word processor packages. If the system is only being used for these types of applications, and coloured text and graphics are not required, then a monochrome monitor would probably suffice.

## What size?

Monochrome monitors are usually limited in the range of sizes available from 12 to 14 in. While VGA/SVGA colour monitors range in size from 14 to 28 in. The screen size usually refers to the diagonal distance from the bottom left corner to the top right corner of the tube face, but this is not necessarily the actual visible screen distance, for example, a typical 17-in. monitor has a diagonal viewing distance of only 15.8 in. and a 15-in. monitor only 13.8 in.

The aspect ratio of most modern monitor tubes is $4 \times 3$ (i.e. the width is 4 units and the height is 3 units) in length. So from simple geometry, the display width is 4/5 and the height is 3/5, of the diagonal distance. On a 14-in. tube this gives 11.2 in. horizontally ($0.8 \times 14$) and 8.4 in. vertically ($0.6 \times 14$).

The standard size for a colour monitor is 14–15 in. For most applications this is adequate but for serious computer-aided design (CAD) and DTP applications a 17-in. or larger monitor is essential. For professional 'on-screen' presentations a 28-in. monitor or an image projection system is required.

Generally, the higher the graphics card resolution in use, the larger the required screen size. For example, running a GUI such as Windows at a resolution of $1024 \times 768$ or above really requires a 17–20-in. monitor, otherwise text and icons become too small to see comfortably. It is worth noting that a 20-in. monitor shows two adjacent A4 pages in portrait mode, or one A3 page in landscape mode, whereas a 15-in. monitor can just display two adjacent A5 pages in portrait mode, or one A4 page in landscape mode.

## What resolution?

Resolutions of $1024 \times 768$ pixels and greater, are necessary for serious CAD, DTP and graphics work. Most games and multimedia applications are best run at the lower resolutions of $640 \times 480$ pixels @ 256 colours, and $800 \times 600$ pixels @ 256 colours. However some modern computer games support palette ranges up to 16.7 million colours and resolutions up to $1024 \times 768$.

Most colour monitors today, regardless of size, have a phosphor dot pitch of 0.28 mm or better. The smaller this value the better the resolution. It is simply not worth trying to save money by going for cheap monitors with a dot pitch greater than 0.28 mm. For example, the text and graphics on a 14-in. monitor with 0.39-mm dot pitch look fuzzy compared to one with a 0.28-mm pitch.

The graphics card and size of monitor should really be considered together, when deciding on the maximum screen resolution. It is pointless having a high-resolution card with a small monitor. A good rule of thumb is to aim for a total horizontal pixel pitch of 80 per in. or just less. Any more than this and text becomes very thin and difficult to read.

The minimum diagonal dimension ($D$), for the monitor to achieve 80 pixels per inch, is found by dividing the maximum required horizontal screen resolution '$H_r$' by 64. For example, to calculate the minimum monitor size to display 1024 × 768 pixels:

$$D = H_r/64 = 1024/64 = 16\,\text{in}.$$

So a 17-in. monitor is ideal and a 15-in. monitor will barely suffice.

### What vertical refresh rate?

The vertical refresh rate is the number of complete frames that appear on the screen in 1 s, measured in Hertz. The faster the rate, the less objectionable screen flicker becomes. The monitor should be capable of operating at whatever refresh rate the graphics card puts out. High-performance monitors can usually cope with refresh rates from 50 to 160 Hz. For example at a resolution of 1024 × 768, a refresh rate >80 Hz is desirable.

## Choosing a graphics card

One of the advantages of running 'Windows' is its ability to operate in several screen resolutions and colours from 640 × 480 pixels at 16 colours to 1680 × 1200 pixels at 16.7 million colours. The normal default screen setting is 640 × 480 pixels by 16 colours (standard VGA). However with a suitable graphics card and driver software, higher resolutions and colour ranges are available.

For graphics intensive applications like DTP and photo-realistic image editing, a 'true colour' card providing 16.7 million colours is desirable. These high-resolution screen modes available in Windows require large amounts of graphics memory. For example, 1 MB is required for a screen showing 1024 × 768 pixels by 256 colours and 2 MB for 1024 × 768 pixels in 65,536 colours.

The amount of graphics memory required for a particular screen resolution and colour range can be found by multiplying the required horizontal resolution ($H_r$) and vertical resolution ($V_r$) by the number of bytes of colour depth ($n$), required per pixel. The colour depth is as follows:

- 1 byte (8 bits) per pixel, provides 256 colours (i.e. $2^8 = 256$),
- 2 bytes (16 bits) provides $2^{16} = 65,536$ colours per pixel and
- 3 bytes (24 bits) provides $2^{24} = 16.7$ million colours per pixel.

The size of the graphics RAM for a particular resolution and colour depth is therefore calculated as follows:

Graphics RAM size $= H_r \times V_r \times n$

For example to run a monitor at 1280 × 1024 × 16.7 million colours, the graphics RAM size must be at least, 1280 × 1024 × 3 = 3.9 MB. So 4 MB of graphics RAM needed. Most cards have more than the required minimum of memory for a particular resolution so that the extra memory can be used as a cache to store frequently used image artefacts. On a 3D card extra memory is also required to store information for the *Z* plane.

The large amount of graphics memory also calls for fast graphics processing. To meet this demand, modern graphics engines clock at amazingly high speeds. The example accelerator card shown in Figure 3.138 clocks at 350 MHz. These top-end cards speed up Windows screen updating considerably and give a 'snappy feel' when switching from one application to another. They also support the fast frame rates required for video playback using MPEG software, and Direct 3D games.

For 3D games, DTP and CAD applications, a 3D accelerator card, supporting Z buffering, texture mapping and alpha blending, are essential.

*Note*: These terms are explained briefly in the glossary at the end of this unit.

## Practical work SAP3.7: setting display properties in Windows

To see what graphics resolution setting you have on your Windows 95/98/2000/ME PC system click **Start**, highlight **Settings** then click **Control Panel**. In control panel double click on the Display icon. Now select the **Settings** tab.

**Warning:** If your graphics card does not have sufficient memory and you change the settings to a higher specification the screen could go blank and you will have to use your Windows startup disk to boot up the system in **safe mode** to change it back (Figure 3.139).

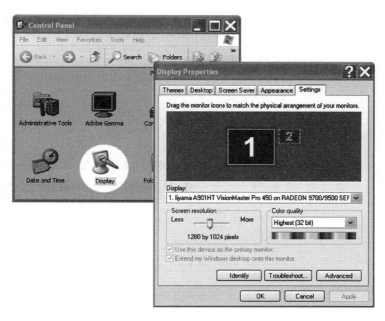

**Figure 3.139**

First chose a resolution suitable for your size of monitor remember the 80 horizontal pixels per inch rule. On a normal 4/3 aspect ratio screen simply multiply the diagonal distance of the screen by 64 to get the ideal horizontal resolution. You can also get the ideal vertical resolution by multiplying the diagonal distance of the screen by 48.

For a 17-in. monitor this works out at $1088 \times 816$ pixels. So select the $1024 \times 768$ option using the 'Screen area' slider.

Now set the colour depth from the 'Colors' slider (American spelling). You should ideally check the graphics card manufacturers installation guide to see what options you have. However in the end it all comes down to how much graphics memory you have. To display 256 colours – this is not much good for photo-editing, etc. – each pixel requires 1 byte. So for our $1024 \times 768$ resolution we need $1024 \times 768$ bytes = 786,432 bytes (i.e. nearly 1 MB) of memory. For 16-bit colour (65,535 colours) you need 2 bytes per pixel so you need 2 MB of graphics memory. Finally for True colour you need three bytes per pixel which corresponds to nearly 3 MB of graphics memory. You should also have some surplus memory for the card to store frequently used graphics elements so for true colour at this resolution 4 MB would be better. Some modern cards come with 8, 16 or 32 MB of graphics memory so running in true colour on a modern PC system is almost standard. Remember that 3D applications require considerably more memory, so do make sure you have enough memory for the most demanding application you are likely to run.

## Digital graphics ports

You can now see why the idea of an analogue monitor system makes a lot of sense. Regardless of the colour depth set by the graphics card, because the signal channels are analogue, the monitor can handle any colour depth thrown at it without a problem.

Flat screen displays using liquid crystals are inherently digital in nature. So ideally they should connect to a digital interface on the graphics card rather than the normal analogue port. The reason for this is that some signal degradation is bound to take place if the digital signal at the graphics card is first converted to analogue form and then it is converted back to digital form inside the display. Some graphics cards now incorporate a special digital output especially for LCD displays.

## Practical Work SAP3.8: monitor installation and adjustment

Equipment requirements:

- PC with DOS and/or a GUI such as Windows installed.
- A pre-installed graphics card compatible with the monitor below.
- A CGA, MDA, EGA or VGA monitor with instruction manual.
- Diagnostics and Monitor line-up software. A free copy of the MONTEST utility can be downloaded from KemTec Training on http://www.kemtec.com

For instructions on installing a graphics card and driver software see section on 'Installing a graphics card'.

(i) If the monitor is packed in a box, remove the monitor carefully and lift it by bending your knees and keeping your back straight (see unit on Health and safety).

(ii) Choose the correct mains lead and fuse for your monitor, by referring to the manual or the ratings label on the back of the monitor. It is a modern practice to provide a mains lead plug and correct fuse with each new monitor. The installation technician however must be aware of correct fusing, as leads often become mixed up (e.g. the monitor lead becomes mixed up with the printer lead). Fit the mains lead to the monitor and wall socket but do not switch it on yet. *Remember even though the socket is not powered up, the earth connection helps remove static.*

(iii) Connect the monitor signal cable to the 'D'-type socket on the ADC. Take care when you do this to prevent damage to the delicate pins.

(iv) Power up the computer. The system should boot and display the normal DOS prompt or start Windows. If you hear *one long and two short beeps* this indicates 'no graphic card fitted' or a graphics adapter installation error. If this is the case, check to see if the card is fitted in the slots correctly.

If you have a system with windows 95/98/ME: click on the START button, then click on '*Shut Down ...*' and then click '*Restart the computer in MS-DOS mode?*'. This will restart the computer and give you the MS-DOS prompt.

## Line-up

(v) When the DOS prompt appears, run the 'MONTEST' program available free from www.kemtec.com. To do this, place the disk in the A: drive and type MONTEST then press the ENTER key. This utility provides several patterns to test both text and VGA graphics screens.

### Purity check

(vi) From the MONTEST menu select PURITY. This function allows you to turn on the RGB guns of the CRT by pressing '1' for 'ON' or '0' for 'OFF'. Select just the red gun by entering '1' for red and '0' for the other two guns. The screen should show an even hue of red across the whole screen, there should be no colour smears. See the two example screens as shown in Figure 3.140.

Purity errors are caused by stray magnetic fields interacting with the three electron beams, causing them to impinge on the wrong phosphor dots. The earth's magnetic field although weak, can severely upset the screen purity of a colour CRT. To see this effect, carefully turn your monitor onto its side while the red raster is being displayed. You should see some spurious colouration appearing over the even red hue.

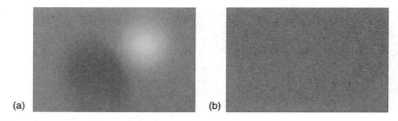

(a)                                                    (b)

**Figure 3.140**  *(a) Purity errors appear as coloured patches. (b) A pure screen shows an even hue*

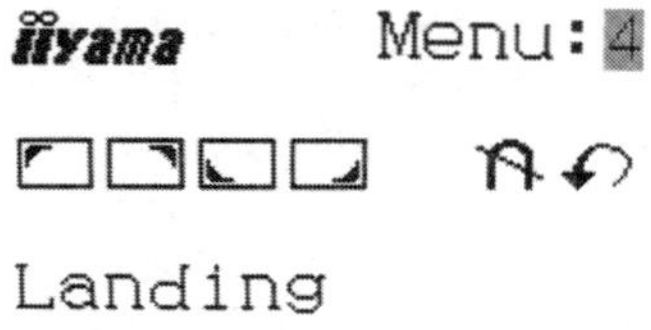

**Üyama**      Menu: 4

Landing

**Figure 3.141**  *A typical monitor OSD menu facility*

Monitors often have a manual degauss facility on the on-screen menu control system on the front of the monitor. If you have one like this then simply select the degauss function from the menu, there is no need to switch the monitor off. Some monitors have a separate degauss button.

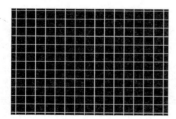

**Figure 3.142**  *Cross hatch pattern*

Purity errors are common with multimedia PCs because they incorporate stereo loudspeakers. These have large magnets which produce quite intense magnetic fields. It is therefore wise to position the speakers as far away from the sides of the monitor as possible.

To combat purity errors the monitor has a built-in *degaussing coil*. This is positioned around the sides of the CRT, close to the screen area. When the monitor is first switched on a diminishing AC voltage is applied across the coil. This creates a decaying alternating magnetic field around the front of the tube. This has the effect of removing any static magnetic fields around the tube and surrounding area. This process is known as '*auto-degaussing*'.

Do this small experiment:

(vii) With the red raster displayed, place a multimedia loudspeaker against the side of the monitor, close to the screen. Coloured patches will appear near the loudspeaker side of the screen.

(viii) Leave the MONTEST program running, and turn off the monitor. Wait for three to five minutes to allow the degauss circuit inside the monitor to cool down (Figure 3.141).

(ix) Switch on the monitor and the auto-degauss will return the monitor back to a normal pure red. The degauss cycle sounds like a low-frequency hum or buzz lasting for 2 or 3 s just after power on, you may also hear the aperture grill inside the tube vibrate under the influence of the magnetic field. This is a useful test if you suspect a faulty degauss circuit.

(x) Return to the menu by pressing ESC and re-select PURITY, this time turn on the green gun only and check for an even hue of green. Finally repeat this for the blue gun.

### Electron beam convergence

(xi) Return to the MONTEST menu and select 'CROSS HATCH'. Select the red and green guns (blue off) and press the space bar until the medium sized yellow cross hatch pattern is displayed as shown in Figure 3.142.

The red and green guns produce a yellow grid pattern across the whole screen area. Look closely at the extreme edges of the screen. The red and green grids should converge accurately to form yellow lines. Slight convergence errors at the edges may cause the grids to separate into red and green lines. This is usually tolerable, however errors near the centre of the screen are highly objectionable and must

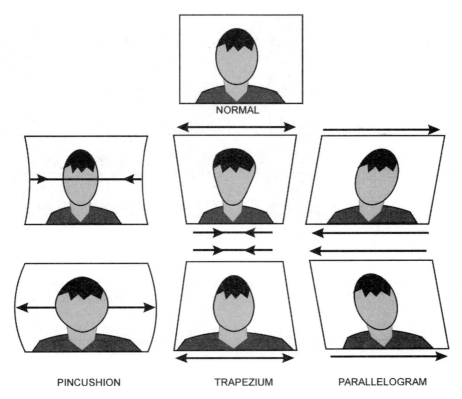

**Figure 3.143** *Types of linearity distortion and the effect it has on the screen*

be corrected by adjustments to the convergence rings on the CRT neck. These align the three electron beams so that they converge at all points on the screen area. Severe convergence errors could also be caused by component failure in the scan circuitry.

Finally the blue grid should also be turned on as well as the red and green grids to give a white cross hatch, again check for convergence errors. This will show up as paired blue and yellow lines instead of a single white line. Some technicians prefer to turn off the green grid and converge the blue grid over the red grid to show magenta lines.

Some high-quality monitors have external convergence adjustments brought out to a front control panel for ease of adjustment.

### Screen linearity

The white cross hatch is also useful for checking and adjusting the screen linearity and centring. Most monitors allow some external adjustment of screen linearity.

(xii) In the MONTEST program, select TEST CARD (H) from the main menu and check the overall screen linearity and focus. If you are working on a VGA or SVGA monitor then also check the screen using the appropriate higher-resolution test cards.

Types of linearity distortion (Figure 3.143)

- Trapezium distortion
- Pincushion
- Parallelogram

**Figure 3.144** *Horizontal grey scale*

Some modern high-quality monitors incorporate front controls to correct pincushion, trapezium and parallelogram distortion. As well as the usual height, width, and horizontal and vertical positioning controls. These are often digitally controlled adjustments where a button must be selected and held down while the correction is applied.

## Grey-scale tracking

(xiii) If you are testing a VGA colour monitor, select 'EXTRA VGA TESTS' (J) from the menu.

Select 'HORIZONTAL GREY SCALE' (A) (Figure 3.144).

The grey-scale pattern can be used to accurately set the contrast and brightness of the screen in graphics mode. Adjust the contrast and brightness controls so that all bars are clearly visible. If the brightness is too low the darker bars will merge into one black bar. If the contrast is set too high, the lighter bars will appear too bright and overpowering.

A correctly set grey scale should show a gradually increasing shade of grey from black to white with no colour tinges. If bars appear with a slight colour cast then the RGB gain and bias controls need adjustment. A failing CRT (i.e. low emission) will show smearing and colour tinges on the brighter bars.

## Colour fidelity

(xiv) Select COLOUR PALETTE (Figure 3.145) and the screen above should appear. The screen shows gradually increasing saturation and brightness for each colour. The brightness increases from right to left and the saturation from bottom to top. The standard 16 colour bars and grey scale are shown on the bottom of the screen. A correctly set up VGA monitor should show all colours and saturations clearly at each brightness level without smearing or loss of colour.

*Note*: saturation refers to the amount of white light added to a particular colour. For example, a fully saturated red contains red light only

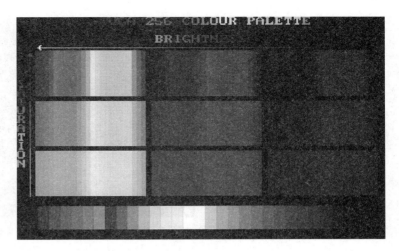

**Figure 3.145** *The VGA 256 COLOUR PALETTE*

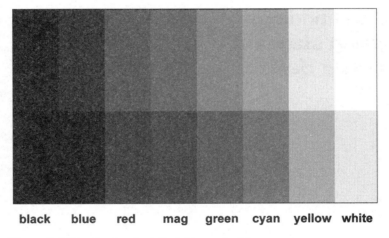

| black | blue | red | mag | green | cyan | yellow | white |

**Figure 3.146** *Contrast and brightness test bars*

with no blue or green light added. A de-saturated red contains red light with smaller but equal amounts of blue and green light added. The greater the amount of blue and green light, the more de-saturated the red light becomes until with equal amount of red, blue and green, it is completely de-saturated to white light.

## Contrast and brightness (text mode)

(xv) Adjust the contrast and brightness controls until the colour bars appear as in Figure 3.146. The bottom bars should be clearly discernible from the top bars, in particular the black, brown and grey bars should be clearly defined. The brown bar might seem odd but the human eye perceives low intensity yellow light as shades of brown.

Standard colour bars (Figure 3.147) are useful for checking the individual R, G and B channels for faults.

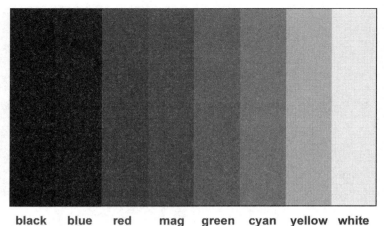

black   blue   red   mag   green   cyan   yellow   white

**Figure 3.147**   *Standard colour bars*

# Fault finding using standard colour bars

From the MONTEST main menu, select 'COLOUR BARS' by pressing the 'A' key, then select bright by pressing '1' and then '1' again for each of the three guns R, G and B. The colour bars as shown in Figure 3.148 (a) will appear.

The voltage waveforms shown opposite will be present on the RGB video channels. These signals could be viewed using an oscilloscope.

(xvi)   Now press ESC to return to the menu and re-select 'COLOUR BARS' but this time press '0' for the red gun and '1' for green and blue, thus turning on the green and blue guns and extinguishing the red gun.

The bars will appear as those shown in Figure 3.148 (b). The red bar as well as the red component of the magenta, yellow and white bars is missing.

This simulates a fault on the red channel.

Under real fault finding conditions the red signal could be traced along the video path using an oscilloscope, until the point is reached where the signal disappears.

### Weak red signal

The colour bars in Figure 3.149 (a) show the effects of a weak red signal. In this example red is only 50 per cent of its normal intensity.

The magenta bar which is normally 100 per cent red and 100 per cent blue now shows a predominance of blue. Likewise the yellow bar now has a predominance of green.

The white bar which is normally 100 per cent of each colour, now has 50 per cent red, 100 per cent green and 100 per cent blue, creating a predominantly cyan hue.

### Weak blue signal

With a weak blue signal (50 per cent in this example), the blue bar looks too dark the magenta bar has a predominance of red (100%R + 50%B) and the cyan bar takes on a green appearance (100%G + 50%B).

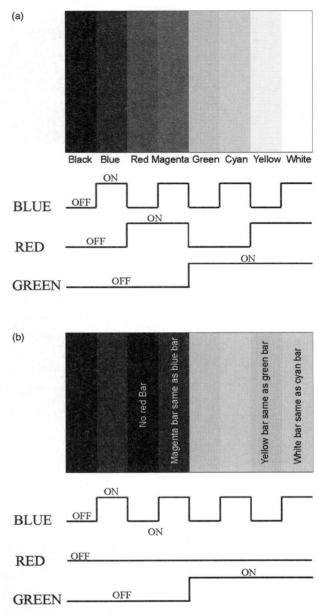

**Figure 3.148** *(a) Normal colour bars and (b) resulting bars with no red signal*

The white bar has a bright yellow appearance (100%R + 100% G + 50%B) (Figure 3.149 (b)).

### Weak green signal

When the green signal is reduced in amplitude (50 per cent), the Cyan bar is predominantly blue (100%B + 50%G), and the yellow bar is predominantly red (100%R + 50%G), giving an orange hue.

The white bar lacking enough green signal, has a light magenta hue (100%R + 50%G + 100%B) (Figure 3.149 (c)).

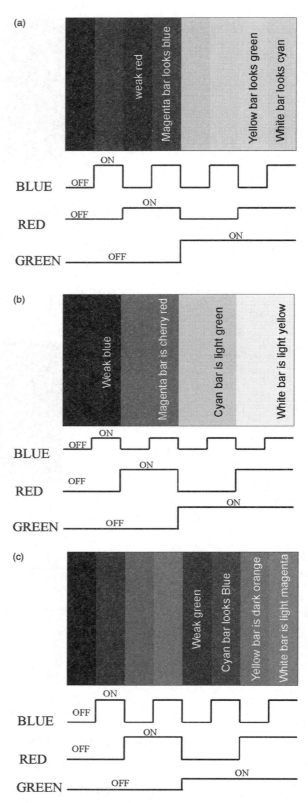

**Figure 3.149**   *The effect on the colour bars with weak red, blue and green signals respectively*

# Flat panel displays

The dominant display technology at present is based on the CRT. However, a relatively new type of display technology is becoming more and more popular as prices fall and the quality and performance begins to rival that of the CRT. This is generically known as flat panel display (FPD) technology. There are several types of FPD available. The most successful type that will eventually replace the conventional CRT-based computer monitor is the thin film transistor (TFT) LCD.

## Liquid crystals

An LCD utilizes the properties of certain nematic substances known as liquid crystals. The molecules of a liquid crystal can be orientated in a different direction to their normal position by the application of a small electric field. Some types of liquid crystal have molecules arranged in a helix. This helical structure has the effect of twisting polarized light through 90°. However when an electric field is applied to the crystal, the molecules untwist and the light retains its original polarization.

Using this remarkable property, a simple on/off picture element (pixel) can be constructed. To achieve this, the liquid crystal is sandwiched between two light polarizing filters placed at 90° to one another. Normally two polarizers placed in this manner block out light, but the liquid crystal in its non-energized state, twists the light from the first filter through 90°, thus matching the polarization of the second filter (see Figure 3.150a). The device therefore allows light to pass through. On application of an electric field however, the LCD molecules straighten out and the light is blocked by the two polarizers.

The LCD pixel therefore behaves like a tiny window letting light through when it is non-energized and blocking light out, when an electric field is applied.

As the device works on an electric field, there is virtually no current flow. So the power consumption is very small. Even large LCD arrays only consume minute amounts of power. LCDs are therefore ideal for battery-operated devices like digital watches, calculators, mobile phones and other hand-held devices, where low power consumption is a very desirable feature.

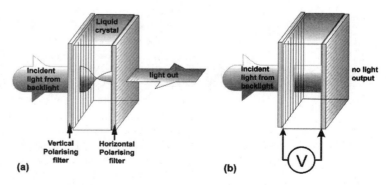

**Figure 3.150** *(a) Non-energized LCD pixel allows light to pass through. (b) When a pixel is energized by an electric field the molecules untwist and the light is blocked*

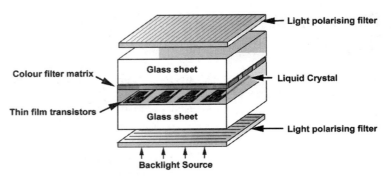

**Figure 3.151** *An idealized sectional view of an Active Matrix TFT*

Unfortunately screen displays for laptops, etc. require a backlight and this does consume power, but generally it is substantially lower than comparable CRT displays.

The early types of LCD are really only suitable for digital watches and calculators as the response time of each pixel is too slow for moving displays. However as the technology matured, better LC materials were discovered. The most important of these was super-Twisted Nematic (STN) liquid crystal. This material contains molecules that can be made to twist through 270° and it has a faster response time than the conventional material. This has allowed the development of large high-resolution displays comparable to those of CRT monitors.

Unfortunately STN liquid crystal on its own, is not responsive enough for today's SVGA resolutions. At the fast scanning rates required, the display becomes very de-saturated and low contrast. It was soon realized that one way to overcome the slow response was to hold the electric field on each pixel for the duration of one complete frame scan period. This would allow the full contrast of each pixel to be realized. To achieve this, each pixel is controlled by a field effect transistor (FET) similar to those used in today's super-large-scale ICs. The FETs are fabricated directly onto one of the glass panels forming the display, so they must be transparent to light. As an example of the technology required, an SVGA colour display of 1024 × 768 pixels, requires 2,359,296 FETs! This is due to the fact that three FETs per pixel are required – one for each of the primary colours RGB. To make the FETs as small as possible and transparent, they are fabricated directly onto the glass panel of the screen using TFT. The FETs are therefore known as TFTs.

To control each FET, an *X, Y* matrix of thin film conductors are also fabricated onto the glass. These select the individual rows and columns and therefore each pixel. This highly successful display system is known as Active Matrix TFT (Figure 3.151).

All LCDs are passive, in other words they do not generate light, they merely act as light shutters or filters. To achieve an acceptable display brightness for computer monitor or TV use, a light source is placed behind the LCD and this must produce even illumination across the whole screen area (Figure 3.152). The backlight uses a miniature florescent tube and a fairly sophisticated screen consisting of a reflector, light spreader and diffuser. The reflector is usually a white polished metal sheet and the light spreader is a flat wedge-shaped glass block, ground on one face and polished on the other. This spreads the light entering at the thick end of the wedge by internal light scattering and

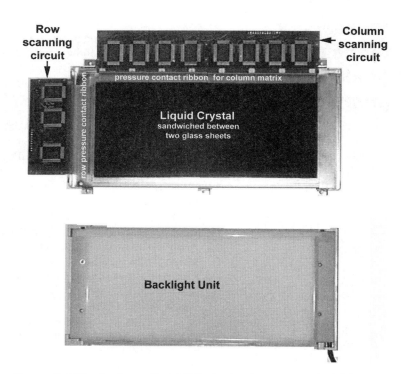

Figure 3.152 *A dismantled LCD display from an early LCD panel showing the key parts*

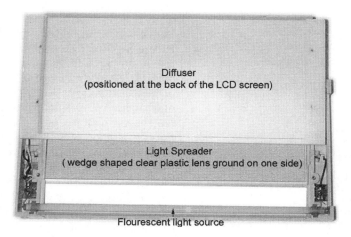

Figure 3.153 *Exploded view of a typical back light unit*

Figure 3.154

reflection from the reflector. The diffuser – a sheet of translucent white plastic – further scatters the light from the glass block, producing an even white glow across the whole screen area (Figure 3.153).

Generally the only serviceable part of a TFT LCD display is the florescent tube. As with most high-tech items nowadays, a new laptop is usually backed by a 3 years on-site or return-to-base warranty so these will normally only be replaced by a PC bench technician.

Figure 3.154 shows a typical Pentium-based laptop computer with an Active Matrix TFT display capable of displaying 65,536 colours @ 1024 × 768.

# Glossary of terms used on modern graphics cards

**AGP**   Acronym for accelerated graphics port, a new interface specification developed by Intel Corporation. AGP is based on the PCI bus, but is intended specifically for 3D graphics. Rather than using the normal graphics card/PCI bus interface, the AGP is a dedicated channel that gives the graphics engine direct access to main memory. The AGP 1.0 specification provides the data bus width and maximum speed implementation of the PCI bus standard (32 bits @ 66 MHz). The AGP 2X supports data transmission on both the rising and falling edges of the clock cycle. This produces an effective data transfer rate of 508.6 MB/s. AGP 4X and 8X is also available providing over 1 and 2 GB/s, respectively.

**Alpha blending**   Combines background information and the colour and transparency of an object to make it look realistic. For example an image of a frosted glass window could be placed over a background scene. The transparency factor and colour of the window is combined with the colour of the background to form a natural looking lighting effect.

**Alpha channel**   An extra channel of information in addition to the RGB channels, used to define a graphic object's transparency or edge characteristics.

**Anti-aliasing**   The smoothing out of the jagged edges of a graphic object, particularly diagonals, by blending the edges of the object with the background.

**Bitmap**   An image stored pixel by pixel, with every pixel represented by individual locations in the memory device, be it in RAM or an auxiliary storage device. Bitmaps take up large amounts of memory or disk space. However, unlike Vector-based graphics, a bitmap requires no mathematical calculations to regenerate the image. Images from a scanner or digital camera can only be sourced as a bitmap.

**Direct X (Direct Draw, Direct Video, Direct 3D)**   An application programming interface (API) for handling D and 3D graphics objects. It was developed by Microsoft as part of the Direct X suite of multimedia APIs. They provide a standard set of software routines, to help programmers develop 2D and 3D applications utilizing the special features provided on today's graphics cards. Direct X runs on Windows 95/98/2000 and ME.

**Frame buffer**   An area of video memory used to store a complete frame of an image while it is being updated.

**Gouraud shading**   A form of shading which blends together colours based on the general direction of a light source, to form a realistic shading on the surfaces of an object. This gives the object a realistic illusion of depth.

**Graphics engine**   The chipset on the graphics card that takes raw image instructions from the CPU and processes and writes image data to the frame buffer. Most modern graphics cards can apply many of the imaging techniques listed in this glossary.

**MIP-mapping**   A texturing technique which adds more detail to an object as it approaches the observer and loses detail as it recedes from the observer.

**Open GL**   Open graphics language. A free 3D graphics programming language developed by Silicon Graphics.

**Pixel**   From picture element. The smallest element on a screen that can be controlled by the graphics card, in terms of its colour and light intensity. The greater the number of pixels in the image, the better the overall quality of the picture on the screen.

**RAM DAC**   The RAM DAC converts the data stored in the frame buffer into the separate RGB signals required by the monitor.

**Texel**   Similar to a pixel but used to define a single element in of a texture map.

**Texture mapping**   In 3D graphics, texture mapping adds realistic looking textures such as sky, sea, grass, etc., to a basic scene.

**Vector graphics**   Graphics in which images are created by calculating the position of certain points in an image using a mathematical algorithm. The position of these points is calculated as a 'vector', that is a value which has magnitude and direction. For example a straight line can be defined by a start position, $X_s$, $Y_s$; angle; and end position $X_e$, $Y_e$.

**Video memory**   The RAM used in the frame buffer. It is also used in the Z-buffer and the texture memory fitted on of some top-end graphics cards. The most common types of RAM used on a graphics card are SGRAM, WRAM and VRAM.

**Z-Buffer**   Used on 3D graphics cards, the Z-buffer stores the depth and position of objects. In collaboration with the graphics engine, it can decide the parts of an object to show on-screen, and the parts to be hidden behind other objects.

## Sound cards

A sound card (Figure 3.155) provides the user with quality sound, for games, multimedia applications and general music playback. It can also process and store incoming audio data from the onboard musical instrument digital interface (MIDI) connector, line input jack and microphone jack. In office applications, sound cards are being used with speech recognition software. Typists no longer have to use just their fingers to type documents. They can now speak to the word processor using a microphone and *speech recognition* software. This converts the spoken word directly into text.

The first sound card to become popular in the PC games world, was the *AdLib Music Synthesizer*. The AdLib system was adopted as the standard sound system by games companies mainly because it was one of the first and least expensive sound systems available for the PC. Then in 1989 a company called Creative Labs introduced their *games blaster* sound card. This was followed a few months

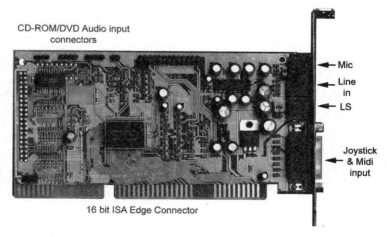

**Figure 3.155**   *A legacy 16-bit ISA, PnP stereo-sound card*

later by the *sound blaster*. The sound blaster was compatible with the AdLib and the games blaster card. It also offered some useful extras, including a MIDI connector, a *microphone* input and a *line* input for analogue sound sources such as a CD player, or radio.

The first PC games to incorporate 'proper sound' – as distinct from the very poor, 'buzzer-like' quality of the PC speaker – used AdLib sound routines. These games also ran quite happily with the sound blaster. Due to the games market, the sound blaster began to grow in popularity, partly due to effective marketing and partly due to the extra facilities offered and AdLib compatibility. Soon it became more popular, than the AdLib card itself and it was gradually adopted as the default sound card for the PC.

There are many PC sound card manufacturers around today, all offering advanced features such as wave table synthesis and 3D sound but they all maintain downward compatibility with the sound blaster specification.

### Multimedia

A PC can be made into a *multimedia compatible* system by adding a CD-ROM, SOUND CARD and VGA graphics adapter and monitor. 'Multimedia', is the term used to describe a system capable of using more than one type of media simultaneously, such as music, speech, text and moving images.

Originally for a microcomputer to qualify as a multimedia machine it had to conform to certain standards laid down by the Multimedia PC Council (MPC) – a group of prominent PC software and hardware manufacturers. To comply with the MPC specification, a multimedia PC had to meet the following minimum requirements:

- CPU 80386SX CPU
- 2MB of RAM
- VGA
- *Sound blaster* compatible sound card
- Single speed CD-ROM drive.

A single speed CD-ROM drive has a data transfer rate of 150 kbs, a double speed drive 300 kbs and a quad speed drive, 600 kbs. Nowadays, 12X speed CD-ROMs are classed as entry level, with a maximum contiguous data transfer rate of 1.8 Mbps.

As you can see from the list technology have moved on from these early systems.

### Installing an ISA PnP sound card in Windows 95/98

In our example PC system, we have already installed a modern PCI PnP sound card. However as a further example here we will go through the process of adding a legacy ISA PnP card to an existing PC system.

ISA PnP sound cards were quite common a few years back so there are many still around in older machines. Incidentally PnP works quite well on ISA cards. The only slight drawback is that the ISA bus does not have the facility to reassign IRQs like PCI. However this is not a problem as the Windows PnP system will juggle the PCI IRQs to accommodate the less adaptable ISA card configuration.

Sound cards are relatively slow devices so they do not really require the high bandwidth of the PCI bus. The 8 MHz ISA bus is more than adequate for sound and music. The reason most sound cards are now

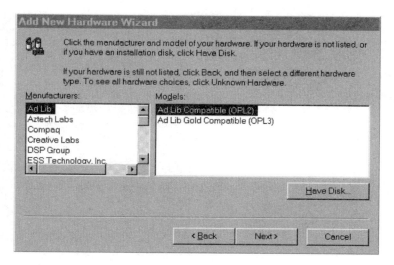

**Figure 3.156**

PCI compatible is because more and more motherboard manufacturers are removing the ISA bus from their motherboards.

*Step 1*   Ensure the PC is switched off. Wearing an antistatic wrist band, carefully remove the sound card from its antistatic bag. Examine the board and locate the correct connector for the CD-ROM audio cable. Notice that there are not many jumpers on the card as it is PnP compatible. See the sound card in Figure 3.156. If there are any jumpers on the card, check that they are set to an appropriate setting. Our example sound card has a jumper to disable the on-board audio power amplifier, if powered speakers are used. It also has a jumper to alter the microphone sensitivity. As we are using ordinary speakers and a normal dynamic-type microphone in this example, we will leave each jumper in the default position.

*Step 2*   Insert the card into a spare ISA slot and fit a retaining screw. Locate the stereo audio cable and plug one end into the analogue socket on the CD-ROM drive. Connect the other end to the audio-in connector on the sound card.

*Step 3*   Ensure that the sound card is seated properly in the slot and then power up the system and let Windows boot up. During the boot process the *Windows Hardware Wizard* will automatically detect the sound card and prompt you to select a manufacturer and model or select a driver from disk. Select *Have Disk*.

Insert the manufacturer's driver disk in the A: drive and type the path to the Windows drivers. On our driver disk it is **A:\win_98**. Then click the OK button and the system starts copying files from the source drive. During the installation you may be asked to insert the original Windows installation disk in the CD-ROM drive so that special drivers and files can be loaded.

*Step 4*   Now the main drivers have been installed, you can add the other software features supplied with the sound card. Most include a virtual, CD player, Sound Recorder and Sound Mixer.

The usual way to add these features to Windows is to place the CD in the CD player and let it auto-run. A menu will then prompt you to select your country and what applications you want to install.

You will probably have to reboot the system to let Windows accommodate the changes (Figure 3.157).

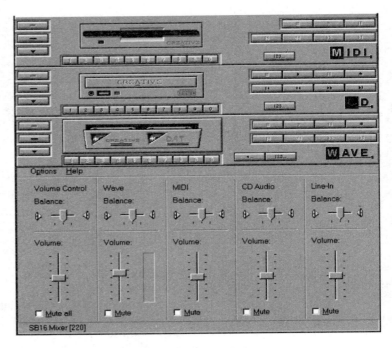

**Figure 3.157**    *A typical Windows audio rack*

*Step 5*    Connect the speakers to the back of the sound card. This is usually the jack socket above the MIDI/joystick connector (often colour coded green). Now test the sound card as follows:

Place a music CD in the CD-ROM drive and wait a few seconds. A virtual CD player will automatically appear on screen or on the task bar and commence to play the first track. Change to a different track and click on the volume control to increase and decrease the sound level, to confirm that the CD-ROM and sound card are working satisfactorily.

As a further test, remove the music CD and place a multimedia CD into the CD player. An ideal CD for this purpose is the type supplied free with reputable PC magazines. Most of these automatically run when placed in the CD and usually have sound accompaniment as well as graphics.

## The keyboard

Since the PC was launched there have been four different keyboard types. The original PC keyboard, with 84 keys; the AT keyboard, also with 84 keys; the enhanced AT keyboard, with 102 keys (Figure 3.158); and the Windows 95 keyboard. The four keyboards differ mainly on the key layout, and number of keys fitted.

All ordinary keyboards have a standard alphanumeric keyboard layout known as the QWERTY layout. This was originally designed over a century ago for mechanical typewriters. Rumour has it that it was designed with this layout, to slow the typist down, in an attempt to avoid jamming the type hammers. An alternative keyboard layout designed to improve typing speed is the DVORAK keyboard, but so far it has proved to be nowhere near as popular as the QWERTY type.

Beneath the keys on the keyboard is a switch membrane made up from a laminate of three plastic sheets. The top and bottom sheets have

**Figure 3.158** *An enhanced 102-key AT keyboard*

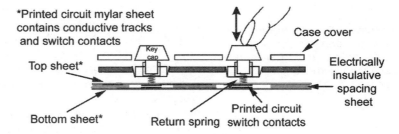

**Figure 3.159**

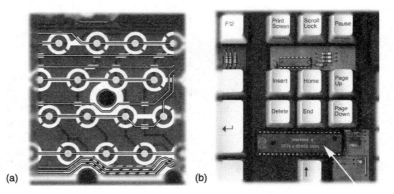

(a)

(b)

**Figure 3.160** *(a) Bottom cover removed showing detail of the contact membrane. (b) Top cover removed showing the keyboard controller*

a matrix of conductive tracks and key contacts printed on their inner surface. The middle sheet insulates the two sets of tracks and has a series of holes stamped in corresponding to each pair of key contacts. When a key is pressed down, a spring loaded plunger presses against the top sheet and bends it in slightly, causing the contacts to come together. On releasing the key the top sheet springs back into shape and the contacts separate (Figure 3.159).

More expensive keyboards often use individual keyboard switches, which generally give a more tactile feel to the keys. The photo below shows part of the membrane from an inexpensive keyboard alongside part of a keyboard using individual keyboard switches (Figure 3.160).

When a key on the keyboard is pressed or released, instead of generating a single on/off pulse it generates several pulses. This is due to the mechanical nature of the switch and is termed *switch bounce*. To

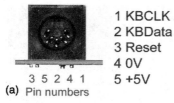

**(a)** 3 5 2 4 1
Pin numbers

1 KBCLK
2 KBData
3 Reset
4 0V
5 +5V

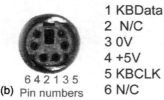

**(b)** 6 4 2 1 3 5
Pin numbers

1 KBData
2 N/C
3 0V
4 +5V
5 KBCLK
6 N/C

**Figure 3.161** *(a) AT-style keyboard connector (5-pin 180° DIN socket). (b) ATX-style mini-DIN connector (6-pin mini-DIN socket)*

remove *switch bounce* an electrical delay can be used or a software delay can be incorporated in the keyboard program. The later method is used on all PC keyboards.

When a key is pressed it is first debounced and then a scan code and break code is assigned to it. The codes are then sent serially, to the computer via a four or five cored screened cable to the keyboard connector on the back of the system-unit. The keyboard socket connections at the back of PC AT and ATX-style systems are shown in Figure 3.161.

# How the keyboard works

The AT-style keyboard contains an intelligent controller IC which performs all the keyboard interfacing functions. The chip contains a built-in RAM, ROM and CPU as well as the I/O ports for the keyboard. It therefore behaves like a tiny computer system dedicated to serving the keyboard. Such a device is called a microcontroller (μC).

The keys are arranged in a matrix of eight rows and thirteen columns, giving a maximum of 104 keys, connected via 21 wires to the controller. The rows connect to the data bus (scan inputs) and the columns connect to the address bus (scan outputs), of the μC. Using the matrix arrangement, only 21 wires are needed plus one to the positive supply rail, to sense all 104 keys. Without a matrix, the μC would require 104 connections plus one common connection for the keyboard alone!

A key scanning algorithm permanently stored in the μC's ROM, continually sends a signal to each column in turn by holding it at Logic 0 for a few microseconds, before returning it to Logic 1. The scanning rate of this action is purposely made many times faster than the maximum possible typing rate of the human operator. The whole 104 key matrix, can be scanned in less than a couple of milliseconds.

When a single key is pressed, it connects a unique row of the matrix with a unique column. Now as the columns are being continually strobed by the μC, the 'lucky' row receives a Logic 0 pulse.

The μC program detects the Logic '0' pulse on one of its scan inputs and by knowing what scan output line (column) is active, it can determine what key is being pressed. It then goes through a short delay of a few milliseconds to allow the key contacts to settle down mechanically, to remove switch bounce. Then it stores the location of the newly pressed key in its internal RAM – the keyboard buffer. It then continues on to complete the current scan.

Up to 10 key presses are stored in the buffer in a first-in-first-out arrangement (Figure 3.162).

After each complete scan of the matrix, the μC runs a data validity check on the stored key presses and then sends two bytes in serial form to the PC. The first byte is called the scan code – a unique number indicating the precise location of the key in the matrix. The second byte is the break code, this indicates if the key has been released. The break code is usually identical to the scan code, apart from the state of the most significant bit which is '1' when the key has been released and 0 if it is still held down.

The scan code and break code is transmitted on the KB data line to the PC. When no data is present on the line, it defaults to the Logic '1' state.

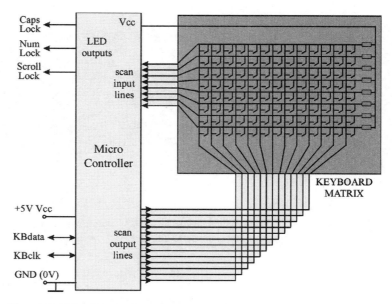

**Figure 3.162** *Block diagram of a 102 key enhanced AT keyboard*

If a key is held down for more half a second or so, the keyboard controller will carry on sending the same scan code and break code at a rate of 10 CPS, until the break code changes. This is useful as it allows the typist to auto-repeat certain characters.

For example the '_' character could be auto-repeated to quickly form a continuous line thus; '_____'.

The KBData signal sent to the motherboard is synchronized by a common crystal-controlled clock (KBCLK). At the motherboard end, the serial signal must be converted back to parallel form using a serial in parallel out (SIPO) shift register. It is then stored in a data buffer ready to read by the whatever program is currently active on the PC.

Most AT/ATX motherboards use an intelligent keyboard controller in place of the simple dumb keyboard arrangement that was used on the old XT system. The main advantage of an intelligent interface is that it can send serial data back to the keyboard as well as receive data. This allows the keyboard to be configured from software.

Special features like specific Windows 95 keys can be accommodated by simply running a driver program supplied with the keyboard. A typical Windows keyboard has extra keys taking the total to 105 or more. Two special *Windows key*s activate the Windows Start menu, and the *Applications key* activates a pop-up menu, which has the same effect as clicking the right mouse button.

The special keys used on the enhanced AT keyboard and the keyboard scan codes are included in Appendix D.

## The mouse

The first mouse is reputed to have been invented by Doug Englebart at Stanford Research Institute in 1963. It used two thin metal wheels at right angles to one another. Each wheel is connected to a variable resistor to sense movement in either the '*X*' or '*Y*' directions. The whole assembly was encased in a wooden box that was barely small enough to fit an average male's hand. Many improved versions of the original mouse have subsequently appeared.

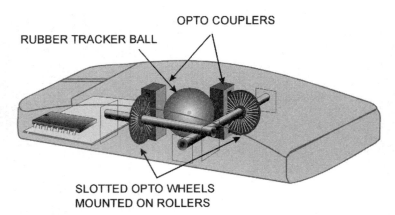

RUBBER TRACKER BALL

OPTO COUPLERS

SLOTTED OPTO WHEELS
MOUNTED ON ROLLERS

**Figure 3.163**  *A skeleton drawing showing the internal parts of an opto-mechanical mouse*

In 1983 Microsoft introduced their first mouse, *Mouse 1.0* specifically designed to operate with the IBM PC. It was supplied with its own 8-bit ISA adapter card. The mouse was known as the *BUS mouse* because it connects via a card to the ISA expansion bus. Subsequently in 1984 a serial version of MOUSE 1.0 was introduced. This version used a spare serial port instead of a separate bus card and was therefore less expensive and easier to install. After several alternative mouse designs, Microsoft introduced their third-generation mouse called the *new mouse* in 1987. This was available as a *BUS mouse* or *SERIAL mouse*. At the same time IBM introduced their PS/2 mouse which uses a similar interface to the keyboard, with clock and data lines. It also uses the neater mini DIN connector rather than the 9-pin D connector found on the serial mouse.

The Microsoft serial mouse and the PS/2-style mouse are now the most popular types in use. In addition to these main makes, there are many clones and grey import versions.

The main method used to detect the position of the mouse, as it is moved around on a flat surface, is a ball and roller system. The rubber coated ball protrudes below the base of the mouse sufficiently to make a friction contact with the surface it is placed on, usually a mouse pad or the none shiny face of a desk. As the ball rotates, it moves two horizontal rollers set at right angles to one another. On axis with each roller is a slotted OPTO wheel that rotates in the path of an infrared beam running between an IR emitter and IR receiver unit. As the wheels rotate, a pulse train is generated in the receiver unit of each OPTO device. The pulses contain both positional, and direction information. A simple digital circuit combines the pulse trains from each OPTO device, into a single data stream.

In the obsolete BUS mouse, the data from the OPTO circuit is conveyed via the connecting cable to an ISA adapter card, which converts the bit stream into parallel form. In the serial mouse system the positional data from the OPTO circuit is converted to RS232, by a built-in chip. This allows the mouse to be connected directly to one of the PC's COM ports (Figure 3.163).

All ATX motherboards, have a PS/2 mini DIN mouse interface. This utilizes the circuitry built into the keyboard controller chip, on the motherboard. If you compare the PS/2 mouse signals with those of the keyboard, you can see the similarity. Some mouse types have

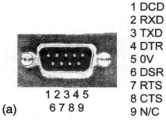

```
1 DCD
2 RXD
3 TXD
4 DTR
5 0V
6 DSR
7 RTS
8 CTS
9 N/C
```

```
1 2 3 4 5
6 7 8 9
```

(a)

```
1 Mouse Data
2 N/C
3 Gnd
4 Vcc
5 N/C
```

```
6 4 2 1 3 5
```

(b)  Pin numbers

**Figure 3.164** *(a) Serial mouse COM1/2 connections (9-pin 'D' connector). (b) PS/2 mouse connections (6-pin mini-DIN socket)*

a switch to select PS/2 or serial mouse mode. A normal serial mouse will not work on a PS/2 port and a PS/2 mouse will not work on a COM port. USB mice are now quite common but there is no real advantage gained. The mouse is a relatively slow speed device and the PS/2 port is perfectly adequate for this purpose. The same reasoning applies to the USB keyboards.

One of Microsoft's latest mouse design is called IntelliMouse. It looks similar to the standard Microsoft mouse but it has an additional wheel between the two buttons. The wheel is easily turned by the tip of the first finger, allowing the user to quickly scroll and navigate around the screen. Several Windows applications make full use of the special features available on an Intellimouse.

The most common problem with a ball-type mouse is the continual build up of grease and grime on the ball and the pressure wheels. One of the support technician's less glamorous tasks is to regularly clean the ball and inner mechanism of debris.

### Optical mouse

However, this may soon be a thing of the past as fully optical designs are now quite common. The optical system does away with the rotating ball arrangement and uses a solid state led/OPTO-sensor array and lens. There are two common types of optical mice: a cordless type and a type that connects directly to the PS/2 mouse port (Figure 3.164). The cordless variety use an Infrared link to a hub connected to the PS/2 port. The only real drawback to the cordless design, is the use of batteries to power the mouse and these will need replacing every so often.

# Self-assessment questions

**SAQ 3.1**   State one advantage a dot-matrix printer has over modern ink jet and laser printers.

**SAQ 3.2**   State two methods used to squirt ink from an ink-jet nozzle.

**SAQ 3.3**   What is the main advantage of a dye-sublimation printer?

**SAQ 3.4**   Why do some current ink jet printers have to be switched on and connected to the computer before an ink-cartridge is replaced?

**SAQ 3.5**   State two types of printer that use a photoconductive drum to form an image.

**SAQ 3.6**   Why is fuser oil used in a colour laser printer?

**SAQ 3.7**   What type of printer uses tractor feed paper?

**SAQ 3.8**   State the typical sequence of steps required to open the Add Printer Wizard in Windows.

**SAQ 3.9**   What are TrueType fonts?

**SAQ 3.10**   What active device forms the digital image on a scanner?

**SAQ 3.11**   Complete the sentences below:

- A modem converts serial RS232 signals into an … signal for transmission on the PSTN, in a process known as. …
- A modem also converts incoming data from the PSTN into a … signal at the PC, in a process known as. …

**SAQ 3.12**   What ITU (CITT) V standard does a current 56 kbs modem usually support?

**SAQ 3.13** When setting up a dialup Internet and e-mail account manually, name at least five pieces of information you need from the ISP to perform the installation.

**SAQ 3.14** What is the digital camera equivalent to the film emulsion of a conventional camera (i.e. the device where the image is formed prior to saving it to memory)?

**SAQ 3.15** What is Flash Memory?

**SAQ 3.16** True or false? When connected to a computer a USB digital camera is regarded by Windows as an extra disk drive.

**SAQ 3.17** State the typical final anode voltage of a colour CRT monitor.

**SAQ 3.18** What is the purpose of the RAM DAC on the video card?

**SAQ 3.19** What speed is the standard AGP 1.0 port clocked at?

**SAQ 3.20** State one advantage Direct X offers since it was introduced into the Windows OS.

**SAQ 3.21** What colour would be produced in the CGA system if the intensity signal is '0' and the RGB signals are '1', '1'and '0', respectively?

**SAQ 3.22** If the green channel failed on a colour monitor (i.e. no green signal), what colour cast would show on the screen, effectively giving away the defective channel?

**SAQ 3.23** Estimate the *approximate* relative RGB colour intensities while the screen is showing a light violet colour on an SVGA monitor?

Red ...%   Green ...%   Blue ...%

**SAQ 3.24** How much memory does a 2D display set to $1280 \times 1024$ by 16.7 million colours consume?

**SAQ 3.25** As a little exercise in setting screen resolutions, what screen size (diagonal) would be required to run in a resolution of $1600 \times 1200$ pixels, to maintain the recommended 80 dpi requirement?

**SAQ 3.26** What is a TFT (e.g. as used in LCD screen technology)?

**SAQ 3.27** What special property is exhibited by a nematic crystal?

**SAQ 3.28** What PC sound interface standard preceded the generic Sound Blaster standard?

**SAQ 3.29** State in simple terms, two methods used to eliminate keyboard switch bounce.

**SAQ 3.30** What is the maximum number of key presses that can be stored in the keyboard buffer?

**SAQ 3.31** State the main reason for using a key matrix on the keyboard controller rather than using separate lines to each switch.

**SAQ 3.32** What is the purpose of the scan code and break code?

**SAQ 3.33** State the main advantage of an optical mouse over an OPTO/mechanical mouse.

# Unit 4 Networks

*Introduction to PC networks*

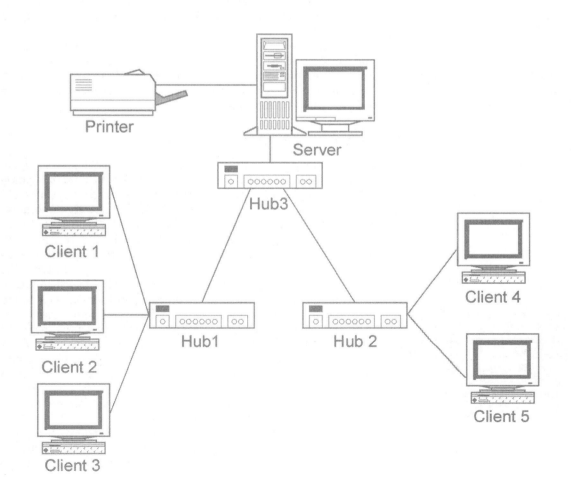

Printer

Server

Hub3

Client 1

Client 2

Hub1

Client 3

Hub 2

Client 4

Client 5

# Communication fundamentals

Networks encompass two of the greatest technological achievements in man's evolution, digital computing and electronic communication. Before we look at the practical aspects in building a small network system, we will see how telecommunications, software and hardware come together to make a network.

To set up a communication link there are four essential elements, namely the *sender*, *message*, *transmission medium* and *receiver*.

How the message reaches the receiver depends on the transmission medium. With speech communication, both the sender and receiver rely on the properties of air pressure waves. The sender's vocal cords vibrate the air released from the lungs and the receiver picks up the tiny air pressure vibrations via the ear drums.

Speech is fine over short distances, as long as the participants have normal hearing. Over distances of more than a few metres however communication becomes more difficult. This is because unwanted sound (i.e. noise), mixes with the signal and the signal strength weakens with distance. In a noisy environment such as that found in a nightclub, the useful communication distance drops down to a few inches. Whereas in a quiet room this can be several tens of metres.

Noise and signal attenuation are a problem with all communication systems. One way to combat these factors is to increase the amplitude of the signal. In a noisy room we do this by shouting. If the noise is too loud however, shouting has no effect and the communication breaks down. The noise has completely swamped the signal.

The words 'signal' and 'noise' in electronic communication are analogous to those of speech. The signal or *data* – as it is known in networking terminology – is the wanted part of the signal and noise is the unwanted part. Usually the word signal is used loosely to mean both the wanted and unwanted parts.

### Signal-to-noise ratio

The ratio of signal to noise – usually measured as the ratio of signal power to noise power – is a useful parameter to the communication systems designer but is only of passing interest to the network technician/engineer. Suffice to say that the higher the signal-to-noise ratio, the better the chance of receiving the signal uncorrupted.

When information is conveyed down a cable, there is a small amount of noise already present in the signal from where it originated. Extra noise is also added as it propagates along the cable. The signal-to-noise ratio at the receiving end is always less than that at the sending end. How much noise a cable picks up depends on the environment and on the technology used to convey the signal.

Noise in cables and other electronic gear is caused by magnetic and electrical interference. In a network environment the major cause of noise is from nearby equipment. One of the skills of a network wireman is to know where cables can and cannot be laid to minimize noise pickup.

Computers and monitors, for example, radiate electromagnetic interference in the form of (radio) waves.

Another major interference source is the mains supply. This generates low-frequency electromagnetic radiation (EMR) which induces a corresponding signal in network cables running close to the mains

You can experience this yourself during a phone conversation with a friend. Ask your friend to move the receiver up close to a working monitor! You will experience noise and signal break-up as a result.

wiring. It is therefore best to keep network cable runs as far away from computers and mains wiring as possible.

The signal and cabling systems used on networks are specially designed to reduce interference but it can still be a problem in some environments.

### Screened cable (coaxial)[1]

Screened cable (Figure 4.1) – as used on bus networks[2] – is composed of an inner insulated cable (the core) and an outer braid of fine copper wires which completely surround the core. In use the outer screen is electrically connected to the metal case of the network equipment and this in turn is earthed via the mains supply. The earth connection acts like a giant sponge soaking up electrical and magnetic interference. The earthed outer screen therefore shields the inner core from interference.

**Figure 4.1** *Screened cable*

### Unscreened twisted pair cable

Physically more straightforward than coaxial cable, twisted pair cable is used on 10BaseT/100BaseT network topologies (to be discussed later).

Twisted pair cable (Figure 4.2) is so called because it contains four pairs of insulated wires loosely twisted together and enclosed in a plastic sheath. There is no screen surrounding the wires, hence the name used to describe this type of cable is *unscreened twisted pair* (UTP).

**Figure 4.2** *UTP cable*

### Differential signal transmission

You may be wondering how a pair of unscreened wires can operate satisfactorily on a network when interference is such a problem. As they stand, twisted pairs of wires are nowhere near as effective as coaxial cable at reducing interference, even if one of the wires is earthed. The secret of the success of this cable system is the way the signal itself is designed. Instead of transporting a single signal along one wire and using the other as an earth return as in a coaxial cable connected network. A mirrored pair of signals is transmitted. One signal is a positive-going voltage and the other is an exact replica but negative-going voltage. This is known as a *differential signal*. The majority of network systems use this system.

The differential signal system works basically as follows (Figure 4.3). At the receiving end the mirrored signals are subtracted from one another, that is, the receiver only passes the difference in voltage between the signals. Let us see the result.

[1] The two conductors share a common central axis, so it is known as coaxial cable.
[2] The various network topologies bus, star and mesh are covered later in this unit.

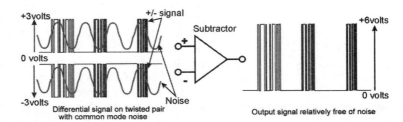

**Figure 4.3** *Using differential signals to reduce noise and improve performance*

If we assume for this description that the mirrored signals have a maximum amplitude of $+3$ and $-3$ V, respectively, the difference signal (i.e. subtracting one signal from the other) is:

$$(+3\,V) - (-3\,V) = 6\,V$$

*(i.e. minus a negative quantity is always a positive quantity)*

The result is a signal with double the amplitude. Now let us consider what happens to the interference induced along the cable. The wires are twisted together, so there is a high probability that they will both pick up identical noise voltages. This is called *common mode interference* because it is of the same polarity and waveshape in both wires. Let us assume the noise amplitude is $+1$ V in each wire. On subtracting the interference signal at the receiver we are left with:

$$(+1.0\,V) - (+1.0\,V) = 0\,V$$

As if by magic, the interference has been nullified. This is of course an idealized example but the principle still applies to real signals and noise. A differential signalling system can therefore handle considerably faster data transfer rates than that obtainable on a single core system because of the excellent common mode interference rejection. The differential mode signals are let through and amplified at the receiver, whereas the interference is largely cancelled out.

Differential signals are also used on ISDN[3] and ADSL[4] lines which are in essence glorified telephone lines using differential signals. This greatly increases the available bandwidth. The universal serial bus (USB) and Firewire™ bus used on computer equipment also use differential signals.

## Fibre optics

Optical cable network systems (fibre optics) are increasingly used as replacements for conventional copper wire topologies. The Ethernet 100baseFX system is a popular example. Instead of using electrical signals on UTP cable as 10BaseT and 100BaseT, for example, extremely rapid pulses of light are used on glass fibre cable. The light pulses emanate from a light emitting diode (LED) or laser diode.

Fibre optic networks offer potentially higher bandwidths and immunity from electromagnetic interference. Being made of glass, the cables are electrically non-conductive, so they can be placed alongside electrical wiring or in areas where electrical isolation is difficult to maintain. They are also far more difficult for a hacker to tap into.

In terms of resources, the raw copper ore used in copper cables is only mined in one or two countries. Glass on the other hand is made from silica, the world's most abundant mineral. Considering future technologies, glass has a rosy future.

Fibre optic cables do have some disadvantages: optical cable is considerably more difficult to join and terminate than copper cable. Light is guided down a fibre optic cable in a slightly analogous way to that of water in a pipe. If the joints are imperfect, leakage occurs and less light reaches the destination. Without special jointing equipment it is virtually impossible to make a perfect optical joint.

Light propagates down the fibre in a series of straight lines, reflecting from wall to wall in a zig-zag fashion. To propagate efficiently, the

[3] Integrated services digital network.
[4] Asymmetric digital subscriber line.

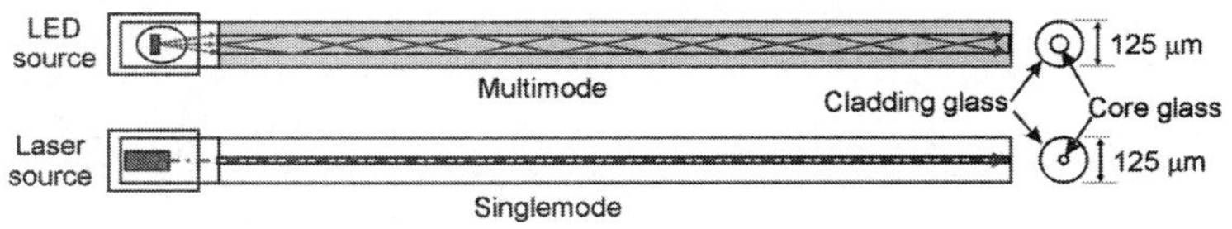

**Figure 4.4** *Idealized drawing of multimode and singlemode light transmission*

light source must be as monochromatic[5] as possible. LEDs and laser diodes are not truly monochromatic but they do emit a relatively narrow spectrum of light compared to conventional light sources. LEDs are the most widely used light source due to their relatively low cost. They are not ideal, however, as the light waves spread out in many directions and are incoherent.[6] This results in much of the light being scattered as it propagates along the cable. However they are perfectly adequate for cable runs up to a few hundred metres.

For much longer runs over several miles, laser diodes are used as the light source. Laser light is coherent and has a much narrower spread and shorter spectrum, than LED emission. Consequently, the losses are far less. The LED system is known as *multimode* and laser diode system as *singlemode*.

With reference to Figure 4.4, the light from the emission source propagates mainly along the central axis or core of cable. The core glass usually has a different refractive index to the cladding glass which helps to confine the light to the central core area.

Multimode cable has a relatively large core diameter and although this allows more light rays to enter, the light source is incoherent and scatters more easily and is therefore not so efficient as singlemode cable. Multimode systems are used for shorter runs, less than 2 km in length, for example inside buildings.

Singlemode cable has a smaller core diameter and allows the coherent light from the laser emitter to propagate through the cable with less scatter and therefore travels a greater distance before a repeater is necessary. It is therefore ideal for long cable runs beyond 2 km.

The individual light-carrying fibres are usually 125 or 250 μm in diameter. The cable itself holds one or more pairs of fibres together with a strengthening cord and an outer sleeve.

Two fibres are necessary per network connection to allow a two-way (duplex) communication.

## Signal types and parameters

Signals fall into two main categories: *analogue signals* and *digital signals*.

### Analogue signals

Our hearing and other senses are based on analogue stimuli. We can perceive a seemingly infinite range of sound pressure levels from a pin dropping, to the ear splitting noise of a jet fighter aircraft passing low overhead.

These two extremes in sound pressure level represent the lower and upper limits of the average human aural response system. Between these limits there is an infinite range of possible sound pressure levels.

Training lead bodies in networking and telecommunications require that network technicians have a fundamental understanding of the signal types and parameters used in telecommunications.

[5] Monochromatic light is light of one frequency or wavelength – virtually impossible to achieve in practice.
[6] In incoherent light all waves of the same wavelength rise and fall in unison, that is, they are in phase.

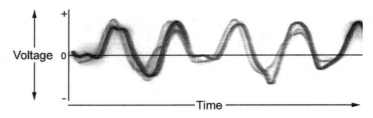

**Figure 4.5**    *A typical analogue waveform*

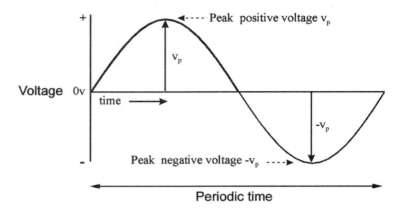

**Figure 4.6**    *One cycle of a sinusoidal voltage waveform*

If we were to examine the sound intensity from a microphone in a quiet room with just one person talking, we would see a signal waveform similar to that shown in Figure 4.5. The microphone converts the incident sound pressure waves into a voltage analogue. The louder the voice, the higher the output voltage.

Notice how the waveform varies smoothly from one level to the next. An analogue signal is one that varies at the same rate and amplitude as the source. It matches the nuances of the original signal using a different physical phenomena (e.g. a corresponding variation in voltage, magnetism, light, etc.). An analogue signal varies smoothly in amplitude with time. It therefore has an infinite number of levels between a defined upper and a lower *limit*.

Using the average human aural system again as an example, all the sounds we hear can be analysed in terms of sine waves. These can be regarded as the basic building blocks of all waveforms. In fact all repetitive signals regardless of their overall type or structure, can be analysed mathematically using groups of sine waves.[7] We will examine a sine wave more closely by referring to the voltage waveform of the UK mains supply. The shape of the UK mains waveform adheres quite closely to an ideal sine wave under normal conditions (Figure 4.6).

## The parameters of a sine wave

The three distinguishing parameters of any repetitive voltage waveform are (1) the shape (in this example the shape is *sinusoidal*), (2) the *peak voltage* ($V_p$) usually referred to as the *amplitude* and (3) the *periodic time*, usually abbreviated to *period* ($T$). These parameters can usually be seen and measured on an instrument called an oscilloscope.

The *periodic time*, also called the *period*, is the time it takes for the basic waveform to go through one complete cycle (Figure 4.6).

[7] This is a mathematical technique known as Fourier analysis.

The number of complete cycles that occur in one second is known as the *frequency* of the waveform.

The SI unit of frequency ($f$) is Hertz (symbol Hz), named after *Heinrich Rudolf Hertz* (1857–1894), the German physicist. He was the first to reproduce electromagnetic waves (radio waves) artificially.

To find the frequency ($f$) in Hertz of any repetitive waveform of period ($T$), simply take the reciprocal[8] of the period ($T$) as shown below:

$$f = \frac{1}{T} \tag{4.1}$$

*Example:* The periodic time of the UK mains AC supply is 20 ms. The frequency is therefore:

$$f = \frac{1}{20 \times 10^{-3}} = 50\,\text{Hz}$$

If the frequency is known then the period of a repetitive waveform can be found by rearranging Equation (4.1), that is, the period $T$ is the reciprocal of the frequency $f$:

$$T = \frac{1}{f}$$

Periodic electronic signals can be monitored and measured on an oscilloscope, where the waveform is usually a varying voltage.

For example, in the three voltage waveforms shown in Figure 4.7, we can observe the following:

1. Waveforms 1 and 2 have the same periodic time and therefore the same frequency.
2. Waveform 3 has a slightly longer periodic time and therefore a lower frequency than waveforms 1 and 2.

If each horizontal division on the grid represents 1 ms, then waveforms 1 and 2, have a periodic time of 4 ms, while waveform 3 has a slightly longer period of 5 ms.

The frequency ($f$) of waveforms 1 and 2 is therefore as follows:

$$f = \frac{1}{4 \times 10^{-3}} = 250\,\text{Hz}$$

and for waveform 3:

$$f = \frac{1}{56 \times 10^{-3}} = 200\,\text{Hz}$$

[8] 'Reciprocal' is a mathematical term meaning divide a quantity into 1, that is, the reciprocal of 4 is a quarter (1/4). So 4 multiplied by its reciprocal is 1. The same applies to any number.

## Amplitude

Regarding the same three signals of Figure 4.6, if each vertical division on the grid represents 1 V, then waveforms 1 and 3 have a peak-to-peak voltage of 2 V. Waveform 2 has a peak-to-peak voltage of 4 V.

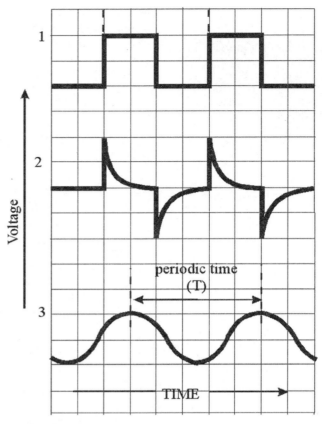

**Figure 4.7**   *Voltage waveforms*

## Root mean square amplitude

As mentioned previously, the AC[9] mains supply waveform – under good conditions – is sinusoidal. This means that the instantaneous voltage of the mains varies cyclically, following the sine law. The voltage swings smoothly from 0 V through to a positive peak value ($V$p), and then back to zero again, to complete one half cycle. It then swings smoothly to a peak negative value ($-V$p), before returning to zero, again for the start of another cycle. The peak mains voltage ($V$p), in the UK is approximately 340 V. The peak-to-peak voltage is, of course, twice this value at 680 V.

You can see from this why AC mains cables are capable of generating significant interference in other signal-carrying cables including network cables. The packets of data on the network cable are just a few volts in amplitude typically less than 5 V. The peak-to-peak voltage swing on a nearby 240 V mains cable is 680 V! So it is no wonder that some mains interference is induced in all network cables.

All AC sine wave supplies are usually rated by their root mean square value (RMS), not their peak or peak-to-peak value. The RMS value of a sine wave is the peak value ($V$p) divided by the square root of 2, that is, $V$p/$\sqrt{2}$. For the UK mains the RMS value $V_{(RMS)}$ is therefore:

[9] AC: alternating current; DC: direct current.

$$V_{(RMS)} = \frac{340}{1.414} \approx 240\,\text{V}$$

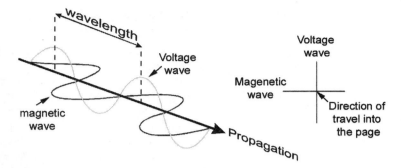

**Figure 4.8** *Electromagnetic waves*

This voltage value is more familiar to us than the peak or peak-to-peak value.

## Electromagnetic waves

Another important group of repetitive waveforms used in electronics and networking are the electromagnetic waves. Various forms of radiation including light, radio waves and atomic radiation fall into this category. These are electromagnetic signals that propagate through space at the velocity of light '$c$', which for practical purposes can be regarded as 300 million metres per second ($3 \times 10^8$ m/s).

Electromagnetic waves are sinusoidal in shape and have a voltage component and a magnetic component at right angles to one another. These are known as *transverse waves* because the electromagnetic variation in the wave is perpendicular to the direction of travel (Figure 4.8).

As electromagnetic waves travel through space, it is useful to know the length between each cycle of the wave. This is known as the *wavelength*. The *wavelength* (symbol $\lambda$, Greek lambda), in metres, is found by dividing the velocity of light '$c$' by the frequency '$f$' of the wave, as shown below:

$$\lambda = \frac{c}{f}$$

For example, to calculate the wavelength ($\lambda$) of the BBC world service medium wave radio broadcast, which transmits at a frequency of 648 kHz, simply divide the speed of light by the frequency:

$$\lambda = \frac{3 \times 10^8}{648 \times 10^3} = 463 \text{ m (approximately)}$$

Several companies use wireless network systems, particularly in situations where the network is only temporary. Wireless networks use radio waves or infrared rather than conventional cables for the transmission medium.

# Digital signals

The word *digital* relates to the fingers of the hand from the Latin word 'digitus'.

A digital signal uses discrete levels rather than the infinite range of values of an analogue signal. The digital signal shown in Figure 4.9(a) has five discrete levels. Two of the levels are a minimum and maximum value, and there are three intermediate levels. Computers use a binary digital system, so there are only two possible states, a minimum and a maximum. The two states can easily be represented by the presence or absence of a variety of physical phenomena including voltage, current, electric charge and light.

If we display digital signals on an oscilloscope they would appear something like those shown in Figure 4.9.

### Transistor/transistor logic signals

The *Transistor/transistor logic* (*TTL*) series was the first successful range of digital integrated circuits (ICs) used in computer manufacture. Launched in the early 1970s, they are still used today in some parts of the PC system. Their strange sounding name is derived from the circuits used in their manufacture. The ICs are fabricated using bipolar transistors throughout rather than the more primitive diode/transistor logic (DTL) used in earlier logic circuits. Data in TTL format is based on a +5 V system. A logic 0 input signal is defined as a voltage level between 0 and +0.8 V and logic 1 as any voltage in the range +2 to +5 V. The LPT1 port on the PC uses TTL signalling.

### High-speed CMOS signals

Today's high-speed complementary metal oxide semiconductor (CMOS) ICs use a different range of voltage levels based on a lower maximum voltage range. Typical chipsets operate at +3.3 V or lower.

For example, using a +2 V maximum signal level. Logic 0 is defined as any voltage in the range of 0 to +0.3 V and logic 1 is any voltage in the range +1.5 to +2 V. The majority of digital ICs used on modern computer systems use high-speed CMOS technology or derivatives.

# Modulation methods

Sending data along a telephone cable or through space as radio waves requires some form of conversion process that matches the signal to the transport medium. The public switched telephone network (PSTN), for example, was originally intended to convey speech

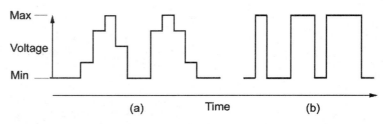

**Figure 4.9** *Digital waveforms: waveform (a) has five states and (b) has two states (i.e. a binary digital signal)*

signals in the form of a varying voltage from a handset microphone. The range of signal frequencies required for intelligible speech is approximately 300 Hz–3.4 kHz. This is known as the speech bandwidth. Although the PSTN handles speech signals admirably, it has no provision for higher bandwidth signals. For example, you cannot successfully receive hi-fi sound via the telephone line as this requires a considerably higher bandwidth of 20 Hz–20 KHz. If you do so, it would sound quite muffled (lo-fi)! Video signals need more than a thousand times the available bandwidth of the PSTN. Without high compression (e.g. MPEG coding) and other processing techniques, video transmission would be impossible on the PSTN. Another problem with the PSTN and incidentally radio frequency transmission, in general, is their inability to transport DC signals. For example, it would be impossible to send binary data (e.g. TTL signals, directly from a computer over the telephone network). Therefore the signal must be converted to a form that can be conveyed on the transmission medium. One such method is to use the original signal to modulate 'carrier' waves. These are usually sinusoidal in shape and fall within the bandwidth of the system.

### Amplitude modulation

One of the simplest modulation methods used in communication systems is amplitude modulation (AM) (Figure 4.10). In this technique the instantaneous voltage level of a carrier waveform of a frequency, several times higher than the maximum signal frequency, is varied in sympathy with the signal using a device called an amplitude modulator. The signal input to the modulator is known as the *modulating signal* and the resultant output is an *amplitude modulated carrier (AM signal)* (Figure 4.11).

It is often useful to know the frequency spectrum of an amplitude modulated system. For example, when a sinusoidal carrier of frequency $f_c$ is amplitude modulated by a sinusoidal modulation signal $f_m$, the modulated carrier wave contains three frequencies. One is the original carrier wave $f_c$ and the other two are sum and difference frequencies of the carrier and modulation waveforms, that is, $f_c + f_m$ and $f_c - f_m$, respectively. These are called the upper and lower side

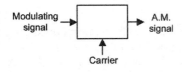

**Figure 4.10**   *AM technique*

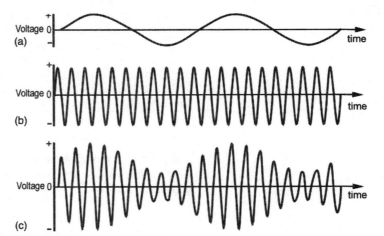

**Figure 4.11**   *(a) Modulating signal* (sine wave used for clarity); *(b) carrier; (c) AM output*

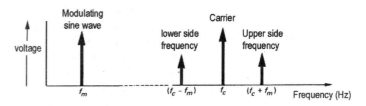

**Figure 4.12** *Frequency spectrum of amplitude modulation system*

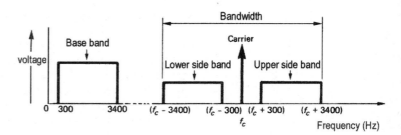

**Figure 4.13** *Example of sine wave modulating signal*

frequencies. The corresponding frequency spectrum of this simple AM system is shown in Figure 4.12.

As a further example, if we replace the sine wave modulating signal with the full range of speech frequencies (i.e. 300–3400 Hz), upper and lower sidebands are created around the carrier frequency. The maximum bandwidth of the modulated speech signal is 6800 Hz. The bandwidth of the original signal is known as the *baseband* (Figure 4.13).

### Frequency modulation

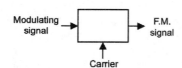

**Figure 4.14** *FM technique*

Frequency modulation (FM) (Figure 4.14) as the name suggests is a system where the modulating signal varies the frequency of a carrier wave. The amplitude of the carrier therefore remains constant (Figure 4.15).

The frequency variation of the carrier waver is directly proportional to the amplitude of the modulating signal. There are several advantages of FM over AM. The most important one is its inherent immunity from noise spikes (a form of AM). In the receiver the FM 'signal' is passed through an amplitude limiter which chops off the peaks of the 'signal' and thus removes much of the noise (Figure 4.16).

In an AM 'signal' the noise once present, cannot be removed successfully.

### Pulse modulation

This form of modulation uses voltage pulses rather than continuous sine waves. Modulation can be achieved in a number of ways. Two popular methods are:

1. *Pulse AM*, where the amplitude of the pulse is directly related to the amplitude of the modulating signal.
2. *Pulse width modulation*, where the width of the pulse is varied in direct proportion to the signal amplitude.

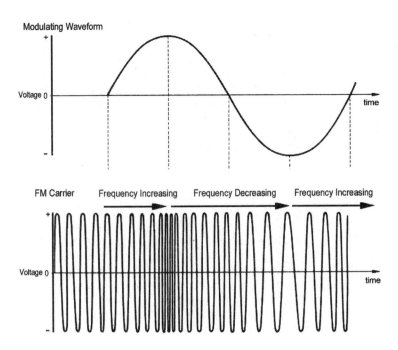

**Figure 4.15** *A carrier (bottom) frequency modulated by a sine wave signal (top)*

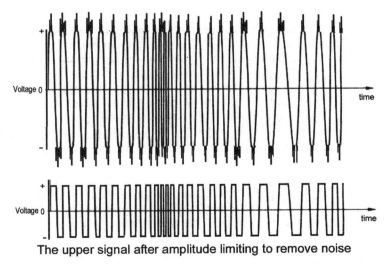

The upper signal after amplitude limiting to remove noise

**Figure 4.16** *A Frequency Modulated signal amplitude limited to remove noise*

To accurately reproduce the original signal (shown here as a sign wave, Figure 4.17) it must be sampled at a pulse rate at least twice as high as the maximum frequency present in the original signal.

For example, to reproduce speech the pulse rate needs to be at least 6.8 kHz. For hi-fi sound a sample rate in excess of 40 kHz is necessary. Uncompressed TV quality video would require a sample rate greater than 10 MHz.

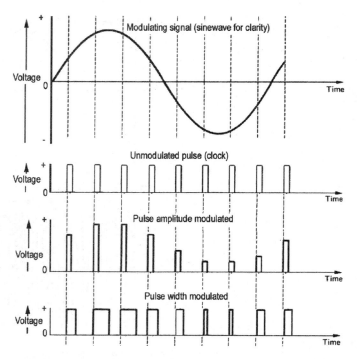

**Figure 4.17**   *Pulse modulation technique*

### Pulse code modulation

Pulse AM is part of an important modulation technique called pulse code modulation (PCM). In this form of modulation the original signal is first pulse amplitude modulated to produce a series of regular voltage pulses whose height it directly related to the amplitude of the signal. Then each pulse is quantized to a fixed voltage level close to the original voltage level and then converted to binary code.

In the simple 3-bit PCM system shown in Figure 4.18, the signal (a sine wave in this example), is first pulse amplitude modulated and then the height of each pulse is quantized to the nearest of eight possible levels. The corresponding 3-bit pattern for the pulses in this example is 100, 111, 110, 010, 000, 010. This forms a serial pulse train output representing the original sine wave.

At the receiver, each group of 3 bits is converted back to a pulse of height commensurate with the binary value. The result is then passed through a low-pass filter to produce a facsimile of the original analogue signal.

## Multiplexing

In a computer network system one cable must share information from many computers. This is particularly essential for wide area networks (WANs) and the Internet where multiple cables between computers would be totally impractical. For a successful network system there are two essential requirements:

- No computer must be allowed to hog the network.
- If data becomes corrupted only the corrupted data should be resent, not the whole file.

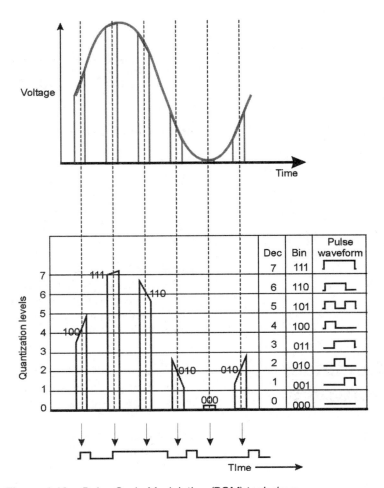

**Figure 4.18** *Pulse Code Modulation (PCM) technique*

Network systems do this by breaking files sent from each computer into smaller self-contained *packets*. Each packet contains a slice of data from the full file, the address of the source and destination computer and other essential data. The addresses are unique to each computer and are stored as a media access control (MAC) address in the computer's network interface card (NIC) (more on this later).

### Time division multiplexing

The process of sending data from many sources in packets one after another in time is called *packet switching* and is a form of multiplexing known as *time division multiplexing* (TDM).

As each packet on a particular cable occupies a unique time slot, the more packets on the network from different sources, the longer it takes to receive all the packets needed to reconstitute the original file. If you are familiar with the Internet, then you have first hand experience of the time it can take to receive data from a remote computer, especially during busy times.

### Frequency division multiplexing

Another form of multiplexing known as frequency division multiplexing (FDM) assigns a unique frequency band to a series of channels.

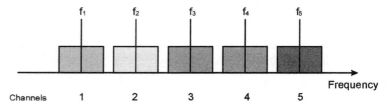

**Figure 4.19**  *Frequency Division Multiplexing (FDM)*

The TV/radio broadcast system is a good example. This is achieved by amplitude or frequency modulating a series of carrier frequencies. A guard band is provided between each channel to prevent channel cross-over. All the channels are sent simultaneously over one link (Figure 4.19).

To receive a particular channel at the receiver, the user 'tunes' (filters out) all but the wanted channel. One inherent limitation of an FDM channel is that it can only send data one way (simplex). To achieve two-way communication two channels are required. FDM can be used to convey multiple network channels. Such a system is known as *broadband*.

# Computer networks

A computer network in a nutshell is a system that allows two or more computers to share information and resources quickly, reliably and securely using a common set of rules.

### Local area networks
When computers are linked in the same building or across nearby buildings the network is called a *local area network* (*LAN*).

### Wide area networks
When two or more LANs are linked together over a large distance, for example between cities or across continents, the network is called a *WAN*.

## Network properties

To be at all useful a network system must allow each computer connected to the network to communicate with other computers and devices as required. To achieve this no computer or device can monopolize the network.

The major breakthrough in networking came with the idea of splitting data up into smaller self-contained packets. If the individual packets contain enough information and are managed according to a strict set of rules, the following features are offered:

- Data can be reassembled at the intended destination back into its original form.
- Corrupted packets can be detected and reissued.
- Data to and from many computers and devices can be handled in a continuous stream.
- Data can be sent securely.

Each data packet must contain:

1. The MAC address of the source and destination computer.
2. Sequence number.
3. The data itself.
4. Error-detecting code.

### MAC address

> Each computer connected to the network is fitted with an adapter card that connects the computer to the network via cable, fibre or wireless link.
>
> These cards are known as Network Interface Cards or NICs for short.

For the reasons stated above, every computer on a network must have a unique address. Not just unique to the devices on the network but to every other computer ever made! This unique address is built into the NIC. The NIC links the computer to the network and provides a unique 48-bit identification number called a MAC address. Batches of MAC addresses are issued to NIC manufacturers worldwide by the *Institute of Electrical and Electronic Engineers* (IEEE) to ensure that no two cards made have the same MAC address. It is unlikely that the IEEE will run out of MAC addresses in the near future as 48 bits provides over 281 billion different addresses.[10]

In network jargon a computer and its associated NIC is termed a *node*, so the words MAC address and node address are synonymous.

To make MAC addresses easier for the support technician to identify, the 48-bit number is usually expressed in hexadecimal, that is, the binary number is split into six 8-bit blocks and expressed in hex.

For example, the following MAC address in binary is what the network sees as a series of voltage pulses:

000000001110000000011000101110100111111100011101

We prefer to see it represented in hex:

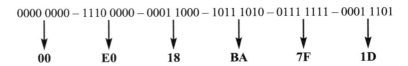

0000 0000 – 1110 0000 – 0001 1000 – 1011 1010 – 0111 1111 – 0001 1101

| 00 | E0 | 18 | BA | 7F | 1D |

> A node can be any device on the network, for example a printer, server and computer with an NIC installed and hence its own unique MAC address.

To check the MAC address of a network PC running Windows 95/98/ME, run the MS-DOS prompt window; | **Start** | **Programs** | **Accessories** | **MS-DOS Prompt** | and type the command **winipcfg** or **ipconfig /all**.

In Windows XP: | **Start** | **All Programs** | **Accessories** | **Command Prompt** | and type the command **ipconfig /all** (Figure 4.20).

### Sequence number

When a file is sent over the network – unless it is extremely small – it will have to be broken down into smaller pieces. In fact the packet size used on the most common network systems is only 1500 bytes in total and this includes essential code in addition to the data itself. The task of breaking down a file from an application, for example a word processor, into smaller pieces is carried out by a special program termed a Transport Protocol. Each piece is given a unique sequence number so that the file can be rebuilt at the destination computer.

So each file fragment along with a sequence number, source and destination MAC address, forms a self-contained data packet but not quite ...

[10] $2^{48} = 281474976710656$ or over 281 billion (US 281 trillion).

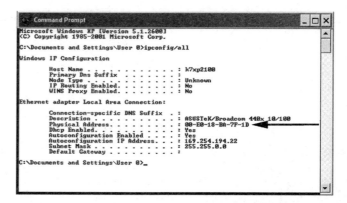

**Figure 4.20** *MAC address (physical address) and other network details*

### Error detecting

If a data packet becomes corrupted during transmission the destination node must be able to detect the error and request a resend. One of the common forms of error detecting is based on a special technique called *cyclic redundancy checking* (*CRC*). Basically sending NIC arithmetically divides the data to be sent by a fixed sequence of bits representing a polynomial. The polynomial is usually 16- or 32-bit long and is generated on the card. The remainder from the division calculation is a unique sequence of bits known as the CRC. This is added to the data packet prior to despatch. On receiving the packet, the destination NIC repeats the calculation and compares it to the CRC, if they match the data is accepted. If an error is detected, a resend of the same packet is requested.

The way data packets are conveyed between the various NICs on a network is determined by the physical and logical properties of the network.

## The physical and logical network

When discussing networks it helps if we segregate the cables, NICs and other connection items, that is, the tangible items you can touch, from the parts of the network that define the protocols, operation and management of the network.

The former part is called the *physical network*, and the more abstract software oriented part of the network is called the *logical network*.

### Physical topology

The *physical topology* is the physical arrangement of the network and in particular the cabling. There are three types of physical topology used in practical networks: *bus*, *star* and *ring*.

### Bus topology

Networks based on the bus topology use coaxial cable to link one computer to another in a daisy chain arrangement. To prevent data corruption the extreme ends of the bus cable must be terminated with a resistor to stop signal reflections. Without a terminator a reflection would be interpreted as a corrupted data packet and no data would ever reach its intended destination.

The first successful network system to use a bus topology was *Ethernet*, introduced by Xerox in the early 1980s. Later it became a standard of the IEEE with the rather unfriendly designation *IEEE 802.3*. Even today people still prefer to call it Ethernet. An Ethernet network system is based on a technique called *carrier sense multiple access/collision detection* (CSMA/CD). Carrier sense refers to the way each computer on the network listens for a carrier indicating that data from another computer is already on the bus. If no data is detected then the computer can gain access to the bus to send a data packet. The words *multiple access* means that all computers have equal rights to use the bus, that is, it is not prioritized in any way. It is a first-come first-served system. Unfortunately this can cause problems when two computers try to send data simultaneously. If this happens each computer reads the data on the bus and detects that the data has indeed been corrupted. This is the *CD* part of the system. Both computers involved in the data collision then wait a random amount of time before attempting to send the data packet again.

The majority of network systems use baseband signalling. Baseband is effectively a single signal conveyed on a cable, as distinct from a *broadband* system that conveys a group of signals simultaneously.

The original Ethernet bus implementation was called 10Base5. The '10' refers to the maximum data transfer rate of 10 Mbit/s. And *base* refers to the signalling method used, that is, baseband. The '5' signifies the maximum allowable cable length of a bus segment, that is, 500 m. 10Base5 networks use RG-8 coaxial cable which is thick and cumbersome so it is bubbed 'thicknet' by technicians.

Later Ethernet 10Base2 (thinnet) was announced. This uses thinner and therefore more manageable RG-58 coaxial cable but the maximum segment length is limited to 175 m not 200 m as one would expect from the name.

An Ethernet bus topology is very easy to set up but the main drawback of all bus systems is that if one of the links in the network chain is broken, the whole network fails. To overcome this limitation, later implementations of Ethernet, for example, 10BaseT and 100BaseT use a star/bus technology. Externally the network is wired in a star arrangement using UTP but inside the central hub the normal Ethernet bus system is implemented in chip form.

### Star topology

As the name suggests, the cables in a star topology radiate out to each computer, from a central hub. If one computer is disconnected the rest are still connected via the hub, the network system will continue to function. This topology is the most popular today.

### Ring topology

In a ring topology all the computers connect to a central cable ring. The main network system that uses a ring topology is called IEEE 802.5 or *token ring* named after the way each computer in the ring uses a token-passing technique to prevent data collisions.

# Ethernet 10BaseT/100BaseT

The practical network system we will set up later is based on the Ethernet 10BaseT and 100BaseT star topology. 10BaseT and 100BaseT use identical cabling arrangements based on UTP cable. UTP cables use RJ-45 connectors (Figure 4.21). These have eight pins and are connected to the twisted wire pairs to conform to the EIA/TIA 568A or EIA/TIA 568B standard. It is wise to stick to one or other of these two standards throughout the network.

For normal computer-to-hub cables, both ends of the cable must be wired the same, to the EIA/TIA 568A or 568B standard (Figure 4.22). However to link two hubs together via the normal hub ports, a so-called 'cross-over cable' must be used. This changes the transmit and receive pairs over so that the transmit pair forms one hub link to the receive pair in the other hub and vice versa. A cross-over cable uses the EIA/TIA 568A arrangement on one end of the cable and the EIA/TIA 568B arrangement on the other end. Most hubs nowadays have a 'cross-over' switch or a dedicated 'cross-over' port, eliminating

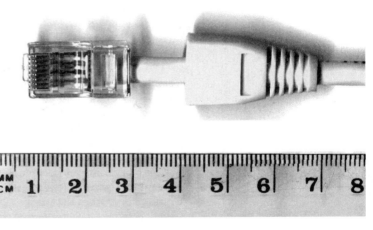

**Figure 4.21**  *The RJ-45 connector*

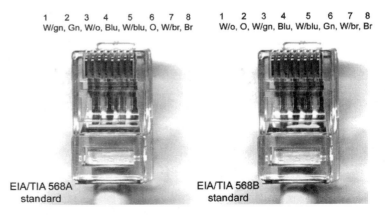

**Figure 4.22**  *The EIA/TIA 568A and 568B twisted pair connection standards*

the need for a cross-over cable. So the same cable type can be used throughout the network.

The wiring for a basic 10BaseT or 100BaseT network in a single room is very straight forward. The NIC on each computer is connected to the central hub via a standard UTP cable (Figure 4.23).

For larger networks spanning several rooms a distribution rack and patch panel arrangement is best. It keeps the whole network interconnection system in one central location and allows the network administrator to make changes quickly and easily without too much disruption to the network users. The distribution rack should be located in a secure lockable wall-mounted cabinet or separate locked room. Figure 4.24 shows a rack of RJ-45 sockets (110 blocks) wired to the various nodes around the building and a hub. In this example the hub is also linked via optical fibre, to a network in an adjacent building. Telephone patching is also conveniently placed in the same cabinet.

The 110 blocks in the distribution rack are wired through the building to numbered RJ-45 wall sockets in each room. Each wall socket can then connect directly to a computer via a standard UTP fly lead.

10BaseT is a baseband system operating at 10 Mbit/s, the 'T' in the name refers to the fact that it uses twisted pair cables. The maximum length of cable between the hub and a node is 100 m. Only two pairs are used in the cable but all four pairs should be wired to the RJ-45 connector to conform to the EIA/TIA 568 standard. This allows the network to be upgraded at a later date without changing the cables. Incidentally UTP cable is available in several grades or categories. The best cable is category 5 and this should be used on all new installations thus making it as future proof as possible.

As long as the network cabling conforms to the EIA/TIA 568 standard and uses category 5 cable, a 10BaseT network can be upgraded to

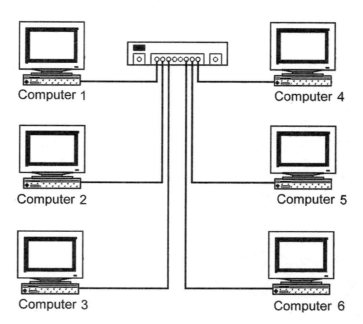

**Figure 4.23**   *A basic star network*

**Figure 4.24**   *Distribution rack and patch panel*

Ethernet 100BaseT simply by changing the NICs and HUBs. The cable will be fully compatible. 100BaseT runs at 100 Mbit/s and the maximum length of cable between the hub and a node is 100 m the same as 10BaseT. Some hubs and NICs claim to support both 10BaseT and 100BaseT but remember that the network will only work at the speed of the slowest device on the network. Mixing 10BaseT and 100BaseT items on the same network will give the same performance as a normal 10BaseT system.

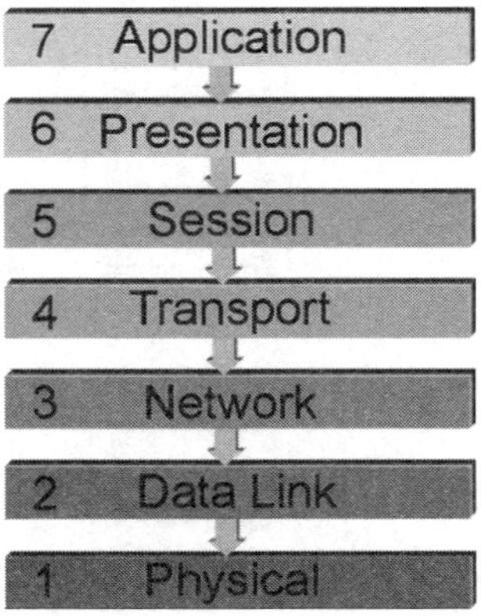

## The Open Systems Interconnect model

We have so far discussed the physical topology of the network and in particular on the popular Ethernet 10BaseT system. To understand the logical topology it is time to introduce the *Open Systems Interconnect* (OSI) model.

In the 1980s a group called the OSI introduced a general seven layer model (Figure 4.25) to help in the design and understanding of

| 7 | Application |
| 6 | Presentation |
| 5 | Session |
| 4 | Transport |
| 3 | Network |
| 2 | Data Link |
| 1 | Physical |

**Figure 4.25**   *OSI seven layer model*

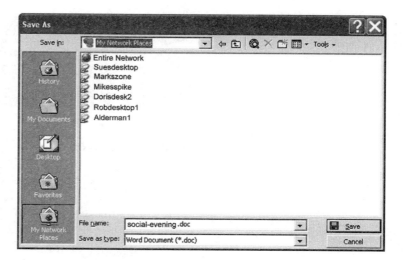

**Figure 4.26**

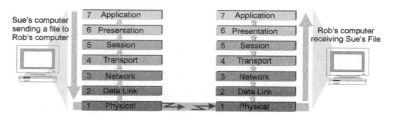

**Figure 4.27** *Example of OSI seven layer model*

networks. The OSI model is not a set of standards but a useful frame work for network designers, administrators and students.

To help illustrate the role of each layer we will take a simple scenario using a group of imaginary users connected on the same network.

Sue is currently sending a word processor file directly to Rob's 'My Documents' folder. In her word processor application Sue clicks | **File** | **Save As** | **My Network Places** |[11] and then chooses Rob's computer | *robdesktop1* | **My Documents** |. While this is taking place other users could well be sending files to Sue. We will follow the role of each layer in this relatively basic process (Figures 4.26 and 4.27).

### *Application layer*

This layer provides applications with network access, that is, protocols called application program interfaces (APIs) that allow applications such as word processors, e-mail software, file management software, etc. to use the network. It allows programmers to write applications that may need to use the network without having to write specific network routines – they just use the networking APIs built into Windows.

[11] Named | My Network Neighborhood | in Windows 95/98/ME.

When Sue saved to Rob's computer, the word processor program used a series of APIs in the application layer to link the file to the network software. At Rob's computer his application APIs perform the reverse process – the file from the lower layers is passed to the application layer where it is saved via the file handling APIs to Rob's My Documents folder.

### Presentation layer

This layer provides a common format for data presentation effectively allowing different computer systems to handle the same data. It also performs data encryption or compression as required by the network operating system (NOS).

Sue's file was stored as 16-bit unicode as used in Windows XP. The file could also be encrypted and compressed depending on the programs configured to operate at this layer. At Rob's computer which is a Windows 98 system the file is converted to 8-bit ASCII and passed to the applications layer.

### Session layer

The session layer establishes connections (sessions) with other computers. It effectively acts like the old fashioned telephone operator system connecting and managing the transfer of files by controlling who connects and for how long.

For example, Sue is sending a document to Rob but at the same time Mark is trying to send one to Sue and so is Doris. The files to-ing and fro-ing from the transport layer are managed here. The session layer establishes a connection between Sue and Rob and validates passwords if required. It also monitors and regulates the data flow. Again the session software is an API. The usual Windows protocols used at this layer are NetBIOS and transport control protocol (TCP).

### Transport layer

The transport layer ensures error-free data delivery or quality of service (QOS). It checks for errors and also has the job at the sending computer of breaking files up into smaller pieces and adding a sequence number. At the destination process the reverse process ensues, building up the composite file from the individual pieces.

At Sue's computer her file is broken down into smaller pieces and sent down through the lower layers to Rob's computer. At Rob's end, the transport layer checks each piece for errors and informs Sue's transport layer software that the data arrived safely or if an error occurred. It also rebuilds the pieces into the original file before passing it on to the higher layers.

### Network layer

The last three layers of the OSI model from the network layer down to the physical layer are termed the *subnet*. The network layer acts like the subnet manager keeping data flowing through the lower layers by routing data through the network's routers (a device used to connect multiple network segments on larger networks), to find the most convenient data path. It works out to be the best route to send packets and add address information to each packet to allow it to take that route. Routers will only work on protocols that support routing. For example, Window's Netbeui protocol does not support routing but other protocols such as the Internet protocol (IP) does.

> On the basic network that Sue and Rob uses no router is fitted.

### Data link layer

The data part of the name refers to the fact that the data link layer at the sending computer adds information to the data to allow it to pass along the physical layer to the correct node. It is also responsible for acknowledging the receipt of error-free packets and for resending corrupted ones.

> Sue's data link layer software adds the source and destination MAC addresses to the data and links the packet to the physical network. At Rob's computer it accepts the packet from the physical layer, error checks it and acknowledges it before passing it to the higher layers.

Network devices that operate at the data link layer are called bridges. A bridge can filter data packets by building a table of MAC addresses for all the nodes on either side of the bridge. So two network domains separated by a bridge will operate more efficiently than if the two networks were linked together as one. Data packets frommachines communicating on one side of the bridge are not passed across to the other side of the bridge, so the chance of a data collision is reduced. If a node on one side needs to communicate to one on the other side, the bridge will detect the destination MAC address and let it pass over to the other side.

### Physical layer

This is the electronic part of the network, the part that turns the binary data in a packet into a stream of voltage pulses along a cable. The cabling, repeaters,[12] hubs and NICs are all part of the physical layer.

[12] An electronic device that regenerates data packets, that is, increases the strength of the signal on long cable runs. A hub also acts as a repeater.

> Sue and Rob's NICs and those of the other members of their group and the cables linking their NICs to the hub and the hub itself form part of the physical layer of their network.

## Peer-to-peer networks

The network that Sue and Rob's team use is known as a peer-to-peer network. For a small group of users this is probably the best type of network as all users are peers that is, equals. Individual users can share their files and resources with the rest of the team. For example, Rob has a colour laser printer attached to his computer and Mark has a duplex printer (prints both sides of the paper). Doris has a high-resolution scanner attached to her computer. These items can all be shared via the network effectively giving everyone access to the facilities. Each member is responsible for backing up their own data onto CD-ROM on a daily basis. This type of system works well in a close knit group of dedicated users. Its main weakness is that there is no overall central management and security is virtually non-existent.

## Client–server networks

A client–server network (Figure 4.28) divides the computers up into servers and clients. A server is a powerful dedicated computer that shares its resources with other computers. It is able to quickly move data back and forth between the client computers and offers a secure and reliable data storage and backup system. Clients are less powerful computers that only connect through a server. They cannot communicate directly to other clients. Such a network offers a far more secure centralized management structure to the network, taking much of the responsibility away from the client. The clients sent files to the server where they are stored on a fail-safe hard disk storage system known as a redundant array of inexpensive disks (RAID) array.

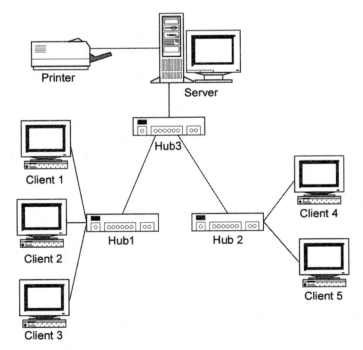

**Figure 4.28** *A small client–server network*

### RAID array

A RAID array, is a group of two or more hard disk drives used collectively to improve the size, data transfer rate and reliability compared to that of a single disk drive. The concept of an RAID array was published in 1987 and suggested several different architectures. The two main types (RAID 0 and RAID 1) use the basic techniques of *striping* and *mirroring*, respectively.

### Striping (RAID 0)

In the disk striping technique a file is saved to the array by splitting it up into several smaller pieces and distributing it over the drives. This greatly increases the apparent data transfer rate. The drawback with this method is that if one drive fails everything stored in the array is lost.

### Mirroring (RAID 1)

The mirroring system saves a copy of the file to two or more drives so that if a drive fails the data remains intact. RAID systems support hot-plugging and automatic data recovery so that the faulty drive can be pulled and replaced and its contents automatically replenished.

# Installing your own windows network

Some otherwise keen computer enthusiasts find the concept of networking all too daunting and tend to steer clear of the subject. A good way to build confidence and open eyes to what is an amazing step up from single stand-alone computing, is to build a basic peer-to-peer network from the ground up. The following practical step-by-step set of instructions for building a 10BaseT network will help both the sceptical and keen take the first step.

What you need for this practical is as follows:

- Two or more Windows 95/98/ME/XP PCs preferably with a Pentium 200 or higher CPU.
- A set of 10BaseT or 100BaseT NICs. One for each PC (preferably PCI but ISA jumperless plug and play types are fine).
- A drum of category 5 UTP cable and at least two modular plugs per PC plus an RJ-45 ratchet crimping tool and wire cutters/strippers. Alternatively several pre-prepared UTP category 5 patch leads can be used. These can be purchased from most computer stores.
- 1 × 8-way hub (not needed if only two PCs are to be linked, in which case a cross-over category 5 UTP cable must be used to connect the PCs).
- 1 × antistatic wrist band and earthing cable.

### Step 1: Installing the NICs (Figure 4.29)

- Switch off the PC and disconnect it from the mains. Remove the case cover from the PC system-unit. Find an empty PCI or ISA slot depending on the NIC type you are using and remove the appropriate cover plate by unscrewing the retaining bolt.
- Wearing an earthed wrist band, remove the NIC from its antistatic bag and insert the card firmly into the correct slot making sure that the back plate cover locates correctly. Fit the back plate screw.

**Figure 4.29**  *Installing an NIC in a PCI slot*

- Refit the case cover and reconnect the mains lead. Repeat the above on each PC.
- Now fit a patch lead between each PC and a port on the hub.
- Switch on the PCs and the hub.

### Step 2: Installing the NIC drivers

It is always best to follow the card manufacturers instructions for driver installation. As we are using a jumperless NIC – the only type supplied nowadays – it will automatically be recognized by Windows. On booting-up into Windows a dialogue box will appear stating that new hardware has been found and after a few seconds the message 'Building Driver Database' appears. The add new hardware wizard dialogue box then prompts you to search for the new driver to install. If you have a popular NIC then Windows will probably have a built-in driver. If this is the case, it will search for the driver and install it automatically. If your card is not on the list of built-in drivers, Windows will prompt to choose from a list of drivers. Select the option 'Driver from Disk' provided by manufacturer and click OK. Usually network cards have the drivers on floppy disk, so insert the disk and select the A: drive.

The 'Add New Hardware' wizard sometimes asks you to insert your Windows CD-ROM. Always use the latest drivers that you can find for your newly installed hardware. The best way to ensure you have the latest driver is to check the manufacturer's web site.

When any new hardware and/or drivers are installed, the registry is automatically updated. The registry is vital to the operation of the system. So before you install a new driver or device, it is wise to backup the existing registry settings, just in case something goes awry. To do this, open the registry by clicking 'Start', then 'Run' and type 'regedit' in the text box and click OK. In Regedit, select the icon 'My Computer' at the top left of the list (Figure 4.30), click the registry menu item and then click 'Export Registry File'. Save the file to a floppy disk using a name of your choice. If after installing the card or driver, the existing registry becomes damaged, the problem can be fixed by simply restoring the old working registry from your floppy.

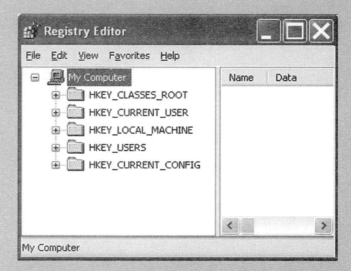

**Figure 4.30**

Windows 98/ME/XP/2000 performs regular backups of the registry automatically each day. In Windows 98/ME&XP you can restore a previous version of the registry in an emergency: (1) Restart Windows in Safe Mode (command prompt) by pressing CTRL during the Windows boot process. (2) At the command prompt type, scanreg/restore. Then press enter to invoke 'Registry Checker'.

You can now select a known good version of the registry (usually the most recent before the problem occurred). (3) Press enter and the registry is restored. (4) Press enter again to restart the computer.

### Step 3: Check if the drivers are installed correctly

- Choose: | **Start** | **Settings** | **Control Panel** | **System** | **Device Manager** |
- Double click **Network Adapters**. You should see your network adapter.
- Double click the line describing your network adapter.
- Click the **General** tab and you will see the name of your adapter card and 'Device Status'.
- If everything is OK it will read, '**This device is working properly**' (see Figure 4.31).
- If there is a hardware conflict with your network card, click on the **Resources** tab to see if there are any conflicting IRQs/DMAs or addresses. De-select automatic settings and click on Change settings. Try some different setting until the confliction is remedied (see Figure 4.32).

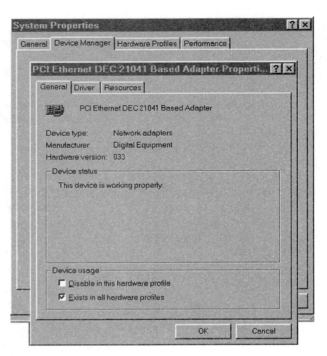

**Figure 4.31**

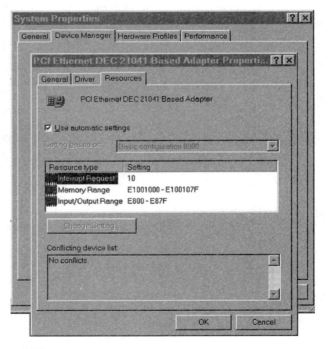

**Figure 4.32**

## Step 4: Setting up the protocol stack Windows 95/98/ME

To set up a network in **Windows XP** see the XP Network setup section at the back of this section.

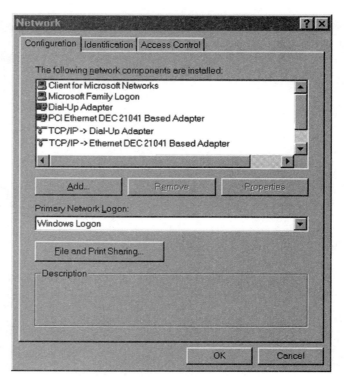

**Figure 4.33**

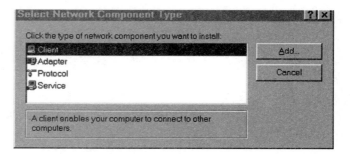

**Figure 4.34**

The protocol stack is the suite of protocols used in a particular network system:

- Choose: | **Start** | **Settings** | **Control Panel** | **Network** | and select the **Configuration** tab. A list of network components appears.
- The list should show your adapter card, 'Client for Microsoft Networks' and various Protocols.
- If the later two components are not shown, click the **Add** button (see Figure 4.33).
- Double click on **Client** in the new window (see Figure 4.34).
- Now select **Microsoft** and then the option **Client for Microsoft Networks**. Click the **OK** button.
- From the network Configuration menu (see the illustration above), change the Primary Network Logon to '**Windows Logon**' (see Figure 4.35).

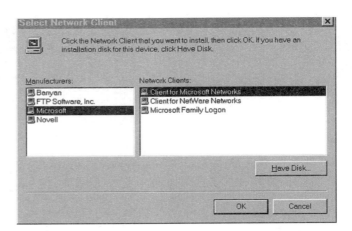

**Figure 4.35**

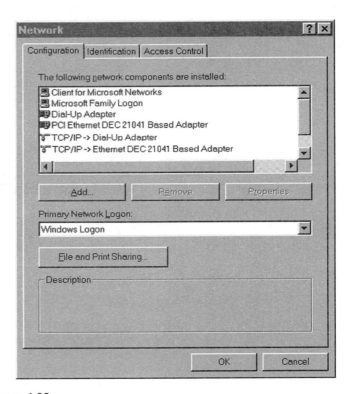

**Figure 4.36**

- For this simple network installation we are using the Microsoft Windows protocols NetBIOS and NetBEUI (*note:* NetBIOS is part of NETBEUI).
- If you wish link the network to the internet instead of using a stand-alone PC for this, then TCP/IP protocol should also be installed.
- Click on **Add** (see Figure 4.36).
- Double click on **Protocol** in the new window (see Figure 4.37).

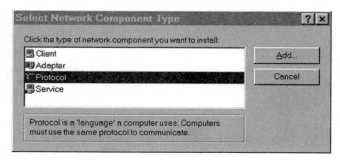

**Figure 4.37**

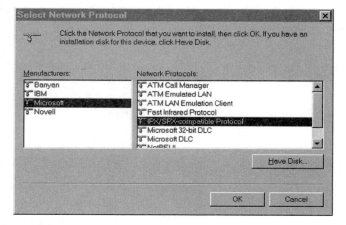

**Figure 4.38**

- Now select **Microsoft** and then **NetBEUI**. Click **OK**.
- Now repeat the above procedure but select **TCP/IP** from the menu opposite.
- Click the **OK** button (see Figure 4.38).

### Step 5: Setting computer and workgroup names

Even in a peer-to-peer network, each computer must have a unique name. It must also be assigned to a particular workgroup. These names were originally entered when Windows was first installed. To change them to suit the proposed new network:

- | **Start** | **Settings** | **Control Panel** | **Network** | and select the **Identification** tab. Your original computer and workgroup names are displayed.
- Change the names if you wish. Then click on the **OK** button (see Figure 4.39).

*Note*: After carrying out the above procedure on each PC, if everything is satisfactory, the indicator lights on the hub showing the status of each port should be lit. If they are not, check the led(s) on the back of each NIC. These should be lit indicating that a link has been established between the NIC and the hub.

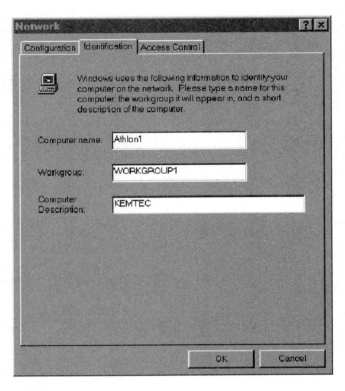

**Figure 4.39**

## Step 6: Setting file and printer sharing options

To enable printer sharing we must inform Windows that we are happy to allow other computers on the network to use the current computers resources, to do this:

- Choose: | **Start** | **Settings** | **Control Panel** | **Network** | and select the **Configuration** tab (see Figure 4.40).
- Select | **Add** | **Service** | (see Figure 4.41).
- Select **Microsoft**, then from the list on the right, select **File and Printer Sharing for Microsoft Networks**. Click **OK**.
- Now click on the **File and Print Sharing** button on the Network main Configuration menu shown above (see Figure 4.42).
- Select, **I want to be able to give others access to my files** and **I want to be able to allow others to print to my printer(s)**. Click **OK**.
- Click **OK** again on the main configuration menu and choose **Yes** to restart Windows (see Figure 4.43).

## Step 7: Sharing files and printers

You have now configured the NIC and installed the necessary network components that link the network to the PC. All that remains to do is to tell the network what resources can and cannot be shared and to what degree. For example, a hard disk can be shared as full read-and-write or as read-only or as a password protected drive. You can assign certain folders for sharing – in which case the hard disk itself must be set as 'unshared'.

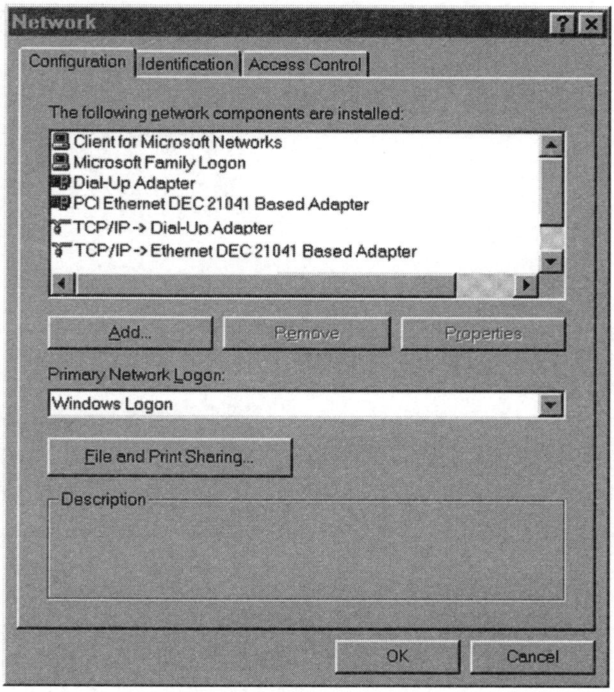


**Figure 4.40**

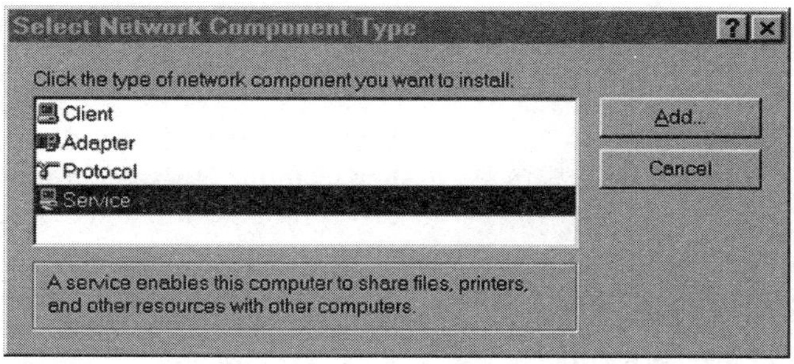


**Figure 4.41**

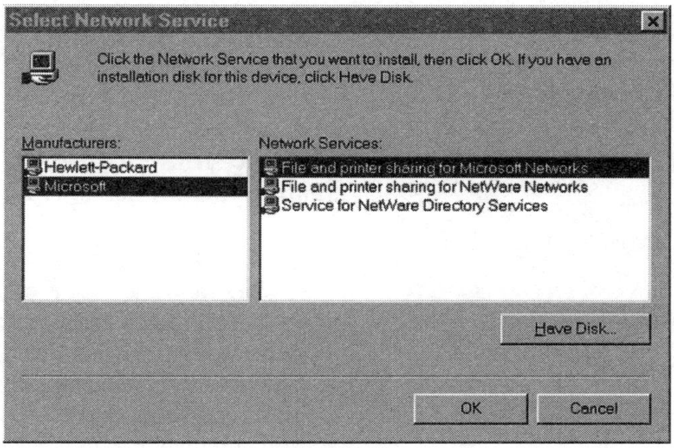


**Figure 4.42**

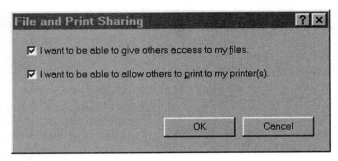

Figure 4.43

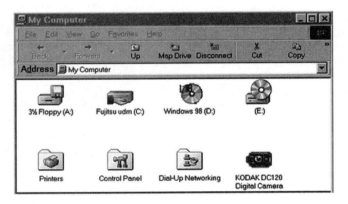

Figure 4.44

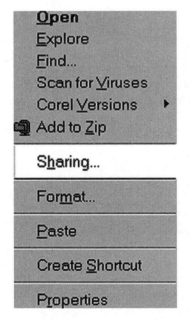

Figure 4.45

## Sharing your hard disk drive

For example, to share your hard disk C: as a password protected drive, for both full and read-only access, do the following:

- Select | **My Computer** | and right click **Drive C:** (see Figure 4.44).
- In the small box, click **Sharing** (see Figure 4.45).
- A drive properties box appears as shown opposite. Click **Shared As** and type a name for the drive as you want it to be known on the network. Click **Access Type Depends on Password** and enter a password for read-only access and another one for full Access and click **OK**. Another box appears asking you to confirm your passwords. Do this then click **OK** to exit (see Figure 4.46).

## Sharing your printer

Sharing your printer, is virtually the same as that for sharing a drive:

- Select | **Start** | **Settings** | **Printers** |. Select the Printer you want to share and then right click it (see Figure 4.47).
- In the small box that appears click on **Sharing** (see Figure 4.48).
- A printer properties box appears as shown in Figure 4.49. Click **Shared As**. Type a name for the printer as you want it to be known on the network.
- You can also assign a password and enter a comment if you wish. Click **OK**.

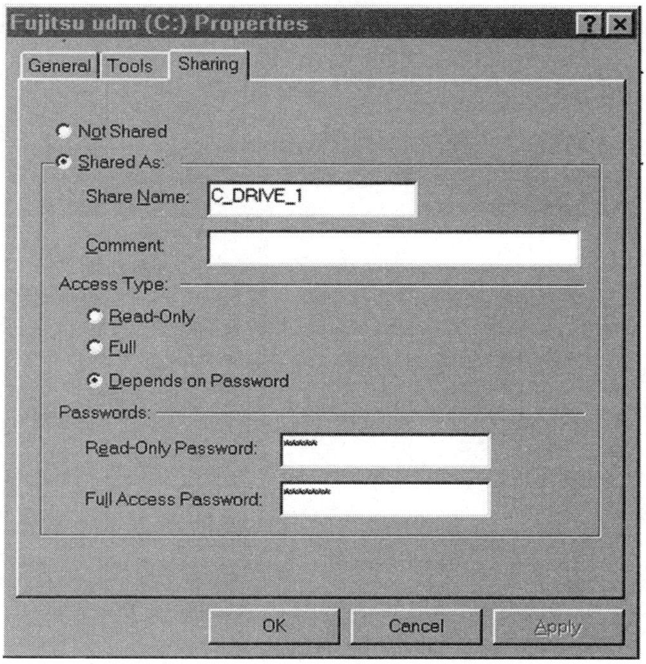

Figure 4.46

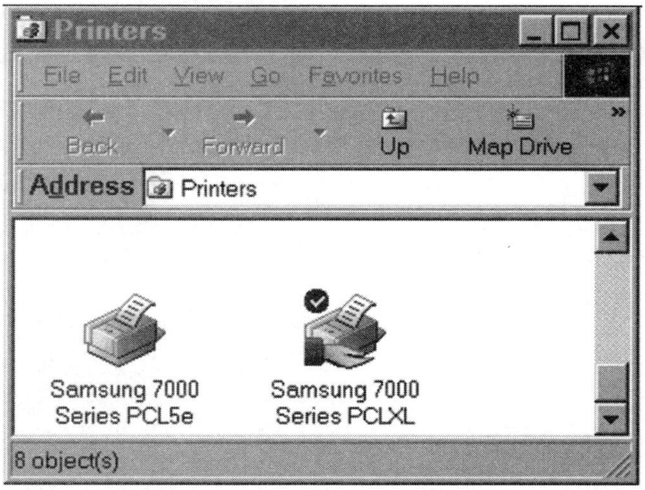

Figure 4.47

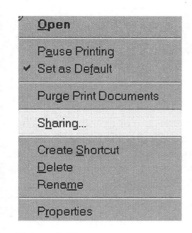

Figure 4.48

## Adding a network printer

The printer has now been declared as a shared item on the network. Now it must be regarded in a similar way to any normal 'local' printer. That is, it must be installed with the 'Add Printer' wizard on each PC that wants access to the net printer.

On each PC in turn:

- Choose | **Start** | **Settings** | **Printers** |
- Double click **Add Printer** to invoke the Add Printer wizard.
- Click **Network printer**. Then **Next** (Figure 4.50).

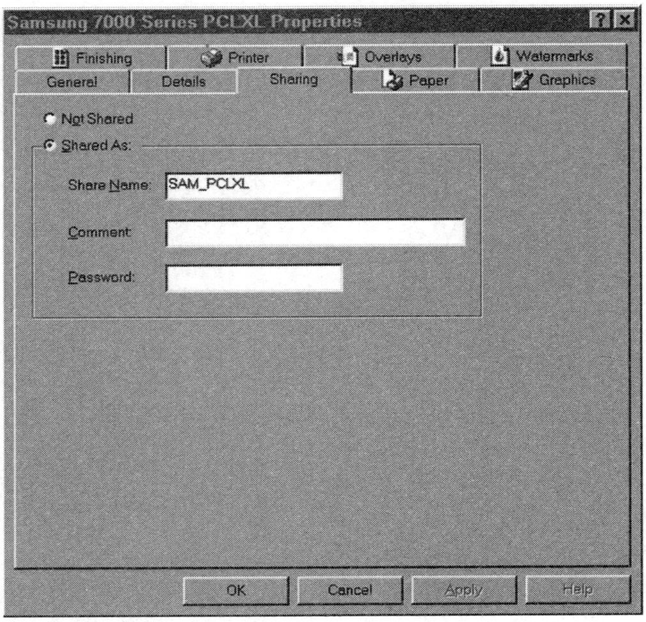

**Figure 4.49**

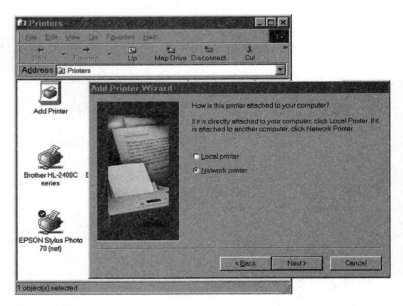

**Figure 4.50**

You are now asked to state the network path or printer name. The best way to do this is to click on the browse button. This will list all the items connected to the network via Network Neighborhood (hic US spelling). In our example (Figure 4.51), we want to add the shared printer from the computer named 'Athlon1'.

- Highlight the printer and then click **OK**.
- You can also click a button to allow printing from MS-DOS applications. If you no need this facility, click **No**. Click **Next>**.

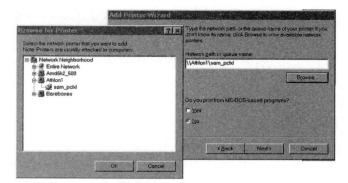

**Figure 4.51**

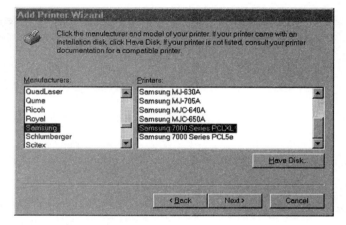

**Figure 4.52**

You must now choose a suitable driver for the network printer.

- Choose the printer manufacturer and select the printer type from Windows built-in printer database or click on **Have Disk**..., if you have a more up-to-date disk-based driver. After choosing a driver click **Next>**.
- You now have options to do a printer test and set the network printer as default (Figure 4.52).

So that in a nut shell how to set up a basic Windows 95/98/ME peer-to-peer network for disk and printer sharing!

# Windows network applications and utilities

Windows has several built-in network utilities and applications. You may have to install some of them from your Windows CD-ROM if they were not already installed.

### Network Neigborhood/My Network Places

This provides the normal access route onto the network for applications. For example, to save a file from WORD onto another computer

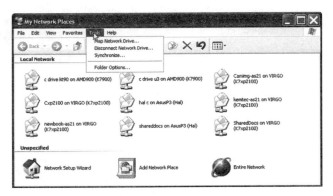

**Figure 4.53** *Windows 2000/ME/XP My Network Places*

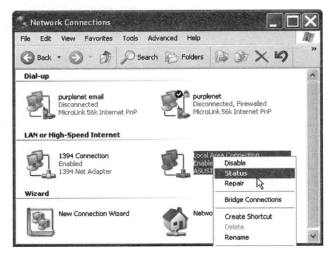

**Figure 4.54** *The Windows XP Network Connections folder*

on the network you click | **File** | **Save As** | to bring up the **Save As** window, then choose **Network Neighborhood** from the **Save in** window and choose the destination computer and folder from the list.

Network Neigborhood (*My Network Places*) can also be used as a utility to view and edit network components. You can also map more convenient names to the various shared network drives and printers (Figure 4.53).

## Netwatcher

Netwatcher is a useful tool for the network administrator, allowing each node to be monitored. For example, it shows all the shared resources on each computer and all the currently opened files on the network. From netwatcher the administrator can add and close resources and also disconnect certain users from a shared resource.

There is no Windows XP Netwatcher. Instead you have a more advanced network setup and trouble shooting utility base than Windows 95/98/ME. You can watch the activity of the network in Windows XP by selecting | **Start** | **Control Panel** | **Network Connections** | and right clicking the **Local Area Connection** icon and select **Status** (Figure 4.54).

# Installing a network using Windows XP

Windows XP excels at making many of the laborious setup, installation and maintenance tasks more pleasant and user friendly. The **Network Setup Wizard** is no exception.

- If you are using **Windows XP Professional** you must be logged on as the network administrator before you can use the wizard. Choose | **Start** | **Control Panel** | **Network Connections** | **Network Setup Wizard** | (see Figure 4.55).
- The Network Wizard welcome window is displayed. Click **Next>** (see Figure 4.56).
- Ensure that the network system is connected to all the items you want to include on the network (see Figure 4.57).
- Select an option that best describes the current computer (see Figure 4.58).

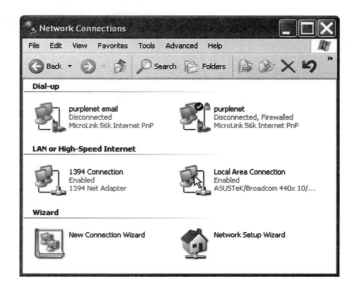

**Figure 4.55**

**Figure 4.56**

*Note:* The Network Setup Wizard can only be run on computers using Windows 98, Windows ME, Windows XP Home Edition or Windows XP Professional.

- This particular computer has a Firewire™ port (IEEE1394) as well as a NIC installed. The IEEE1394 port can be bridged to the LAN if desired. In this example we have decided to let the wizard carry out the bridging. If you do not have a Firewire™ interface then this option will not appear (see Figure 4.59).
- Now you have the option to let the wizard setup the network on the other computers. You can let the wizard make a Network Setup Floppy disk (see Figure 4.60) or you can use the Windows XP CD. The setup floppy is run on each and makes the network setup process very straight forward.
- Before running the network setup wizard on computers running Windows 98 second edition or Windows Millennium Edition, make sure Internet connection sharing (ICS) is disabled.

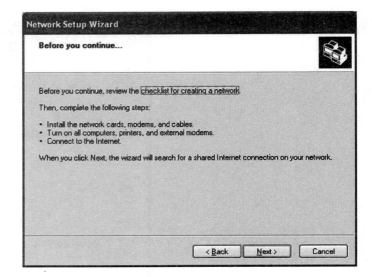

Figure 4.57

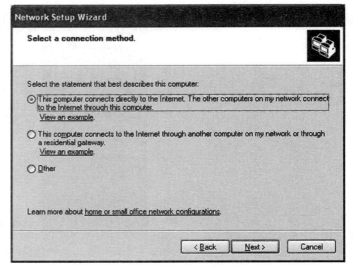

Figure 4.58

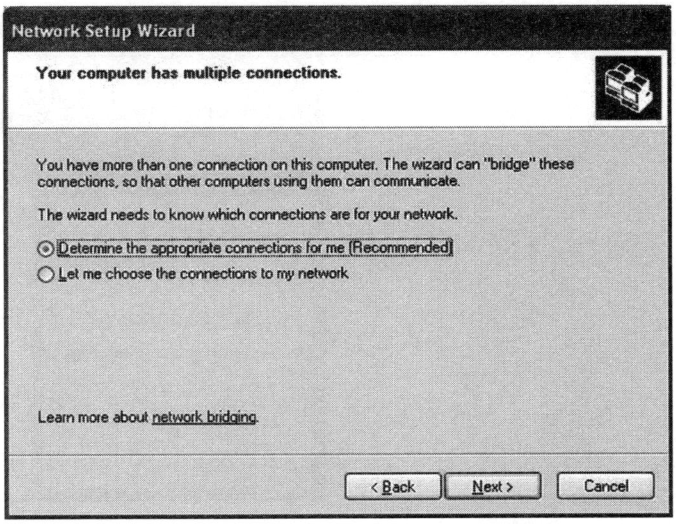

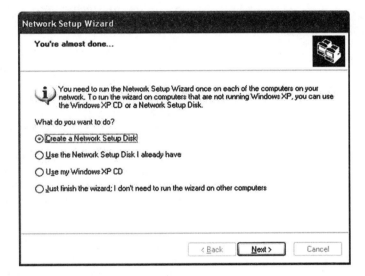

**Figure 4.59**

**Figure 4.60**

- To disable ICS in Windows Millenium click | **Start** | **Settings** | **Control Panel** | **Add/RemovePrograms** | and select the **Windows Setup** tab and click | **Communications** | **Details** |. Now clear the **ICS** check box. Click **OK**.

- To disable ICS in Windows 98 second edition click | **Start** | **Settings** | **Control Panel** | **Add/RemovePrograms** | and select the **Windows Setup** tab and click | **Internet Tools** | **Details** |. Now clear the **ICS** check box. Click **OK**.

- Run the network setup wizard first on the Windows XP computer that will be physically connected to the Internet. Insert it into the A: drive choose | **Start** | **My Computer** | and click on the 3½ **Floppy (A:)** icon. Then double click on **netsetup.exe**.

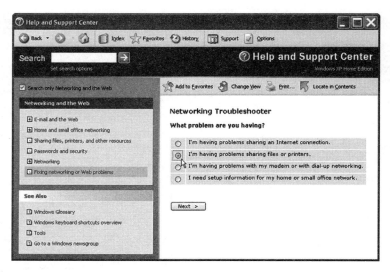

**Figure 4.61**

- Follow the instructions as netsetup configures each computer in turn.

### Windows XP network troubleshooter

Windows XP provides a very comprehensive help and troubleshooting database.

To use the Help and Support centre choose | **Start** | **Help and Support** |

To get help on your network via the Help and Support menu:

- Choose **Networking and the Web** in the Pick a Help topic window.
- Then choose **Fixing networking or Web problems**.
- Select **Home and Small Office Networking Troubleshooter**.
- Choose an option from the list of problems as shown in Figure 4.61:

Having got this far you will certainly want to try out your new network facility. So go ahead! and get more 'hands on'. The more you use a network – even a simple home setup – the more you will gain confidence and appreciate the power available from a group of linked PCs.

If you have not built one yet … do it! … and enjoy!

## Making RJ-45 UTP flyleads

Any self-respecting network technician should be capable of fitting an RJ-45 plug to a UTP cable when the need arises. Acquiring this basic skill will allow you to carry out emergency cable repair work and cut cables to the right length. This will eliminate the untidy coils of excess lead that are inevitable with ready-made leads.

As with any practical skill, it is important to have the right bits and pieces for the job in hand. To make up your own UTP cables, the minimum tool kit you require is an RJ-45 crimp tool; a pair of side cutters and a Stanley knife. You also need a reel of category 5 UTP cable, and a bag of RJ-45 modular plug shells with strain-relief hoods to cover the plugs – the latter are not essential.

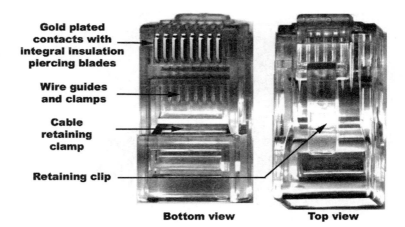

**Figure 4.62**

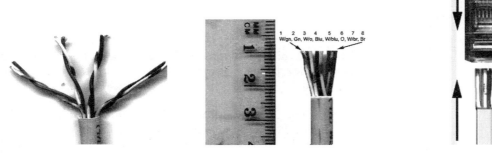

**Figure 4.63**          **Figure 4.64**              **Figure 4.65**

# The RJ-45 modular plug design

An RJ-45 modular plug is cleverly designed to form a secure and reliable connection to category 5 UTP cable using nothing more than a single crimp action. The connector has a row of eight raised gold plated terminals, each with an integral cutting blade. During the crimping process – after the cable end has been prepared and inserted into the shell – the terminals are pushed under force into the shell while the integral cutters pierce through the outer insulation of each wire and make electrical contact with the underlying copper cores. During the crimping process, special clamps moulded into the plastic shell are released onto the individual wires and the cable sheath itself, firmly clamping the cable into place (Figure 4.62):

1. First slide a PVC strain-relief cover over the end of the cable. Now strip off the outer sleeving (about 2.5 cm) using a stipping tool or carefully score around it with a Stanley knife without cutting into the wire pairs (see Figure 4.63).
2. Fan out the wires and lay them out in the correct order to the TIA 568A or 568B standard, as shown in Figure 4.64. Then trim off at right angles, to a length of 13 mm.
3. Keeping the wires in the correct colour order, push the wires carefully into the shell so they line up with the guide channels (Figure 4.65).

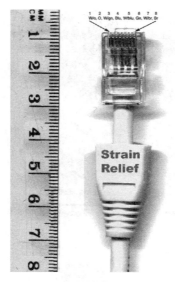

**Figure 4.66**

**Figure 4.67** *RJ-45 ratchet crimping tool*

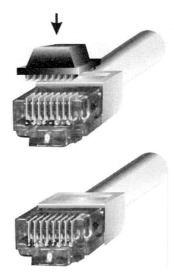

**Figure 4.68** *An inexpensive crimping tool*

4. Then push the cable home firmly until the wire ends butt up against the top wall of the shell. The end is now ready for crimping (Figure 4.66).

In the crimping process, the individual terminals are pushed down through the insulation and into the inner conductors of each wire.

It is advisable to use a properly designed crimping tool that applies the correct amount of pressure.

An inexpensive alternative to the ratchet crimping tool is the diecast crimping head (Figure 4.67). This is shaped in a similar fashion to the jaws of the more expensive ratchet action crimping tool.

The method is to align the tool on the top of the plug so that it fits over the terminals and snugly into the cable clamp recess. The plug and tool are then squeezed together gently in a soft-jawed vice (Figure 4.68).

# Making thinnet RG-58 coaxial leads

Thinnet is not so popular nowadays but it is still in use in many establishments. The beauty of a bus system like 10Base2 is its simplicity and low cost. Cables can simply run from computer to computer in a point-to-point topology.

Thinnet uses special 4.95 mm diameter, 50 ohms coaxial cable known as RG-58 C/U.

Making your own thinnet cables is straightforward enough as long as you have the correct tools. You need a BNC crimping tool, a coaxial cable cutting/stripping tool or a pair of pliers, and a Stanley knife.

For the leads, you need a drum of RG-58 C/U cable, and a quantity of Thinnet BNC connectors, BNC T pieces and cable glands. The latter provide strain relief for the cable ends (Figure 4.69).

You also need terminator plugs for the extreme ends of the network. The exact quantity of these items obviously depends on your particular requirement. For point-to-point wiring you need one T piece per

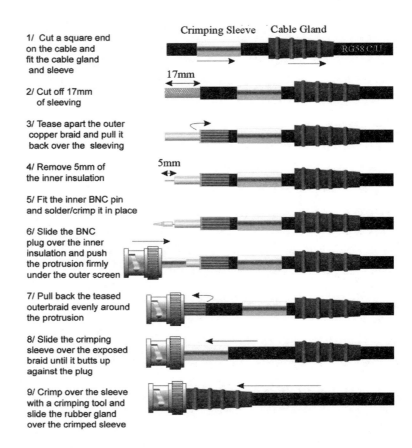

1/ Cut a square end on the cable and fit the cable gland and sleeve

2/ Cut off 17mm of sleeving

3/ Tease apart the outer copper braid and pull it back over the sleeving

4/ Remove 5mm of the inner insulation

5/ Fit the inner BNC pin and solder/crimp it in place

6/ Slide the BNC plug over the inner insulation and push the protrusion firmly under the outer screen

7/ Pull back the teased outerbraid evenly around the protrusion

8/ Slide the crimping sleeve over the exposed braid until it butts up against the plug

9/ Crimp over the sleeve with a crimping tool and slide the rubber gland over the crimped sleeve

**Figure 4.69**   *Fitting BNC connectors*

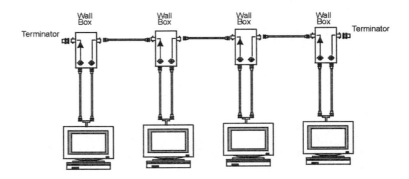

**Figure 4.70**

station and two BNC connectors. If you decide to use wall boxes then you will need six BNC connectors per computer. Four on the box itself (wall box in, computer in, computer out and wall box out) and two on the BNC T connector at the computer (Figure 4.70).

Figure 4.69 shows the step-by-step process required to make a BNC thinnet cable end.

**Part 2** **Assembling, configuring and upgrading**

# Unit 5 Building a complete PC system

*Building a new system-unit and installing the OS*

# Handling and transporting equipment

The delicate parts of a PC system are easily damaged by careless handling during installation, maintenance and transportation. The following tips on good handling methods should therefore be followed at all times:

- When moving a computer and peripherals from one location to another, unplug all connectors and carry each item separately. Do not try and carry two items at once to save unplugging the joining lead, such as a monitor and base unit. If the computer system is to be moved over any great distance, place the monitor system unit and peripherals in their original packing.
- Hard drives are particularly delicate due to their delicate head mechanism. It is therefore essential to protect them from severe bumps and jolts. When transporting a hard drive, wrap it in an anti-static bag and place it in a box with plenty of shock absorbing material.
- CD-ROMs and floppy drives are sometimes supplied with a head restraining device. Fit these back on the drive before they are transported (Figure 5.1).
- New printers, scanners and other peripherals with moving parts are packed using some form of restraint to arrest the mechanism during transit. This varies from a special arresting mechanism to pieces of foam. The restraints must be removed before use and refitted when the equipment is being transported over any great distance. For items with a *Return to Supplier Warranty*, keep the original packaging for the duration of the warranty period, in case they need to be returned for replacement or repair.

## Transporting potentially dangerous equipment

Monitors and laser printers have circuits that operate at several thousand volts. When moving these devices, it is important to allow time for the internal voltages to decay to a safe value prior to transportation.

For example: a monitor that has been abruptly disconnected from the mains, could hold an electrical charge of several thousand volts. This happens because the internal high voltage circuits have not had sufficient time to discharge to earth via the mains socket. A person carrying the monitor could therefore receive a sudden electric shock, as the high voltage finds another convenient path to earth through his or her body. Even if the resulting electric shock is not

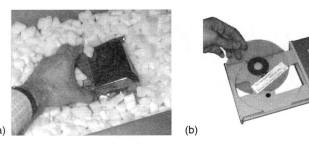

(a)                                        (b)

**Figure 5.1**  *Handling of peripherals (a) Hard drives and (b) CD-ROMs*

directly harmful, the sudden jolt of the shock could cause the monitor to be dropped, resulting in possible CRT implosion and injury.

Here is the correct way to disconnect all mains operated peripherals:

1.  After closing down the operating system (OS), switch off the power to the computer and peripheral at the mains socket but keep the power leads plugged in to allow any built up charge to dissipate safely to earth through the earth pins of the mains plug and socket.
2.  Wait at least 30s to allow high voltages to discharge to earth.
3.  Now disconnect the signal cables first followed by the power leads. This prevents any remnant static charge from dissipating through the computer input/output (I/O) ports.

Carrying out this simple procedure will prevent accidents due to electric shock. It will also eliminate the chance of damaging the associated I/O circuitry when the device is removed. One of the biggest causes of damaged adapter cards, motherboards and peripheral devices, is careless handling – particularly during connection and disconnection.

## Static sensitive devices

Most integrated circuits used in a modern computer system are fabricated using complementary metal oxide semiconductor (CMOS) technology. These devices are very easily damaged by static electricity picked up on our bodies going about our normal daily life. For example, when we walk across a carpeted floor, touch a CRT monitor screen, or push a supermarket trolley, we charge ourselves up to several thousand volts. We have all experienced this at some time or other when we inadvertently discharge ourselves on a large metallic object like a central heating radiator or the metal work of the shopping trolley.

The important point to realize here is that, all the electronic parts making up a computer system are easily damaged with very low levels of static electricity – much lower in fact than that required to draw sparks or make us jump. It is almost impossible not to gain a low level static electric charge on our bodies under normal daily activity. So it is important that we remove the charge before we handle static sensitive electronic devices. This is best accomplished by wearing an anti-static wrist band.

Some technicians who should know better, are very blasé about static electricity precautions. They often come out with comments like, *'I never wear a wrist band and I've never had a device failure'*. The point they are missing is that static damage does not always show up immediately, it could take a month or so before a damaged device finally fails.

### CMOS devices

A typical CMOS IC (e.g. a CPU or RAM device) is fabricated on a silicon chip from tens of millions of transistors called MOSFETS (metal oxide semiconductor field effect transistors). The fundamental building block of all digital CMOS circuits is a complimentary pair of MOSFETs forming an electronic switch. One of the pair is a P channel MOSFET, and the other is an N channel MOSFET. (Basically this means they are exact opposites of each other. When one device is 'on' the other is 'off' and vice versa.) Each MOSFET has three electrodes called the gate, source and drain. The gate electrode is electrically insulated from the other electrodes by an ultra thin layer of silicon dioxide, just a few molecules thick.

In operation, a control voltage applied to the gate electrode creates a tiny electrostatic field in the channel between the source and drain electrodes, which directly influences the current flowing across the channel. The device can therefore be made to switch on and off, simply by changing the voltage level at the gate terminal. An important property of a CMOS switch is that virtually no current flows in the gate terminals because the gates are electrically insulated from the channel by the silicon dioxide layer but the voltage level at the gate has a marked effect on the conducting properties of the channel.

In a CMOS device, the P channel and an N channel MOSFETs are arranged, one above the other, as a complementary pair. The gate electrodes are connected together to form the input terminal (see Figure 5.2). The presence or absence of a signal voltage at the input, determines which half of the pair is on, and which half is off.

When the upper device is in the 'on' state, the lower device is in the 'off' state, and vice versa. So when the device is maintained in a high or low state very little current flows. The power consumption is therefore extremely low.

For interest only:

The basic complimentary arrangement shown in Figure 5.2 forms a simple NOT GATE circuit, that is, the output state is always the opposite of the input state. If a logic '1' signal is applied to input (A), the lower FET turns on and the upper one turns off, so the output (Y) is pulled down to logic '0'. Conversely when logic '0' is applied to the input, the top FET turns on and the bottom FET turns off, so the output is pulled to logic '1'.

As mentioned previously, as one device is on while the other is off, there will be no significant current flow in the circuit. For steady DC and low frequency operation the power consumption of a

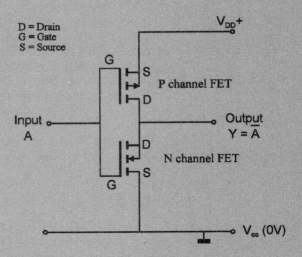

**Figure 5.2**  *A basic CMOS device (NOT Gate)*

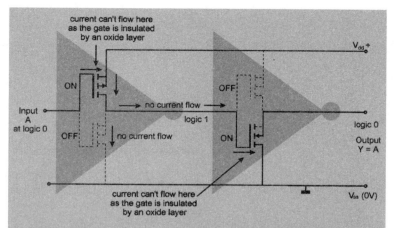

**Figure 5.3**  *A basic circuit showing two CMOS NOT gates connected in cascade. Showing that virtually no current flows when operated at very low frequencies*

CMOS circuit is very low. So CMOS ICs are eminently suitable for use in battery operated equipment. The real time clock (RTC) and CMOS setup memory are good examples of low frequency CMOS circuits – the Lithium battery on the motherboard lasts for years keeping the time date and CMOS information intact. Even when millions of CMOS devices are connected together, as in a modern CPU, no significant power is drawn at low switching rates. This is because only one FET of each switching pair can be on at a time and current cannot flow through a gate terminal because of the insulated layer. Figure 5.3 demonstrates this with a basic CMOS 'NOT' gate circuit. The same applies to far more complex circuits running at fixed or low frequencies.

Unfortunately as the switching rate increases, the current consumption increases. This is due to the tiny inter-electrode capacitance of each MOSFET. The capacitance between the gate region and silicon substrate has a marked effect at high frequencies. Due to capacitance, the current drawn by a CMOS IC increases linearly with operating frequency. This is why the power consumption of a modern CPU increases as the clock frequency increases. One way to reduce power consumption is to reduce capacitance by reducing the size of the electrodes, that is, by using a smaller scale of integration during the chip fabrication process. Modern flip-chip CPUs are fabricated using 13 μm technology and lower so power dissipation is kept to a reasonable level.

## Static damage

As the insulated layer at the gates of each CMOS pair are so thin: a static discharge at the gate terminal, will electrically break down the layer and short circuit the gate to the silicon substrate, thus destroying the device. Alternatively a tiny hole may be punctured through the oxide layer. This could result in device failure, perhaps several weeks or even months later (Figure 5.4).

If static is allowed to discharge onto an adapter card or a memory module such as a SIMM, for example, which contains several hundred thousand MOSFETs, then there is a fair possibility of at least one of them being damaged.

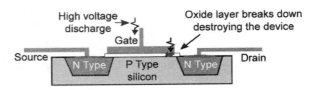

**Figure 5.4**   *An idealized sketch of a damaged MOSFET, magnified several thousand times*

### Preventing static damage

1.  When inserting or extracting printed circuit boards, or handling disk drives and ICs, always wear an earthed anti-static wrist band. An earthed wrist band removes the static from your body and clothes. Adapter cards, memory modules, CPUs and drives, often cost several hundred pounds each, so it is foolhardy to risk damaging them for the sake of a wrist band that costs just a few pounds.
2.  Try not to touch the ICs and tracks on adapter cards. Handle all cards by their edges only. Keep all adapter cards and ICs away from monitors and laser printers. The high static charges present on these units can easily cause static damage.
3.  Always store unused adapter cards, disk drives, memory devices and other static sensitive devices in their anti-static bags. Only remove them from the bag when you are ready to use them and wear an earthed wrist band.
4.  Disregard and put down to ignorance, the often quoted phrase from some complacent members of the PC fraternity, '*I never wear a wrist band …*' What happens to a static damaged card, weeks or months later, they would never think of putting down to their own former careless handling methods (Figure 5.5). Wear one!

Wear an anti-static wrist band when handling all ICs and printed circuit boards. (a) This includes simms/dimms, CPUs, motherboards, adapter cards, and disk drives.

When building a PC: make sure that the mains is switched off at the wall socket and at the system-unit. Then earth the system-unit by plugging in a 3-core mains lead to a properly earthed wall socket. Now wear your wrist band and clip the free end of the earthing lead to the bare metal chassis. (b) The earth wire in the mains lead and socket will ensure that the case and therefore your wrist band, is properly earthed. An alternative method is to use a special dummy plug fitted with a wrist band stud.

**H&S**
*Although most wrist bands incorporate a high value resistor to lessen the wrist of electric shock, for electrical safety, earthed wrist bands must be removed if you work on exposed live circuits that operate at high voltages, such as monitors and laser printers. This of course does not apply when working inside the system-unit case or handling PC components during assembly and repair.*

# Assembling a complete PC system

This section shows you how to assemble and commission a PC.

To configure the individual parts into a working computer, ideally you should have the complete set of installation instructions and driver software. New adapter cards and peripherals are normally supplied with full instructions and driver software. Basic adapter cards are often supplied with a set of instruction sheets, whereas more expensive items like motherboards and graphics cards have full user manuals. It is unwise to attempt to install an unfamiliar part without instructions. If you do not have any, browse the manufacturer's web site for information and driver updates.

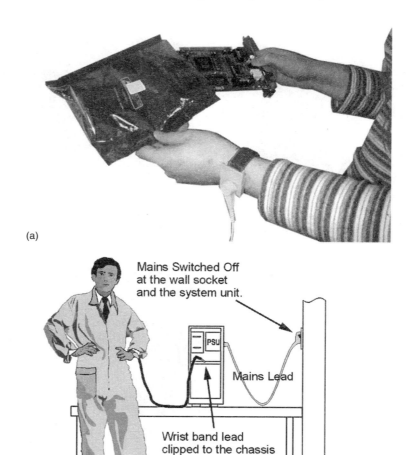

(a)

(b)

**Figure 5.5** *Preventing static damage: (a) earthed wrist band and (b) properly earthed wall socket*

## Before you start

Make sure you have all the items necessary to build your PC. Our example system parts are shown in Figure 5.6, including the OS software. There are several OSs available for the PC. The most popular ones today are Windows 98, Windows Millennium Edition (ME), Windows 2000 and Windows XP. These are known as *Graphical User Interfaces* (GUIs) and are intended for mouse operation. OSs that rely on commands entered via the keyboard are known as *Command Processors*. MS-DOS, UNIX and LINUX are examples of command driven OSs. However, they can run GUI bolt on to support a Windows like environment.

You should choose an OS to match the type of work you want to do on your computer. If popular application software is to be used on the computer, then it only really makes sense to use a modern GUI. For home and general office applications Windows XP (Home Edition) is a good choice for a new PC.

**Figure 5.6** *Typical Microsoft® Operating Systems*

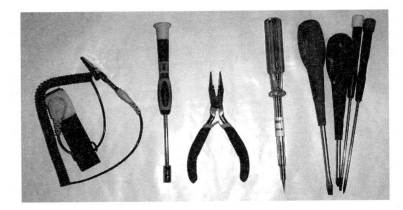

**Figure 5.7** *The basic set of tools needed to build a PC*

For a reliable workstation in a client/server network, Windows 2000 Professional or Windows XP professional are ideal.

Not forgetting the experimenter; if you are knocking up an old system to do control experiments, for example, robotics, practising hardware troubleshooting, etc., then MS-DOS is probably a more suitable OS in view of its relative simplicity and command oriented interface.[1]

## Tools

You will need a basic tool kit to build a PC, the minimum being a set of screwdrivers, a pair of snipe nosed pliers and an anti-static wrist band. A box spanner or socket set containing a 5 mm (~3/16 in.) socket will also come in handy for fitting hexagonal motherboard pillars and tightening 'D' connectors (Figure 5.7).

[1] Information on MSDOS can be obtained from Appendix E on the Elsevier website at http://books.elsevier.com/companions/0750660740

## Step by step guide to building a PC

In the following scenario we will cover the full construction of a modern Socket A ATX system. The skills covered will meet many of the skills required to install an Intel Socket 370 or Socket 478 system.

### Building a system to a customer specification and budget

A regular customer (a local creative design and marketing company) has ordered a new stand alone PC system with monitor, keyboard, mouse, modem, A4 scanner and A4 photo-quality colour inkjet printer. The proposed PC will be used as a workstation suitable for 2D/3D graphics design, presentations and general office and multimedia applications. The cost of the proposed system must fall within the companies maximum budget of £1500 + VAT.

To generate a list of features and arrive at a rough specification the proposed system was discussed with the customer. Typical details gleaned from the discussion were as follows:

- The critical application work carried out on the system will be graphics oriented (Video editing and 3D rendering). The customer will use the system with professional digital cameras so a facility for downloading and archiving large image files and video clips up to 200 MB is required. Initially the system will be used stand-alone but it will be connected to a network at a later date.
- After noting the output signal connections/specifications of the digital cameras and the recommended system specification for their graphics application software, an item list was drawn up.

### Costing

Working to a specification within a restricted budget is not a particularly easy task and compromises may have to be made. It is therefore quite normal when commissioning a purpose built system to discuss options more than once with the customer in order to achieve a mutually agreeable outcome. A good way to cost this type of work is to use a spreadsheet as follows:

1.  First determine an interim price and specification. Do this by listing the essential parts and peripherals without compromising on specification or cost. This involves choosing peripherals and components known from experience to be reliable, properly supported, good value and well suited to the proposed application (columns labelled Model A and Cost A in Table 5.1).
2.  Now produce a second list by choosing items of a slightly lower specification and price you feel that will not compromise the specification adversely (columns labelled Model B and Cost B in Table 5.1).
3.  Taking into account the standard percentage fee to be charged for building and commissioning the system, a satisfactory price can now be arrived by mixing and matching the items from the two lists to produce a third and final list (column labelled Cost C in Table 5.1).[2]

### Choosing the motherboard/CPU combination

To save the cost of a separate sound card, modem and network interface card, a high quality motherboard was chosen with these features built in. Several motherboards by various reputable manufacturers

[2] This basic spreadsheet is available on the Elsevier website at http://books.elsevier.com/companions/075066 0740

**Table 5.1**

| Part | Description | Model A | Model B | Cost A | Cost B | Cost C |
|---|---|---|---|---|---|---|
| Monitor | 19 in. flat screen CRT monitor | Model x | Model y | £393 | £335 | £393 |
| Graphics Card | 128 MB AGP 8X | Model x | Model y | £150 | £85 | £150 |
| ATX motherboard | Soc. 462 (USB2/firewire/ Snd/NIC/modem) | – | – | £92 | £92 | £92 |
| CPU | Athlon XP 2100 (AMD) | XP 2400 | XP 2100 | £105 | £70 | £70 |
| Heat sink/fan | AMD XP compatible fan/heat–sink | – | – | £12 | £12 | £12 |
| RAM | DDR-SDRAM 266 MHz | 512 MB | 256 MB | £53 | £30 | £30 |
| Hard disk drive | UDMA 133 | 80 GB | 40 GB | £75 | £55 | £75 |
| CD-drive | CD-R, CD-RW, drive | With DVD | Without DVD | £50 | £38 | £38 |
| Floppy disk drive | 3.5 in. 1.44 MB (standard) | – | – | £10 | £10 | £10 |
| Case | Midi Tower ATX 400 W PSU | – | – | £37 | £37 | £37 |
| Keyboard | High quality 104 key PS/2 type | – | – | £15 | £15 | £15 |
| Mouse | Ball-less wheel mouse PS/2 type | – | – | £25 | £25 | £25 |
| Printer | Photo quality ink jet printer | Model X | Model Y | £92 | £76 | £92 |
| Scanner | 48 bit 1200 DPI | Model X | Model Y | £165 | £79 | £94 |
| Operating system | Microsoft Windows XP (Home Edition) | – | – | £70 | £70 | £70 |
| | Total less service charge | | | £1344 | £1029 | £1203 |

proved to be equally acceptable. All supported a wide range of state of the art features as well as the essential built-in facilities mentioned above. A useful degree of future proofing was therefore provided as a bonus. Several motherboards also supported IEEE1394 (Firewire™ and USB2 and these are ideal I/O ports for fast transfer of digital images again saving money by obviating the need for separate plug in cards. All the motherboards considered had a fast 266/333 MHz front side bus and memory bus and several supported the AGP 8X specification. A motherboard with the ideal specifications was finally chosen from the usual supplier. The board supports all AMD Socket 462 CPUs, and DDR 266/333/400 DIMMs so an Athlon XP 2400 CPU and 512 MB DDR266 RAM was picked as a good compromise on cost and specification.[3]

A combined DVD/CD-R/CD-RW drive was chosen as the ideal means of archiving large graphics files. The additional DVD capability was regarded as a useful bonus and the bundled software included top end CD writing software thus saving the customer the money for additional software.

To suit the graphics intensive applications a top end AGP 8X mode graphics card was selected that also featured state-of-the-art 3D rendering technology.[4] The OS was chosen on a *cost versus performance and user friendliness* basis. Windows XP Home Edition was considered the best option. The extra facilities in the professional edition are not needed by the customer so the extra cost was not justified.

A proposed specification and price was finally sent to the customer for approval (Table 5.1).

[3] A similar motherboard to the one specified in this scenario is the ASUS A7V8X, featured in Unit 2.

[4] A similar graphics card to the one specified here is the ATI Radeon 9700, featured in Unit 3.

After a few days the customer gave the go-ahead to build the system. The following pages describe the techniques involved in assembling and commissioning the system. Here we put you in charge in a step by step process ...

### Preparing to assemble the system-unit

- Plan your time so you have at least two uninterrupted hours to physically build the system-unit and install the OS.
- Prepare a clean uncluttered work bench or table top.
- Wear an earthed anti-static wrist band.
- Avoid working on nylon carpets as they can generate high charges of static electricity.
- Do not rush around. The more you do so, the greater the static electricity build up on your body.
- Check the parts against the parts list to ensure you have all the PC parts close at hand along with appropriate tools and accessories.

### Stage 1: Prepare the system-unit case

- Unpack the system-unit case and place it on the work bench. If you are using your dining room table! cover it with a few layers of news paper or an old table cloth to prevent scratches.
- Now undo the screws at the back of the case to release the two side covers and place them in a suitable receptacle. Remove the covers and place them out of the way (under the work bench, out of leg reach, is a good place). Remove the bag of screws, washers and motherboard pillars from inside the case.

**PSU rating**

Check the output wattage rating of the pre-fitted PSU and make sure it is suitable for your system. The higher the speed of the CPU the greater its power requirement. The best way to check if the PSU is suitable is to visit the CPU manufacturer's web site for a list of recommended PSUs or ratings for your CPU.

For reliable operation it is essential to have an adequate power supply unit. The maximum available output power in watts, for each DC output from the PSU is found by multiplying the output voltage by the maximum output current.

For example, if the $+5\,$V output of the PSU has a stated max current of $30\,$A, the maximum power output capability of the $+5\,$V output is $5 \times 30$, that is, $150\,$W. However, the total wattage rating of a PSU is usually less than that found by totalling the individual output power ratings together.

For example: The maximum rated outputs for a typical $300\,$W ATX PSU are $+3.3\,$V at $28\,$A, $+5\,$V at $30\,$A, $-5\,$V at $0.25\,$A, $-12\,$V at $1\,$A and $+12\,$V at $15\,$A. The maximum total power according to this is approximately $436\,$W however this must never be consumed in practice. At the absolute maximum, the total power consumed from this PSU must never exceed the rated $300\,$W. This particular PSU manufacturer states a maximum permissible combined power for the $+3.3\,$Vt and $+5\,$V lines of $200\,$W (Figure 5.8).

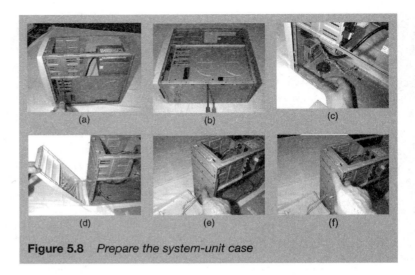

**Figure 5.8**   *Prepare the system-unit case*

- If the case has a removable base plate for the motherboard, unscrew and remove it from the side of the case, see (a) and (b) above. Now remove the plastic front panel. You will find that the fixing method varies. Sometimes it is secured with self-tapping screws (c) and sometimes it uses a simple push-on system.
- The next task is to remove the metal screening plates that cover the apertures for the floppy disk drive and CD drive. Figures (e) and (f) show what plates to remove. Each plate usually has a screwdriver slot to facilitate the gentle backwards and forwards flexing required to break the retaining lugs. **Watch your fingers on the sharp edges!** Once the plates have been removed the front panel can be refitted to the case. Discard the sharp jagged plates carefully.

**Figure 5.9**   *A removable bolt on rear cover plate*

You will also have to remove some of the cover plates from the rear of the case to accommodate the graphics card and I/O panels (see Figure 5.9). This is best done just prior to fitting the motherboard. The respective positions of the plates that need to be removed can then be determined correctly. The better quality cases are usually fitted with proper screw mounted cover plates. Less expensive cases have flex-to-remove plates with a screwdriver slot to facilitate removal.

### Stage 2: Setting up your wrist band
- Before you handle the motherboard and cards, etc., your wrist band must be worn. To make it effective it must be earthed. The easy way to do this is to earth the case and fix the crocodile clip of the wrist band onto the bare metal work.

You can earth the case via a mains wall socket as shown in Figure 5.5. *Before plugging in the mains lead, make sure both the case on-off switch and wall socket switch are in the off position and leave it in this position!!*

### Stage 3: Fitting the motherboard to the base plate
- Open the motherboard box and identify each part and find where it fits on the motherboard and case. Use your motherboard manual

**Figure 5.10**   *Fitting the motherboard to the base plate*

Make certain you have not placed a pillar where there is no hole in the motherboard otherwise a short circuit between the case and the underside of the PCB could result in serious damage to the motherboard and other components.

If you have a case with a non-removable base plate do not fix the motherboard until all the case metal work has been prepared.

Normally it will also be necessary to remove cover plates for sound, modem and network adapter cards. However, in this example they are built into the motherboard and their respective connectors are brought out on the rear ATX I/O panel.

[5] Some cases do not have a detachable base plate so you will have to work within the confines of the case.

if necessary. As well as the normal set of cables and ATX I/O plates, our example motherboard is supplied with a set of back cover panels and leads for two IEEE 1394 ports, two extra USB ports, a midi/joystick port and digital audio ports. The ideal positions for these rear panel items will be established when the motherboard is being fitted (Figures 5.10 and 5.11).

- Lay the case base plate[5] face side up on the table and wearing your earthed wrist band carefully remove the motherboard and anti-static foam pad from the anti-static bag. Holding the motherboard and foam pad in place over the base plate, determine which base plate holes line up with the motherboard fixing holes – you could mark the ones not to be used with a pencil. Now place the motherboard and foam pad to one side.
- Find the set of brass hexagonal stand-off pillars in the bag of screws provided with the case and screw these into position on the base plate in the predetermined fixing positions. A 5-mm box spanner is the best tool to use to fit the pillars (g and k).
- Remove the motherboard from the foam pad – whatever you do, do not leave the conductive foam pad in place – and align it carefully over the pillars and screw it to the base plate using the insulated fibre washers and screws provided (h and i).
- Temporarily place the base plate and motherboard back into the case to determine which of the rear panel cover plates to remove to accommodate the AGP graphics card, and extra I/O port panels.
- After placing the motherboard and base plate to one side, remove the cover plates from the required slots in the case by flexing each one gently back and forth with an old screwdriver until the thin joining tabs shear apart.

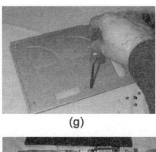

(g)

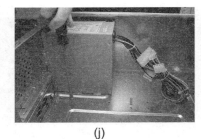

(j)

(k)

**Figure 5.12** *An ATX midi tower case with a fixed motherboard mounting plate*

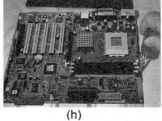

(h)

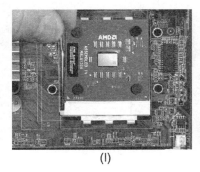

(l)

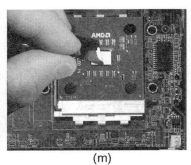

(m)

**Figure 5.13** *Fitting the CPU*

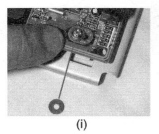

(i)

**Figure 5.11** *Fitting the motherboard on a case with a removable mounting plate.*

Some cases are supplied with removable 'bolt on' or 'clip on' cover plates rather than the type that need to be flexed off. These are far more satisfactory as they can be removed at any time without the fear of damaging the motherboard components.

- Remove the rear ATX I/O panel cover plate and replace it with the one supplied with the motherboard. (e) This is necessary as the connector layout pattern is different to the standard ATX layout due to the extra built-in features of the motherboard. Watch your fingers on the sharp edges!

## Stage 4: Fitting the CPU

- Place your CPU, heat sink and fan ready for use. Wearing your wrist band, remove the CPU from its anti-static wrapping and examine it to locate the reference corner. The reference corner of the CPU is indicated by a small triangle or spot. This usually points to the hinged end of the CPU lever. Pull the lever on the side of the CPU socket upright to open the contacts. Now line the CPU with the socket so that the reference corner of the CPU lines up with the corresponding reference corner of the socket and let it drop gently into place, see Figure 5.13(l). The CPU should fit into the socket with virtually no force except perhaps for a very gentle amount of downward pressure. If you have to force it you probably have the CPU incorrectly orientated. When it is located properly, push down the lever to lock the CPU into place.

### *Heat sink and fan*

Before we continue with the CPU installation it is wise to be aware of the mechanics involved in removing heat from the CPU: current CPUs run at extremely high clock frequencies and consume a lot of electrical power which is dissipated as heat in the chip.

Without a method of removing the heat efficiently, the chip would fry in a few seconds. The standard way of removing heat from a chip is to clamp a large finned aluminium or copper block in intimate contact with the chip surface. The device is known as a heat sink.

To quantify a heat sink's effectiveness at removing heat, physicists use a characteristic called *thermal resistance*. The lower the thermal resistance of a heat sink, the greater the heat dissipation factor and the cooler the chip.

*For the technophobe, thermal resistance is measured as the change in temperature in degrees centigrade for each watt of change in power dissipation (°C/W). A perfect heat sink would have a thermal resistance approaching 0°C/W.*

As every science pupil knows, copper is a better heat conductor than aluminium. A copper heat sink has a significantly lower thermal resistance (almost 50 per cent lower) than an aluminium heat sink of the same size and shape. Unfortunately copper heat sinks are substantially more expensive. Another factor that improves the effectiveness of a heat sink is air flow. If the hot air surrounding the heat sink is exchanged with a continuous stream of colder air by attaching a fan, the thermal resistance is reduced even further. The effectiveness of the fan is influenced quite markedly by the temperature of the surrounding air. So it is essential to maintain an adequate airflow through the case via vent holes. Some manufacturers recommend fitting an extra fan inside the system-unit to ensure a good air flow.

Unfortunately there is a potential weak layer in the heat transfer path. This occurs in the chip to heat sink interface. If the bare metal of the heat sink is mated to the bare surface of the chip, microscopic gaps (air pockets) are present. This occurs because the two surfaces are not perfectly smooth and flat, even though they may look it to the naked eye. These minute air pockets have an extremely high thermal resistance which creates areas of high localized heating in the underlying chip. This is not a very desirable situation so a heat conductive paste with a very low thermal resistance is used between the two surfaces. This ensures an even heat transfer. Some heat sinks are pre-fitted with a wafer thin pad of heat conductive compound on the mating surface, so there is no need to apply a separate quantity of paste. Now on with the installation ...

- Check your heat sink to see if it has a pre-fitted pad of heat conductive compound. If it has, remember to peel off the protective paper backing sheet. If you accidentally leave it in place, the chip will burn out as paper has a very high thermal resistance. Only peel off the protective sheet immediately prior to fitting the heat sink to prevent dust and grit sticking to the pad. If this happens you must use a new pad otherwise the heat conducting properties will be impaired.

  If your heat sink is not pre-fitted with this compound, smear some heat sink paste evenly across the exposed face of the CPU chip with a spreading tool – a thin strip of soft plastic (m). The paste is a bit messy so do not worry about spoiling the appearance of the mirror like surface of the chip. Do not use too much compound, the object is to make a good thermal joint you are not making a jam sandwich!

### *Fitting the heat sink and fan to the CPU*

This process is quite straightforward once you are aware of the hazards and how to avoid them. The exposed silicon chips on current AMD and Intel CPUs are very fragile. This is particularly the case with AMD Socket A CPUs. Socket 7 CPUs and earlier designs used a ceramic substrate and the chip was mounted in a well on the same side as the pins. The upper heat sink side was just plain ceramic. So fitting a heat sink was very straightforward. However, the latest chips dissipate a lot of heat so the manufacturers decided to flip the chip to the top of the substrate to allow the chip and heat sink to form a more intimate thermal bond. The exposed chip is similar in fragility to a wafer thin piece of coal – it is easily cracked or crushed on the corners if the heat sink/fan is not fitted correctly. The correct way to fit the heat sink and fan is as follows (Figure 5.14):

* After applying the heat sink compound – or peeling off the backing sheet in the case of the pre-fitted variety – align the heat sink and fan over the CPU socket with the stepped (rebated) end over the socket lever hinge cover (1). Now gently lay the heat sink assembly on top of the CPU so that it rests on the four foam pads of the CPU (2). Now fit the short end of the spring clamp over the lug on the CPU socket (3). Notice that the heat sink is parallel to the CPU socket at all times. On no account tilt the assembly when you hook on the spring clamp otherwise you will crush the CPU. Now fit the other end of the spring clamp using a screwdriver with a tip that fits snugly into the recess in the spring clamp. To do this; push the clamp down-and-out to clear the lug (4) and then push down-and-in to locate the clip over the lug (5). The heat sink/fan assembly is now fitted! (6) (Figure 5.15).

Note: Some heat sink/fan assemblies have a thumb lever instead of a screwdriver lug. To fit these you press down while pressing the thumb spring in, until the spring aligns over the lug. While maintaining the downward pressure release the pressure on the thumb spring to locate the spring over the lug.

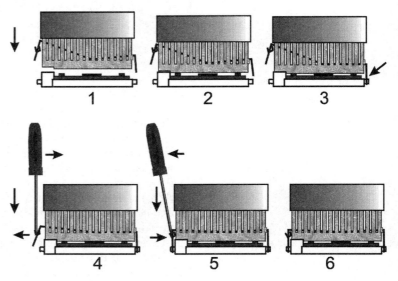

**Figure 5.14**   *Fitting the heat sink and fan to the CPU*

Some heat sink manufacturers have devised a simple lever operated spring clamp system which removes the need to manually depress the strong spring clamp. Even with this improved system you must still keep the heat sink assembly in a parallel plane with the CPU otherwise the chip can still be crushed.

- The final step to fitting the CPU and heat sink/fan is to attach the fan power leads to the motherboard. If you cannot locate the connector on the motherboard refer to the user manual (Figure 5.15(b)).

## Stage 5: installing RAM

The motherboard we have chosen supports up to 3 GB of unbuffered DDR DIMMs in three speeds, PC2100 (266 MHz), PC2700 (333 MHz) and PC3200 (400 MHz). The motherboard has a bank of three 184-pin DDR-SDRAM DIMM sockets available but the manufacturer limits their use as follows:

All three sockets can be used with PC2100 DDRs, a maximum of two can be used with PC2700s and only one can be used with PC3200 DDRs.

This limitation is due to the electrical losses in the PCB tracks and socket contacts at higher operating frequencies. Therefore, if you are going to use PC2100 or PC2700 devices it is best to choose a single high capacity DIMM, rather than using two or three lower capacity devices, to allow for future memory expansion. If you are using a PC3200 device then you have only one expansion option and that is to replace the existing DIMM with one of higher capacity (Figure 5.16).

- Ensure you are still wearing your wrist band and place the motherboard and baseplate in front of you on the work surface. Remove the DIMM device/s from the anti-static bag and notice how the notches line up with the index tabs on the motherboard DIMM slots. If you have a single DIMM device insert it in the first bank usually labelled 'bank 0'. To do this: open up the slot's retaining

(a)                                          (b)

**Figure 5.15**  *(a) Alternative thumb fixing heat sink (b) Fitting the CPU fan lead*

**Figure 5.16**  *Installing RAM*

(a)

(b)

(c)

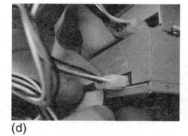

(d)

**Figure 5.17** *Fitting the drives into the case*

**Figure 5.18** *CSEL jumper setting on a typical hard disk drive*

[6] See below for description of CSEL.

clips and line up the DIMM with bank 0 keeping it in a horizontal plane with the socket. Now push the DIMM down firmly until it locates into the socket contacts. The retaining clips either side of the DIMM assembly will automatically close into place. If the clips have not closed fully, ensure that the DIMM is fully inserted into the slot, and flip them into the closed position.

### Stage 6: fitting the drives into the case

The CD/DVD player, floppy drive and hard drive may be fitted into the case prior to fixing the motherboard on the removable base plate, the order of events makes little difference. If you have a case with a fixed motherboard plate then the drives are best fitted before you secure the motherboard into the case (Figure 5.17).

- The drives can be fitted in any order so the following procedure just happens to be the order we decided to use: wearing your wrist band, remove the CD-R/CD-RW/DVD drive from its protective bag (a), and set the rear jumper to CSEL[6] if it not already preset to this position. Now slide the drive through the drive cage until the front of the drive lines up with the front panel and the screw holes are visible through the fixing slots either side of the drive (b). You may have to temporarily clip the case front back on to achieve this. When it is aligned properly, screw in two self-tapping screws either side of the drive to secure it into position. Fit the floppy disk drive using the same procedure.

    Unwrap the hard disk drive from its protective wrapper and read the instruction sheet. Now set the drive select jumper to CSEL. Slide the drive carefully into its bay with the connector side pointing into the case. Take care not to drop the drive otherwise this may damage the head and disk, and render it inoperative. Align the threaded holes in the side of the drive with the slots in the cage and secure it into position, using two self-tapping screws either side of the drive. (c) The ribbon and power cables can now be connected. Picture (d) shows two CD drives, the lower one is automatically set as slave using CSEL.

### The CSEL drive setting

Current hard drive and CD drive ribbon cables incorporate the *Cable Select* (CSEL) system. This allows the drives to be automatically set as master or slave depending on their position on the cable rather than using separate jumper settings. To implement this feature the drive/s must be set to CSEL. This is an extra jumper position in addition to the normal master and slave jumper positions. When the CSEL jumper is in place, the drive on the far end of the cable is automatically assigned as master and the drive in the middle of the cable is automatically assigned as slave. In a single drive system with the CSEL jumper in place, the drive must always sit on the far end connector (Figure 5.18).

The ribbon cable is designed so that the wire carrying the CSEL signal passes directly from the CSEL pin on the motherboard connector straight through to the CSEL pin on the end connector. The CSEL pin on the intermediate connector however is not connected. This is achieved during manufacture by removing the CSEL pin from the connector itself.

The CSEL cable system works as follows: the CSEL wire is grounded at the motherboard end, effectively sending a logic 0 signal to the end drive's CSEL input thus setting it as master. However, the intermediate drive's CSEL input is not connected to ground due to the missing pin. The drive on the intermediate connector therefore automatically pulls the CSEL input pin to logic 1, thus setting it as slave.

Some early implementations of CSEL on old 40-way ribbon cables may use a different arrangement. But the aim is the same – automatic master/slave setting via the cable (Figure 5.19).

Some system builders still prefer to use the old master/slave setting system. The slight inconvenience with this system is that if the drives are changed or a drive is added temporarily, for example, when copying user files across onto a new drive, the jumpers will have to be changed. With the CSEL system the drive is automatically assigned as a master or slave depending on its position on the cable.

- Now the motherboard and base plate can be fitted into the case. Locate the base plate into the guide slots in the case and slide it into position. Now secure in place with the original screws. For cases with a fixed base plate fit the motherboard carefully over the pillars installed previously and insert the fixing screws taking care not to drop any under the motherboard.

### Stage 7: Connecting the wires and cables

*Front panel wiring*

- Connect the front panel leads. These connect the motherboard to the various indicator lights and pushbuttons on the front of the case. These are usually labelled '*Power led, Speaker, Reset switch, Hard Drive led and Power-on/off*'. It is best to refer to the motherboard manual for the actual functions supported. The power-on/off and reset switch connectors can be placed either way round but the indicator leads must be connected with the correct polarity. This is necessary because the indicators are *light emitting diodes*. These have an anode and a cathode. The anode must connect to the positive '+' terminal and the cathode to the negative '−', terminal on the respective motherboard connector (Figure 5.20).

  Some cases have a standby switch which allows the operator to manually place the PC into a power saving or standby mode. This switch connects to the system management interrupt connector (SMI).

- Once the front panel connectors are fitted in their correct locations, tidy up the bundle of wires with cable ties and tuck them neatly away from the motherboard.

- Refer to the motherboard manual and the fit any remaining I/O cables to the motherboard. In our example there are digital audio,

*Note:* The positive (anode) lead is usually the coloured lead whereas the negative (cathode) lead is usually black or white.

**Figure 5.19**   *UDMA66/100/133 IDE cable*

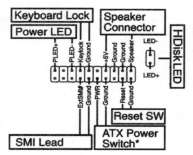

**Figure 5.20**   *Front panel wiring*

(a)

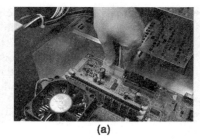

(a)

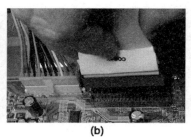

(b)

**Figure 5.22**   *(a) Fitting the power cable and (b) the primary IDE ribbon cable to the motherboard*

(b)

**Figure 5.21**   *Fitting of I/O cables to the motherboard*

USB and IEEE1394 I/O cables and connectors (Figure 5.21(a)). Fit the corresponding rear I/O cover plates in a convenient position leaving space for the AGP graphics card (Figure 5.21(b)).

- Fit the main ATX 20-way motherboard power cable socket to the motherboard. It only locates one way, so after aligning it in the correct direction push it home fully until the locating lug (if fitted) clips into place (Figure 5.22(a)). Some ATX motherboards have extra power connectors for a +12 V feed. If this is the case with your motherboard you should fit these as well.
- The disk drive ribbon cables can now be connected to the motherboard. Observe the correct polarity. There is a stripe marked on the first wire of each ribbon cable and this must go to pin 1 of the drive and motherboard connector. Nowadays most drive cables are indexed and only fit one way into the corresponding sockets but do make yourself aware of the 'leader stripe to pin 1' rule. The hard disk must connect to the primary motherboard IDE connector and the CD drive to the secondary IDE connector (Figure 5.22(b)).

### Stage 8: Inserting the graphics card
- Wearing your wrist band, remove the AGP graphics card from its wrapper and carefully align it with the AGP port on the motherboard (a). Make sure an empty rear panel slot is available for the card. If not, remove the offending screening plate. Now push the card firmly into the AGP edge connector. The card goes into the slot quite deeply, so make certain that you have pushed it home fully. Now secure the mounting plate into the rear case frame with a suitable screw from the set of case screws (b) (Figure 5.23).

**Figure 5.23** *Inserting the graphics card*

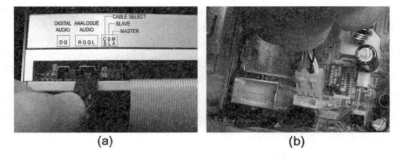

**Figure 5.24** *Fitting sound, modem and network cards*

- Some motherboards have a locking lever at the rear of the AGP slot. This should be pushed fully home when the card is inserted properly (c).
- The graphics card used in this scenario is a high speed AGP 3.0 8X card. Due to its high processing speed it consumes more power than a conventional AGP card. To accommodate the extra power requirement a separate power connector is fitted to the card – the same type as that used on a floppy disk drive. This must be fitted otherwise the card will not function (d).

### Stage 9: Fitting sound, modem and network cards
- Our motherboard has the sound, modem and network subsystems built into the motherboard so separate adapter cards are not required. However, conventional AT and ATX motherboards do not include these items so now is the time to fit these into the spare PCI slots (Figure 5.24).
- Now take the audio cable supplied with the CD/DVD drive and connect it from the drive's analogue audio output connector (a), to the appropriate audio input connector on the sound card. This enables the user to play music CDs directly into the sound system.

You may wish to refer to the CD/DVD manual and sound card manual to do this. On our system the CD audio connector is located on the motherboard itself (b). The analogue audio lead has polarized connectors to ensure that the left and right stereo signals connect to the corresponding L&R channel inputs on the sound card.

### Stage 10: powering up the system

- Check over all the leads, connectors and cards and make sure there is nothing loose or badly fitted. A typical check list is as follows:
  - PSU screwed home fully into case and nothing fouling the fan
  - PSU mains selector on rear (if fitted) set to 240VAC for UK mains
  - Motherboard fitted securely no screws loose underneath
  - Rear ATX I/O panel fitted correctly (all sockets accessible)
  - Motherboard power lead/s fitted correctly
  - CPU fitted correctly and FAN supply connected
  - DIMMs pushed home fully and clipped in place
  - All adapter cards pushed fully into their slots and secured
  - Auxiliary supply lead correctly fitted to AGP card (if one is provided)
  - All extra I/O connectors and panels fitted correctly
  - Front panel wires fitted correctly
  - Drives fitted flush to front and fully secured
  - Drive power and ribbon connectors connected properly
  - No loose screws or other loose items in system-unit case
- Clear a space on your work top area and place the monitor and system-unit at a convenient location close to a mains outlet. Connect correctly fused mains cables to the monitor and system-unit. If the recommended mains plug fuse rating is not specified on the back of the equipment use the suggested fuse rating below:
  - System-unit:          2–3 A fuse.
  - CRT type monitor:    2 A fuse.
  - LCD monitors:        1 A fuse.
- Check that all mains on-off switches are in the OFF position and plug the monitor and system-unit mains leads into the wall socket. These items are now earthed through the mains supply.
- Connect the mouse and keyboard to the rear I/O panel mini din connectors.
- Connect the monitor signal lead into the appropriate socket on the graphics card.
- Switch on at the wall socket and power up the monitor and system-unit. After a few seconds you should see details of the graphics card and/or the power up procedure on screen during the power on self-test (POST). You should see details of the POST RAM check on screen. After waiting a few seconds, if the monitor displays 'no signal', you have probably not pushed the graphics card fully into the AGP socket.
- If the system appears dead with no on screen activity, switch off the computer and carefully check the seating of the CPU, DIMM and adapter cards. Also check all the ribbon cable connections and other wiring. If you still have a dead system, wear your wrist band and carefully remove all the adapter cards and power up the system again. If you get a succession of beeps from the internal speaker, then switch off and reinsert the AGP graphics card,

*Warning:* Never fit a plug fuse higher than the specified fuse rating. Under certain fault conditions mains equipment fitted with a fuse of higher rating than the recommended value can pose a serious fire risk!

remembering to push it firmly down into its socket … Try again. If still no joy, remove the DIMMs completely and then refit them and … try again.[7]

### Stage 11: Configuring CMOS setup

The next step is to enter CMOS setup to set the CPU and RAM parameters. Also while you are in the SETUP menu, the boot sequence, cache, date/time and other options can be set.

| **BIOS manufacturer** | **Key or key combination** |
|---|---|
| AMI | Del |
| Award | Del alternatively Ctrl Alt Esc |
| PHOENIX | F2, Ctrl Alt Esc or Ctrl Alt S |

- Boot your PC and enter CMOS setup by pressing the appropriate key or key combinations during the POST as shown in the above data.
- In the 'Main' setup screen, set the time and date and check to see if the hard disk drive and floppy disk drive have been auto-detected correctly. In this Award CMOS setup utility, you can view the Sub-Menu of the items prefixed with an arrow head by highlighting the item and pressing Enter.
- Now the CPU parameters must be set. The exact method used depends on the system ROM manufacturer. Refer to your motherboard manual for the correct procedure. Current motherboard/CPU combinations automatically detect and set the main CPU parameters, which is just as well, as CPU core voltage and operating frequencies vary widely. The CPU core voltage in particular must be set accurately to avoid burning out the CPU.

All we have to do is set the correct CPU core clock frequency as a multiple of the system clock (external CPU clock). In our example, the system clock is 133.33 MHz and the CPU is an Athlon XP 2400.

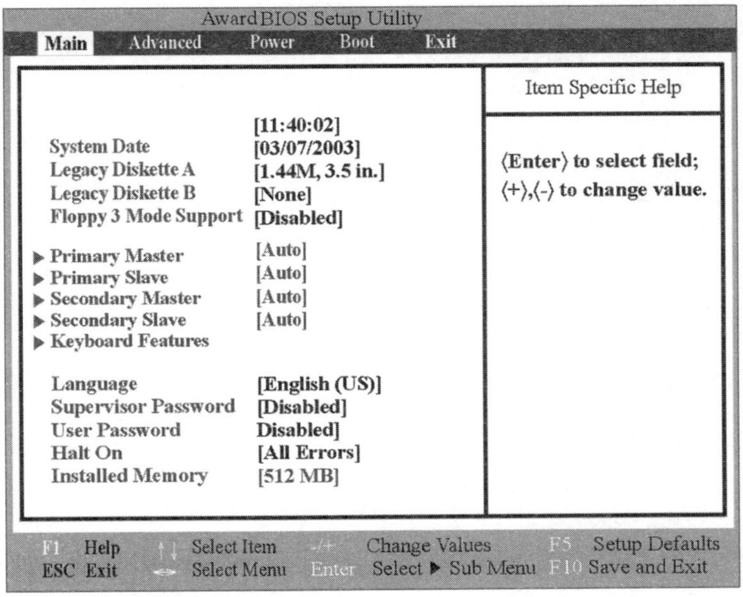

> The following screen shots are for the AWARD CMOS setup but the basic procedure is similar for other system ROM manufacturers (Figure 5.25).

> The CPU automatic parameter detection and setting system works via links on the CPU. The pre-set links set a series of unique binary codes and these are read by the motherboard and used to automatically set the correct parameters in hardware. For example, on an Athlon XP CPU the codes VID4.0, FID3.0 and FSB-Sense automatically set the CPU core voltage, core frequency and front side bus frequency, respectively.

**Figure 5.25**

---

[7] Troubleshooting is covered fully in Unit 6.

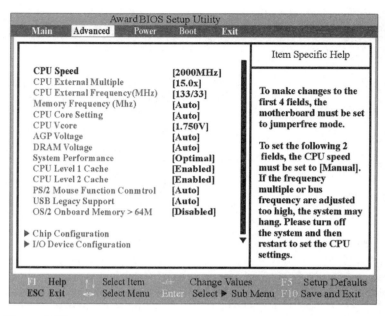

**Figure 5.26**

**Table 5.2**

| Athlon XP | Core frequency (MHz) | FSB frequency (MHz) | System clock (MHz)/multiplier |
|---|---|---|---|
| 1700 | 1467 | 266 | 133/×11.0 |
| 1800 | 1533 | 266 | 133/×11.5 |
| 1900 | 1600 | 266 | 133/×12.0 |
| 2000 | 1667 | 266 | 133/×12.5 |
| 2100 | 1733 | 266 | 133/×13.0 |
| 2200 | 1800 | 266 | 133/×13.5 |
| 2400 | 2000 | 266 | 133/×15.0 |
| 2600 | 2083 or 2133 | 266 or 333 | 133/×15.5 or 166/×13.0 |
| 2700 | 2167 | 333 | 333/×13.0 |

Historically the floppy disk has always been set as the first boot device. This was useful in MS-DOS and earlier versions of Windows up to and including Windows ME as it allowed the user to boot off a utility floppy disk in case of emergency. For example if the hard disk failed with a damaged boot sector it could often be repaired using a disk repair utility. Windows 95, 98 and ME allow the user to create a bootable Startup floppy disk. This contains various tools and repair utilities. Including a CD-ROM boot facility in-case the BIOS does not support CD-ROM booting. However, most BIOS ROMs manufactured from 1999 onwards do support CD-ROM booting.

This requires a core clock frequency of 2000 MHz. We therefore have to set the multiplier to 15.0× (see Figure 5.26 and Table 5.2).

- The Memory (RAM) bus frequency setting has been left on 'Auto' but it can be set manually if you prefer.
- Make sure that the Level 1 and Level 2 caches are enabled (Figure 5.26).
- Now select the boot menu and set the search order for bootable devices. Here we have set the boot order as floppy disk, hard disk and CD-ROM. This the best sequence for most situations (Figure 5.27).
- Select Yes for plug and play OS and then select the 'Exit' menu and save your setting and exit CMOS setup. As we have not installed an OS at this point, in a few seconds you should see the on screen report '*no operating system detected*' or a similar message. This is the next step ...

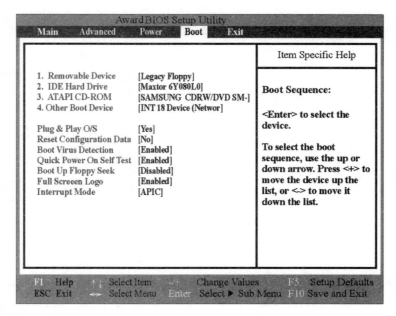

**Figure 5.27**

# Installing Windows XP Home Edition

## Stage 12: Installing the operating system

We have chosen Microsoft® Windows XP Home Edition for our system as it is one of the most stable and user friendly GUIs available at a reasonable price.[8] Also different users can configure the system to their own requirement. Tom an experienced Windows 95/98 user will quickly get used to the new environment and if he chooses he can set it up to look and feel like Windows 98. Jane a relative new comer to computers will find the friendly and informative Windows XP environment easy to learn and use.

- Unpack the Windows XP Home Edition CD from its wrapper taking care not to tear the Product Key and Certificate of Authenticity (COA) labels. These must be kept as they are needed to Activate and Register the software.
- If the system is switched off, switch it on and insert the Windows XP Home Edition CD into the Compact Disc drive. If the system is still powered up displaying the '*no operating system message*', insert the Windows XP CD and press the Reset button on the front of the system-unit. The disk starts up and prompts you to press enter to set up Windows XP. On pressing *Enter* the installation proceeds (Figure 5.28).
- You can see from the screen above that the Windows XP Home CD can also be used to repair an existing Windows XP installation by pressing 'R' to start the Recovery Console (Figure 5.29).
- As the hard disk is brand new, the Windows XP installation program detects the unpartitioned hard disk space and prompts you to press C to create a partition. The program then automatically partitions the space you allocated. We chose to use the whole of the disk as a Windows XP partition (Figure 5.30).
- After choosing a quick or normal high level format, the format procedure proceeds. On a large partition this can take over 30 min to complete (Figures 5.31 and 5.32).

[8] Installation details of other Windows OSs follow later in this unit.

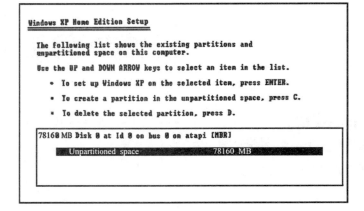

```
Windows XP Home Edition Setup
═══════════════════════════

  Welcome to Setup.

  This portion of the Setup program prepares Microsoft(R)
  Windows(R) XP to run on your computer.


    •  To set up Windows XP now, press ENTER.

    •  To repair a Windows XP installation using
       Recovery Console, press R.

    •  To quit Setup without installing Windows XP, press F3.



  ENTER=Continue   R=Repair   F3=Quit
```

**Figure 5.28**

```
Windows XP Home Edition Setup
═══════════════════════════

  The following list shows the existing partitions and
  unpartitioned space on this computer.

  Use the UP and DOWN ARROW keys to select an item in the list.

    •  To set up Windows XP on the selected item, press ENTER.

    •  To create a partition in the unpartitioned space, press C.

    •  To delete the selected partition, press D.

  78160 MB Disk 0 at Id 0 on bus 0 on atapi [MBR]
        Unpartitioned space              78160 MB
```

**Figure 5.29**

```
Windows XP Home Edition Setup
═══════════════════════════

  A new partition for Windows XP has been created on

  78160 MB Disk 0 at Id 0 on bus 0 on atapi [MBR].

  This partition must now be formatted.

  From the list below, select a file system for the new partition.
  Use the UP and DOWN ARROW keys to select the file system you want,
  and then press ENTER.

  If you want to select a different partition for Windows XP,
  press ESC.

     Format the partition using the NTFS file system (Quick)
     Format the partition using the NTFS file system
```

**Figure 5.30**

Now the hard disk has been formatted the Windows XP OS installation proceeds in earnest. The previous installation screens were based on a rudimentary text only display. Now a more user friendly sequence of screens ensues. Details of the process under way and time to completion is displayed along with other useful information (Figure 5.33).

After a minute or so you are prompted to enter country and language options (Figure 5.34).

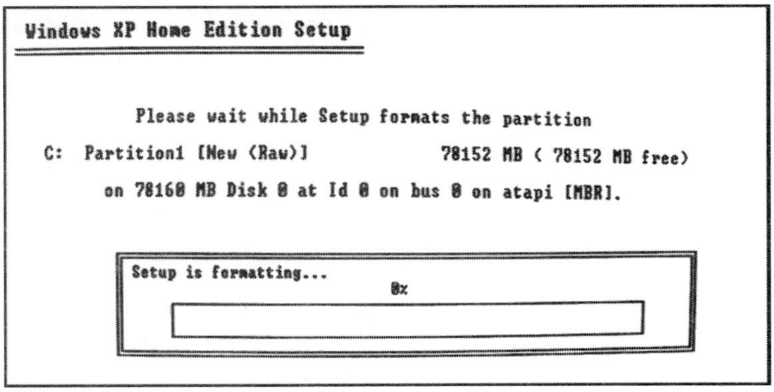

```
Windows XP Home Edition Setup


            Please wait while Setup formats the partition
    C:  Partition1 [New (Raw)]                78152 MB ( 78152 MB free)
        on 78160 MB Disk 0 at Id 0 on bus 0 on atapi [MBR].


         ┌────────────────────────────────────────────────┐
         │ Setup is formatting...                         │
         │                            0x                  │
         │  ┌──────────────────────────────────────────┐  │
         │  │                                          │  │
         │  └──────────────────────────────────────────┘  │
         └────────────────────────────────────────────────┘
```

Figure 5.31

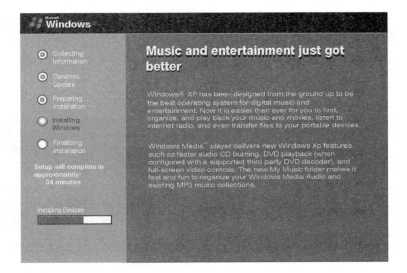

Figure 5.32

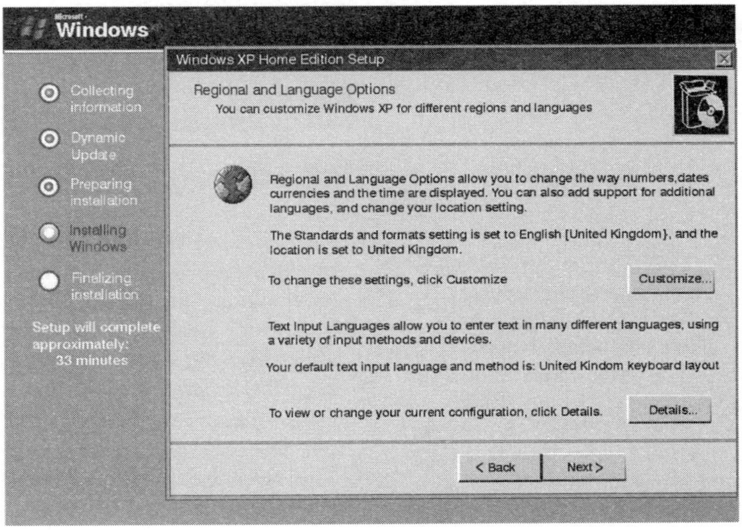

Figure 5.33

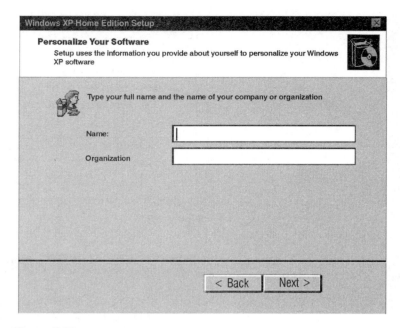

Figure 5.34

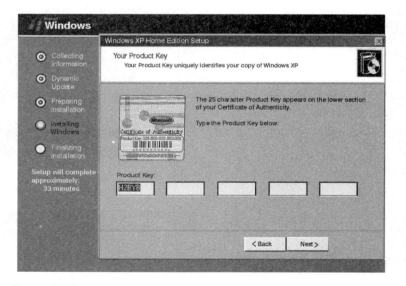

Figure 5.35

[9] The 'product key' is a sequence of alphanumeric characters usually arranged in five groups with five characters in each. It is Microsoft's way of reducing software piracy by assigning individual CD-ROMs or program suites, to a unique ID. Microsoft usually inserts this on the cover of the user manual or wrapping accompanying the CD.

- Next you are prompted to enter the user's name and company or organization (Figure 5.35).
- Now comes the all important authentication process. Here you must enter the unique 25 character long Product Key from the labels attached to the Windows XP package.[9] After completing the process, keep the *Product Key* and *Certificate of Authenticity* (COA) labels in a safe place. Subsequently they should be attached to the system-unit in a secure but accessible location (Figure 5.36).
- The next user prompt requests a unique name to identify your computer. This is particularly important if the system is part of a network of computers. If this is the case choose a name that easily

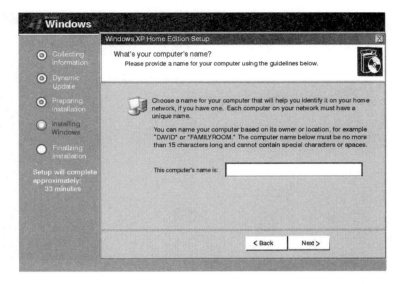

**Figure 5.36**

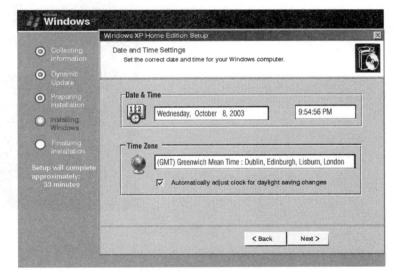

**Figure 5.37**

identifies this system or one that follows the same naming convention as the rest of the network.

If you are installing Windows XP Professional, you will also be prompted to enter the network administrator's pass word. If you are the 'network administrator' on a stand alone PC then choose a password you can easily remember and write it down in a notebook or diary. If you are the network administrator of a LAN system at this point you should enter your usual administrator password and log the computer name, OS, product code and installation date into your network log (Figure 5.37).

* Finally after entering the correct date, time and time zone, the installation program has most of the information it needs to proceed with the main installation process (Figure 5.38).

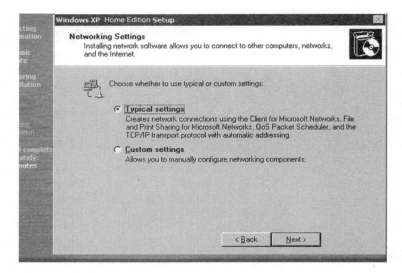

**Figure 5.38**

**Figure 5.39**

- If you have a network card installed, Windows XP setup detects it and prompts you to choose *Typical settings* or *Custom settings* to set up the system to work on the network.

  Unit 3 contains details of the Windows XP *Network Setup Wizard*. After several minutes of disk activity the installation goes through the final phase before rebooting the system (Figure 5.39).

- The Windows XP OS is now installed and the plug and play system automatically detects hardware devices including the graphics card, sound system and modem and prompts you to locate the correct drivers from the disk provided by the manufacturer (Figure 5.40).
- After prompting for user names, Windows XP is ready for use.
- Naturally the next step is to install the required peripherals and applications software to make the Windows XP machine a useful, stable, user friendly companion.

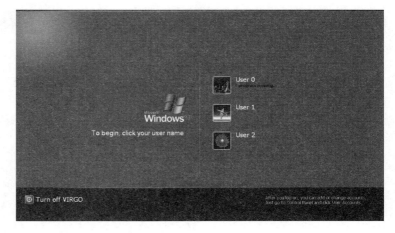

**Figure 5.40**

**Figure 5.41**

# Files and settings transfer wizard

In previous versions of Windows, after replacing your old computer with a new one, it was necessary to manually reconfigure your display properties, folder options, Internet browser and e-mail settings. Thoughtfully, Windows XP provides a utility called the *Files and Settings Transfer Wizard* to help you transfer settings from your old computer to your new one. It also has an option to transfer user files from your old computer too but you will probably prefer to do this manually.

- To invoke this utility run the new Windows XP computer. Click Start, highlight *All Programs/Accessories/System Tools* and click on *Files and Settings Transfer Wizard*. The following window appears (Figures 5.41 and 5.42). Click Next>.

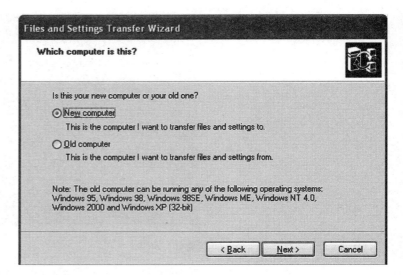

**Figure 5.42**

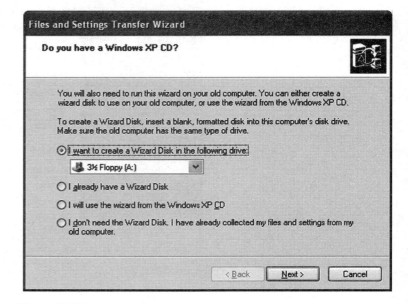

**Figure 5.43**

- Tell the wizard which computer it is currently running on by clicking on New computer (Figure 5.43).
- Insert a formatted 3.5 in. floppy disk in your A: drive to create a wizard disk to use on your old computer. This copies a utility program named FASTwiz.exe (Figure 5.44).
- To run the wizard, insert the disk in the A: drive of your old computer and run it by clicking on *Start* then *Run* and typing *A:\FASTWiz* then click OK (Figure 5.45).
- After closing any other open applications or utilities, click next to allow the wizard to gather information about your old system and prepare for the next step (Figure 5.46).
- There is a short interval while it does the information gathering exercise (Figure 5.47).

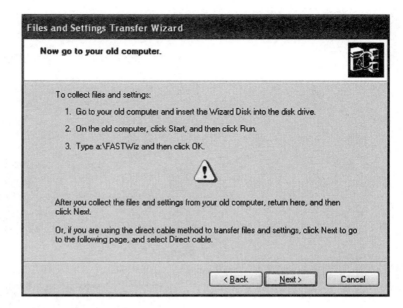

**Figure 5.44**

**Figure 5.45**

**Figure 5.46**

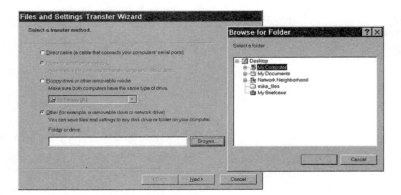

**Figure 5.47**

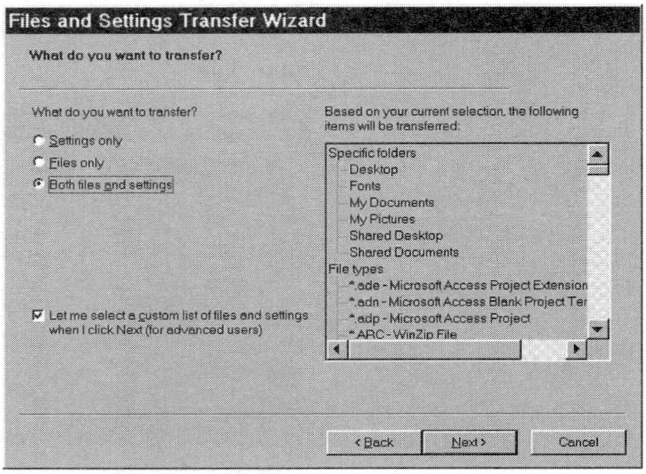

**Figure 5.48**

**Figure 5.49**

- Choose the method you want to use to transfer the files and settings. If you choose the removable drive/network drive option click on *Browse* to select a folder to save the files and settings to. If you want to transfer large files or a large number of small files then a network or removable drive is best. Transferring large amounts of data via floppy disk (Sneakernet) will be extremely tedious and requires a lot of disks (Figure 5.48).[10]
- Select what you want to transfer by clicking on the relevant radio button. Also click the check box to let you select a custom list of files and settings (Figure 5.49).
- Remove all the items in the windows that you do not want to transfer by highlighting the item and clicking on the Remove button. Any folders or items not listed can be added by clicking on the relevant 'Add' button (Figure 5.50).
- The wizard now collects all your selected items (Figure 5.51).

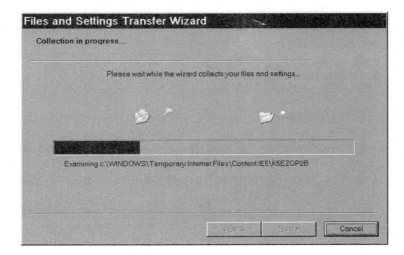

**Figure 5.50**

[10] Sneakernet, colloquial name for floppy disk transfer because the user does a lot of footwork.

**Figure 5.51**

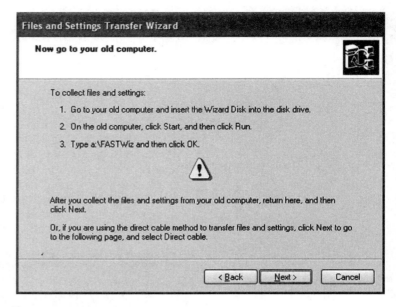

**Figure 5.52**

- Once all the data has been collected the information is saved to the folder/disk you specified previously. Click the *Finish* button to close the wizard (Figure 5.52).
- Now go back to your new computer and complete the *files and settings* transfer by clicking on *Next* in the wizard window you left earlier.
- On completion of the transfer process, click on *Finish* to close the wizard.

# Windows XP registration and activation

The XP product activation system was set up by Microsoft in an attempt to reduce software privacy. After installing Windows XP you have a 30-day period in which to activate the software. After this period you will not be able to access your files and applications but you will still be able to access the activation menu.

Activation is a compulsory non-invasive procedure whereas product registration is purely voluntary. If you wish you can register at the same time as the activation process.

To open the activation menu at any time during the 30-day grace period click Start, highlight *All Programs/Accessories/System Tools* and click on *Activate Windows* (Figure 5.53).

You can activate Windows XP via the Internet or via freephone: The freephone procedure is as follows:

Arm yourself with a pen and paper:

1.  After invoking the activation menu, contact Microsoft by phoning the 0800 number displayed in the menu. A Microsoft customer representative (Mcr) responds and prompts you to enter your Installation ID via the telephone keypad. *Your installation ID is automatically generated from your product ID entered during installation and is displayed in Step 3: of the activation menu.*

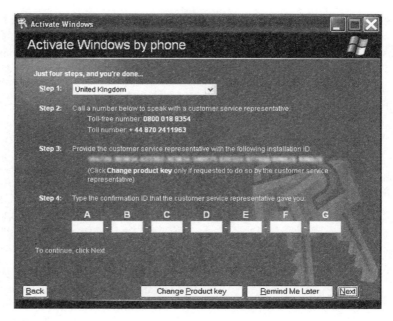

**Figure 5.53**

2. The Mcr then issues a long confirmation ID split into seven segments A to G. Jot these down carefully onto a piece of paper and end the call.
3. Now enter your confirmation ID into the menu in Step 4: and click *Next*.
4. That completes the activation process.
5. Finally destroy the piece of paper.

# Installing Windows 98 on a new PC

## Entering CMOS setup

Whatever system you have built, before you install Windows 98 enter CMOS setup and set the boot sequence and general settings for the cache, 'plug n play' and disk drives.

- Boot your PC and enter CMOS setup by pressing the appropriate key or key combinations during the POST:

| BIOS manufacturer | Key or key combination |
|---|---|
| AMI | Del |
| AWARD | Del alternatively Ctrl Alt Esc |
| PHOENIX | Ctrl Alt Esc or Ctrl Alt S |

- In the main setup screen, confirm that your hard disk has been auto-detected correctly and that the floppy disk is correctly identified as the A: drive. Now select the BIOS FEATURES SETUP MENU (Award) or ADVANCED CMOS setup (AMI) and set 'CD-ROM' as the first boot device. Also ensure that the CPU Internal and External Caches are enabled.
- Check to ensure that the USB function is enabled if this is an option.
- Turn off Ultra DMA (UDMA) to the secondary IDE port if this is an option. Several CD drives will not work correctly if this is enabled.

### DMA and interrupt request settings

- If all your adapter cards are modern ISA/PCI/AGP plug and play devices then enter PNP/PCI setup and enter 'YES' for *PNP OS Installed*. Set all DMA channels to *PNP* and all IRQ channels to *PNP/PCI*. In some systems to enter these settings you will have to change the *Resources Controlled by* option in PNP/PCI setup to 'Manual' instead of Auto.

### Legacy non-plug and play cards

- If you have any non-'plug n play' adapter cards installed such as a legacy ISA Sound Card, write down the DMA, IRQ settings of the card by referring to the on-board jumpers or dip switches and set these in the PNP/PCI SETUP menu to 'legacy ISA'. In some systems to enter these settings you will have to change the *Resources Controlled by* option in PNP/PCI setup to 'Manual' instead of Auto.

If you are installing Windows 98 on a newly built machine with no OS currently installed follow the steps below to carry out the installation: to upgrade to Windows 98 see the subsequent instructions.

*Step 1*  Place your Windows 98 CD-ROM in your CD-ROM drive and reboot the PC. This must be a disk specifically intended for a new machine. It usually states on the CD-ROM 'For distribution with a new PC only'. It must not state 'UPGRADE' on the disk. An UPGRADE disk is intended for an existing PC with Windows 3.1 or Windows 95 already installed.

*Step 2*  The CD-ROM spins up and displays the following message on-screen. Select 2. 'Boot from CD-ROM'.

```
Microsoft Win98 CD-ROM Startup Menu

   1. Boot from hard disk
   2. Boot from CD-ROM                          <
```

The Windows setup program boots and shows the following message:

```
Microsoft Win98 Setup

      Welcome to Setup

The Setup program prepares Win98 to run on your computer

   • To set up Windows now, press ENTER
   • To learn more about Setup before continuing, press F1
   • To quit without installing Windows, press F3

Note: If you have not backed up your files recently, you might
want to do so before installing Windows. To back up your
files, press F3 to quit Setup now. Then, back up your files by
using a backup program.

To continue with Setup, press ENTER.
```

*Step 3*   Press Enter: The following screen shows that Windows 98 Setup has detected the presence of a hard disk and has found empty space. In the case of a newly built PC, the hard disk will be completely empty. Select 'Configure unallocated disk space' and press ENTER.

---

Microsoft Win98 Setup

Setup needs to configure the unallocated space on your hard disk to prepare it for use with Windows. None of your existing files will be effected.

To have Setup configure the space on your hard disk for you, choose the recommended option.

**Configure unallocated disk space (recommended)**
Exit Setup.

To accept the selection, press ENTER.
To change the selection, press the UP or DOWN ARROW key and then press ENTER.

---

*Step 4*   In the next screen select 'Yes, enable large disk support'.

---

Microsoft Win98 Setup

You have a drive over 512 MB in size. Would you like to enable large disk support?

This allows more efficient use of disk space and larger partitions to be defined.

No, do not use larger disk support
**Yes, enable large disk support**                                    <

To accept the selection, press ENTER.
To change the selection, press the UP or DOWN ARROW key and then press ENTER.

---

*Step 5*   You are now prompted to restart the computer, with the Windows 98 CD-ROM still in the drive.

*Step 6*   The System reboots the drive is partitioned and a widow appears informing you that the hard disk is being high level formatted.

*Step 7*   Windows 98 is then installed onto the hard disk. As the process unfolds you will be prompted to respond to various options, including a request to enter the *product key*. The process usually takes 30 min or so on a modern system.

*Step 8*   When the installation has been completed, restart the PC, go into CMOS setup, select the BIOS FEATURES SETUP MENU (Award) or ADVANCED CMOS SETUP (AMI) and choose 'Floppy Disk', 'Hard Disk 1' and 'CD-ROM' for the first, second and third boot devices respectively. Then Save and Exit CMOS setup. Windows then completes the installation procedure by automatically configuring device drivers for the adapter cards and other plug and play peripherals in the system. If Windows cannot find a suitable driver in its extensive driver database you can choose an option to insert a disk provided by the manufacturer. As a general rule, if the OS is more up to date than the card and a driver has not been automatically installed by Windows, then see if a new driver is available for download via the Internet.

# Upgrading from Windows 95

Windows 95 exists in three versions namely Windows 95 (4.00.950), (4.00.950A) and (4.00.950B), the latter was only available as an OEM release. The 950A and 950B versions of Windows 95 correct several bugs found in the first 95 release. They also support the new FAT 32, large disk format as well as FAT 16. The first issue only supports FAT 16.

Windows 98 is also available in two versions; the early 98 (4.10.1998) release and the later (4.10.2222A) *Second Edition*. You can check the version of any Windows 9* installation by right clicking *My Computer* and selecting *Properties*.

## Upgrading from Windows 95 to Windows 98

The upgrading process includes the following steps:

*Step 1*   Insert the Windows 98 CD into the CD-ROM drive and wait for the AutoRun facility to boot the Windows 98 Setup program. If it fails to load, run SETUP EXE by clicking on the *Start* button and then *Run*. In the open window type *d:\setup*. or whatever drive name is applicable.

*Step 2*   Have your *product key* handy and respond to the various prompts as the Windows98 Installation proceeds.

## Installing Windows 98 on an existing machine after clearing the hard disk

*Do not do this with the UPGRADE version of Windows 98 only the full install version.*

You may want to clear all the information off the hard disk before you install Win98. Before you do this, backup any important files to tape or CD-R, as everything will be erased. If the motherboard is quite old it may have a BIOS that does not support direct booting from the CD-ROM drive. If your BIOS <u>DOES</u> support CD-ROM booting go to the next set of instructions overleaf.

If your BIOS does not support CD-ROM booting, you must make a startup disk to support the CD-ROM before wiping the hard disk. To do this:

*Step 1*   In Windows 95: if you do not already have a Windows Startup disk, create a new one by placing a formatted floppy disk in the A: drive. Now click *Start* highlight *Settings* then click *Control Panel* and click on the *Add/Remove Programs* icon. Select the *Startup Disk* Tab and create a startup disk. You will be prompted to enter your Windows 95 Installation disk. Make sure the disk contains FDISK.EXE, FORMAT.COM and the SYS.COM file.

*Step 2*   Copy the CONFIG.SYS and AUTOEXEC.BAT utilities from the hard disk to your startup disk, along with any referenced files or folders that related to your CD-ROM drive.

*Step 3*   Boot your machine with the startup disk and ensure that you can access the CD-ROM drive.

*Step 4*   Run FDISK.EXE, and remove the old partition/s from your hard drive and create a new one. To do this type FDISK then ENTER. The following menu appears:

```
                    Microsoft Win95
                 Fixed Disk Setup Program
           ©Copyright Microsoft Corporation 1983–1995

                       FDISK Options

Current Fixed disk drive: 1

Choose one of the following:

1:  Create DOS partition or logical DOS drive
2:  Set active partition
3:  Delete partition or logical DOS drive
4:  Display partition information
5:  Change current fixed disk drive

Enter choice: [3]
```

If you have a hard disk over 512 MB, when you run FDISK you will see a message and prompt, asking you to specify whether you want to run in FAT 32 rather than FAT 16, and set the drive to one large partition. The message is as follows:

---

Your computer has a disk larger than 512 MB. This version of Windows includes improved support for large disks, resulting in more efficient use of disk space on large drives, and allowing disks over 2 GB to be formatted as a single drive.

*Important*: If you enable large disk support and create any new drives on this disk, you will not be able to access the new drive(s) using other OSs, including some versions of Windows 95 and Windows NT, as well as earlier versions of Windows and MS-DOS. In addition, disk utilities that were not designed explicitly for the FAT 32 file system will not be able to work with this disk. If you need to access this disk with other OSs or older disk utilities , do not enable large drive support.

Do you wish to enable large disk support?

---

This is something you must decide yourself. However, for all modern applications, FAT 32 and large disk support is the best option.

Example 1: If you ever want to boot off the A: drive in MSDOS 6.22 or earlier, and access the hard disk using MSDOS FAT 16 utilities, then do not choose large disk support. If you have a drive larger than 2.1 GB and you do not choose large disk support you will have to partition the drive as one active partition and several smaller extended partitions (logical drives).

Example 2: If you have a large drive and you only want to run in Windows 98/ME/2000 and you do not want to install FAT 16 disk utilities, then your best option is to choose large disk support.

*Step 5* Choose option 3 from the menu to delete the existing partition. In the next screen, choose 1 to delete the primary DOS partition. Then enter the volume label name if there is one, or else press enter.

```
                    Delete Primary DOS partition

Current fixed disk drive: 1
Partition  Status  Type       Volume label  MB    System  Usage
C: 1       A       PRI DOS                   ****  FAT16   100%

Total disk space is ****MB (1 MB 5 1 048 576 bytes)

Warning data in the deleted primary DOS partition will be lost.
What primary partition do you want to delete..? [1]
Enter volume label ..........................................?[_   ]

Press Esc to return to FDISK options
```

*Step 6* The partition is deleted and then you are returned to the main menu. Now select option 1 to create a DOS partition or logical DOS drive and in the subsequent menu select create primary DOS partition.

*Step 7* Now format your hard drive using FORMAT COM from the floppy disk and make it bootable by running SYS.COM on it. Or use Format c:/s.

*Step 8* Copy the CONFIG.SYS, AUTOEXEC.BAT and the CD-ROM support files from your startup disk to your hard drive.

*Step 9* Remove the floppy disk and reboot your machine from your hard drive.

*Step 10* At the command prompt, change the directory to your CD-ROM drive, insert the Windows 98 CD, and run SETUP.EXE. The installation should then proceed as normal. Have your product key handy.

# Installing Windows 98 on an existing Windows 98 system

*Do not do this with the UPGRADE version of Windows 98 only the FULL version*

If you already have Windows 98 installed and you want to start all over with a clean hard disk by wiping everything off it, carry out the following procedure:

*Step 1* Make sure you have backed up all important files as everything will be erased permanently. Also make sure you have a working Windows 98 Startup disk – have at least two to be safe as floppy disks are notorious for failing at the most inopportune moments. If you need to create one, place a blank formatted floppy disk in the A: drive. Now click *Start* highlight Settings then click *Control Panel* and click on the *Add/Remove Programs* icon. Select the *Startup Disk* Tab and create a new startup disk. You will be prompted to enter your Windows 98 installation disk.

*Step 2* Reboot the computer from your Startup disk and notice that a prompt appears asking if you want to (1) *Start computer with CD-ROM support.* (2) *Start computer without CD-ROM support*, choose the second option. Wait for the floppy disk to boot, then type FDISK and repartition the hard disk as shown in steps 3–5 of the previous example.

*Step 3* Now the hard disk has been partitioned you can proceed to reinstall Windows 98. To do this reboot from your Startup disk.

*Step 4* The prompt (1) *Start computer with CD-ROM support.* appears. Place your Windows 98 full installation disk in the CD-ROM drive, highlight (1) *Start computer with CD-ROM support.* option and press enter.

*Step 5* *Windows 98 setup* should now automatically format your hard disk and install Windows 98.

# Upgrading to Windows 98 from Windows 3.X

The upgrading process includes the following steps:

*Step 1* First backup any important files the user may wish to keep.

*Step 2* As a precaution, make a back up copy of CONFIG.SYS and AUTOEXEC.BAT and check your adapter cards and peripherals to see if there are Windows 98 drivers available. You may be able to download new drivers from the manufacturer's Internet site if they are not available in the Windows 98 built-in driver list.

*Step 3* Place the Windows 98 CD in the drive and use File Manager to run SETUP EXE. As the installation proceeds you will be prompted to enter your product key.

*Step 4* Follow the on screen prompts as the upgrade proceeds.

> **HINT 1:** We have noticed that sometimes when performing a Windows 9* or Windows 2000 installation, the CD-ROM becomes inaccessible for some reason. If this happens the installation has to be aborted and restarted. To overcome this problem it makes a lot of sense to copy the full Windows CD onto a directory of the hard disk and run it from there. To do this you will need plenty of spare drive capacity, especially if you are installing Windows 2000.
>    Apart from the advantage outlined above, it also much more convenient, especially when adding new peripherals or installing new applications. You do not have to rummage around looking for the original Windows CD.
>
> **HINT 2:** If want to install the full version of Windows 98 on a newly built (or rehashed PC) without BIOS support for CD booting and you already have a Windows 98 Startup disk, you could use it to provide the initial CD-ROM support needed to start the installation process.

# Installing Windows 2000

Microsoft have gone to great lengths to make the Windows 2000 installation easier than previous versions of Windows. They have achieved this by adding many more devices to the built-in driver database and by improving the Windows Setup Wizard.

There are a few things to check before installing Windows 2000, it must have the right CPU, enough RAM and a CD-ROM or DVD fitted. Also check that your peripherals are compatible, that is, that there are Windows 2000 drivers available either built in to Windows 2000 or as downloadable drivers from the Internet. Windows 95/98 drivers will not work, as Windows 2000 uses a radically different driver interface system known as the hardware abstraction layer (HAL). Many manufacturers only offer Windows 2000 drivers for their latest products, neglecting older products, so beware. If they are not available on the manufacturer's web site or built in to the Windows 2000 driver database, you may have to discard the hardware or use Windows 95/98.

It is also important to note that some built-in drivers are not 'all singing all dancing'. For example, a particular printer driver may only offer rudimentary facilities and produce relatively poor quality output compared to the equivalent Windows 95/98 driver. If this is the case, check the manufacturer's web site to see if a later version is available.

# Windows 2000 minimum system requirements

### CPU

The minimum CPU specification for Windows 2000 is a Pentium 166 MHz. You cannot run it on a 80486 or lower CPU.

### RAM

The minimum RAM size stipulated by Microsoft is 32 MB but we recommended at least 64 MB.

### Hard disk drive

As Windows 2000 has a huge driver base we recommend a minimum hard disk size of 2 GB. It will fit on a drive with a smaller capacity, say 1 GB but that will leave less than a 600 MB for application programs.

### CD-ROM drive

If your BIOS does not support a bootable CD-ROM, you can obtain Windows 2000 with three floppy disks and a CD-ROM. The floppies prepare the installation and set the system to accept the CD-ROM. For newly built systems, the single CD-ROM OEM version is all that is required, as the BIOS will support booting from CD-ROM.

# Installing Windows 2000/ME on a freshly built system

If you are installing Windows 2000 on a new machine with no OS currently installed follow the steps below to carry out the installation:

*Step 1*   Check the system requirements as mentioned above and thoroughly read through the installation instructions. Run the interactive installation tutorial if supplied with your Windows 2000 CD.

*Step 2*   Boot your PC and enter CMOS setup by pressing one of the following key or key combinations during the POST:

| BIOS manufacturer | Key or key combination |
|---|---|
| AMI | Del |
| AWARD | Del alternatively Ctrl Alt Esc |
| PHOENIX | Ctrl Alt Esc or Ctrl Alt S |

*Step 3*   In the main setup screen confirm that your hard disk has been auto-detected correctly and that the floppy disk is correctly identified as the A: drive. Now select the BIOS FEATURES SETUP

MENU (Award) or ADVANCED CMOS setup (AMI) and choose 'CD-ROM' as the first boot device. Then Save and Exit.

*Step 4* Power up the PC and immediately place your Windows 2000 CD-ROM in the CD-ROM drive. After performing some initial system checks, the Windows 2000 setup wizard boots, click *Next* to continue with the installation.

# Windows 95/98/ME/XP architecture

Probably the most important feature of Windows 95/98/ME is its dual 16/32 bit OS. This allows it to run both DOS and Windows 3.1, 16 bit applications, as well as Win32 applications based on the Windows NT. If the user only wants to run 32 bit applications then Windows 2000 or older versions of NT being true 32 bit OSs are the best choice. You can still run DOS applications in a virtual DOS mode as long as they are not disk utilities or programs that edit the FAT.

Windows 95/98/ME/XP however is an excellent choice for people with existing DOS and Windows 3.1 application software who also want to use 32 bit applications. It is also fine for the person new to computing as it offers a very user friendly interface. In this role it is unsurpassed and demonstrates the ingenuity of the Microsoft Systems programmers.

The advantage of 32 bit software over its 16 bit variety, is speed – it handles data in wider chunks without the memory restrictions of DOS and without the need to address memory in 64 KB segments as required in the earlier Intel processors.

Earlier versions of Windows could also use 32 bit code for 32 bit hardware devices by using 32 bit *Virtual Device Drivers* (VXDs). These are drivers that make use of the 32 bit protected mode operation of the 386 and higher processors. The normal DOS drivers are called *Real Mode* drivers as they can operate with the very early 8 bit 8088/6 CPUs as well as the later 386 and higher CPUs.

Windows 95/98/ME/XP has taken the 32 bit coding of earlier Windows programs further by also using VXDs for all hardware devices, including the mouse, keyboard, I/O ports/buses and graphics adapter. The disk filing system – also a VXD – has been upgraded to support long file names. This makes the system more stable as it stops applications from using the hardware directly. Instead they must use the virtual device drivers.

The core of the Windows OS is a 32-bit engine called the *Virtual Machine Manager* (VMM). Surrounding the VMM is a set of program modules, that perform either 16 bit or 32 bit, low level system operations. So the OS can be classed as neither 16 bit or 32 bit, but a 16/32-bit hybrid.

At a higher level, the Windows application program interface (API) handlers, namely the user, kernel, and graphical device interface (GDI), are also hybrids. As each handler has a 16-bit and 32-bit module. The modules are USER16, USER 32, KERNEL 16, KERNEL 32, GDI16 and GDI32. These handle the on-screen windows, and user input and output, graphics and printing functions and allow Windows 3.1, DOS and 32-bit applications to run seamlessly. Data via the API handlers must therefore be converted from 16-bit to 32-bit code and vice versa. The process of converting code backwards and forwards between 16 and 32 bits is called *Thunking*, by the Microsoft program team (Figure 5.54).

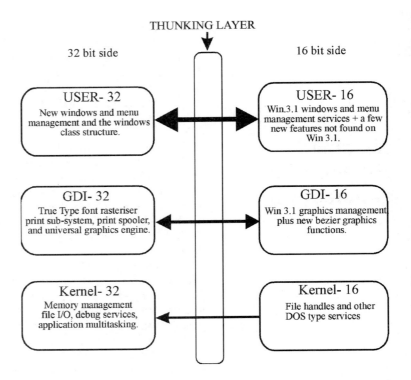

THUNKING LAYER

32 bit side                                  16 bit side

**USER- 32**
New windows and menu management and the windows class structure.

**USER- 16**
Win.3.1 windows and menu management services + a few new features not found on Win 3.1.

**GDI- 32**
True Type font rasteriser print sub-system, print spooler, and universal graphics engine.

**GDI- 16**
Win 3.1 graphics management plus new bezier graphics functions.

**Kernel- 32**
Memory management file I/O, debug services, application multitasking.

**Kernel- 16**
File handles and other DOS type services

**Figure 5.54**    *A representation of Microsoft's thunking diagram showing how the Win95 API-handlers are paired as 32 bit and 16 bit modules. The middle,* thunking layer, *carries out the 16/32-bit code conversion between the two sides*

## Windows 3.1

We have purposely not discussed Windows 3.1 in this book as it was superseded by Windows 95 in August 1995. One of the less impressive aspects of Windows 3.1 was the way application programs could hang and bring the whole system to a halt. This often resulted in the loss of valuable data. The way around the problem was to ensure that regular backups were taken every few minutes. The later versions of Windows are far less susceptible to complete system crashes because programs are handled in a pre-emptive multitasking system.

A pre-emptive multitasking system interrupts one running application to give another application a share of the processor's time. The OS effectively allocates *time slices* to each application program depending on its individual priorities. A calculation intensive application might be allocated a large time slice and a slow application such as a printer driver, might be allocated only a short time slice. Allocating time slices in this way keeps all applications alert, with no single application hogging the processor all the time. The other good thing about pre-empting programs is that if one hangs, due to a bug, etc., the other programs will continue to function normally. Well it should do in theory!

In Windows 95/98/ME/XP, all applications are pre-emptively multitasked including Win16/32 and DOS applications. A hung program can be closed by simply pressing Ctrl Alt Del and closing the offending program.

Windows 95 is not as crash proof as later versions of Windows because part of the system code is based on the Windows 3.1 system. The OS cannot be called by another program before the currently tasked Win3.1 program has completed. If this hangs for any reason, then the system itself will probably lock up.

Windows 95 was superseded by Windows 98 and that in turn has been superseded by Windows ME and Windows XP. Windows 2000 and the later implementations – Windows XP Professional and Windows XP Home, will be the dominant OSs over the first few years of the new millennium. Windows 2000 and Windows XP is not Windows 95/98/ME and it is not Windows NT. It is a hybrid using parts of Windows 98 and NT. The key feature is it incorporates the user-friendliness of Windows 98 with the stability of NT. Current user reports on Windows XP suggest that it really is as good as it is hyped up to be that is, a stable, user friendly, true multitasking system ideal for the home and business user alike.

# Windows 2000/XP architecture

Windows 2000/XP is the most stable, user friendly all singing all dancing Microsoft Operating System you can buy. It differs from Windows 95/98/ME in several major ways. Although the core of Windows 2000/XP is based on Windows NT there are some major changes and enhancements. Microsoft have seemingly done the impossible by integrating the user friendly Windows 98 GUI environment, to a stable NT core with some major changes and enhancements. Apart from increased security and advanced networking features it forces application programs and drivers to be Windows 2000/XP compliant by issuing certificates – any non-compliant drivers or programs will be rejected. In Windows 95/98/ME, protected mode and real mode programs can run side by side and if a real mode program crashes it can hang the system. Windows 2000/XP will only run protected mode programs. This all bodes well in maintaining a stable system.

| Win2000/XP | Win95/98/ME |
| --- | --- |
| Runs in protected mode only | Runs in protected mode and real mode |
| Multithreaded OS | Single threaded OS |
| Full user level security | Limited password security |
| HAL using plug and play | Plug and play |

The full architecture of Windows 2000/XP is quite complex and really only of use to system programmers, so here we will use a simplified model. We can think of the full system architecture being composed of two modes. These are called User Mode and Kernel Mode by Microsoft. These are both made up of several smaller components that each work together to perform a whole variety of different tasks. Despite the obvious complexity, like all OSs it ultimately links user applications to the underlying BIOS and system hardware. How it does it however is quite different to Windows 9* and DOS.

Unlike its Windows 9* predecessors, Windows 2000/XP is not a monolithic OS. It is composed of a collection of purpose built modules which each focus on a specific set of tasks. This makes the system very versatile and easy to adapt and upgrade without the usual interaction between components. One major innovation – first seen in Windows NT – is the (HAL). This is a major component in the Kernel that links the rest of the OS to the BIOS and hardware. As it is modular, it can be replaced with a different version to link the OS to other CPU architectures outside the Pentium X86 series. Currently Microsoft have made an agreement with Intel to keep Windows 2000/XP in the X86 realm – they originally specified that Windows 2000 would run on X86 and Alpha systems.

### User Mode

This is the application execution area where all user interactions are handled. This forms the outer ring of protection keeping user actions and applications programs away from the more sensitive kernel. Software in user mode operates on a non-privileged basis with no direct access to the BIOS or hardware (Figure 5.55).

### Kernel Mode

Kernel Mode is the inner ring where system operations occur. These are a group of highly privileged programs that have direct control over the operating environment. They are written to adhere to the strict protected mode requirement of the underlying CPU architecture accessed via the HAL. This two-tier structure isolates rogue programs, drivers, etc. from crashing the system.

The modules in Kernel Mode can be simplified into a few groups as follows:

- *Executive Services*. This is the interface layer for components in user mode and the underlying kernel mode. It includes inter-process

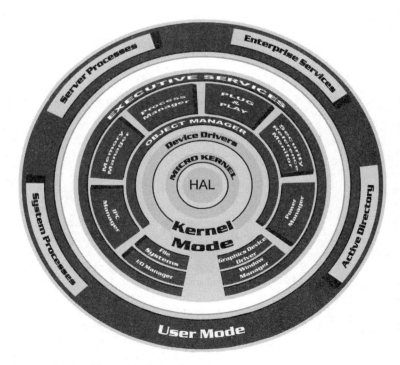

**Figure 5.55**   *The Win2000 architecture*

functions, I/O, file management, plug and play management, power management, communications management and virtual memory management.

- *Device Drivers.* These programs interface general I/O functions and multimedia functions, e.g. sound, video, to the hardware via HAL.
- *HAL.* This is the platform specific component of the system. This layer interfaces the rest of the OS to the BIOS and Hardware effectively isolating and protecting it. It also allows the OS to be easily adapted to other CPU architectures.
- *Microkernel.* This is like the control centre of the whole OS as it manages CPU execution including hardware, interrupts and I/O.

## Windows 2000/XP memory architecture

The memory architecture of Windows 2000/XP is much simpler than that implemented on DOS systems. It utilizes a linear 32-bit contiguous addressing system and does away with the legacy, conventional, upper, extended and expanded memory modes.[11] This is possible because modern CPUs have 32 bit registers and address buses. The old 8088/86 CPUs around at the launch of MS-DOS used 16-bit registers and a 20-bit address bus. To provide downward support for DOS applications, Windows 2000/XP uses a virtual DOS machine that supports this legacy memory structure.

With 32 bits the Windows system can manage memory up to 4 GB and this can be made up of physical RAM and Virtual RAM. Virtual RAM is simply a portion of the hard disk space treated as RAM. When Virtual RAM is being used, the disk area is called the **paging file**. The Virtual Memory Manager moves data a page at a time to the paging file. A page is simply a 4-KB chunk of data. The main draw back of virtual RAM is its operating speed which is around a hundred times slower than normal RAM.

## Power management in Windows 2000/XP

Windows 2000 and ME are new OSs that support the *Advanced Configuration and Power Interface (ACPI)* specification proposed by Compaq, Intel, Microsoft. Phoenix and Toshiba. It specifies how software and hardware components like OS, motherboard and peripherals, manage power usage. Its goal is to integrate power management and plug and play into the OS. Previous BIOS managed power management and plug and play systems proved to be less than ideal. In fact early plug and play implementations were referred to as 'plug and pray' by the fraternity.

The primary goals of ACPI are:

- To be OS independent
- Allow the OS to control all motherboard and peripheral components
- Flexible architecture
- Legacy power management support for non-ACPI systems
- Cost effective implementation

[11] Information on MS-DOS can be obtained from Appendix E on the Elsevier website at http://books.elsevier.com/companions/0750660740.

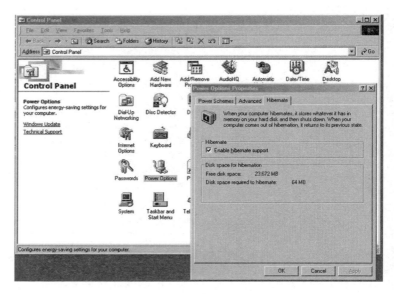

**Figure 5.56**   *The ACPI provides some impressive power saving options*

On a compatible motherboard – one with a BIOS supporting ACPI – Windows 2000 and ME offers some impressive power management features (Figure 5.56).

Apart from a selection of different power down modes and times, Windows 2000 and Windows ME and XP has a *Hibernate* mode. This uses a portion of the hard disk space to save the existing memory contents. When the system is switched on after hibernating it starts up where it left off. This is obviously very useful for portable systems.

### Windows 2000 versions

- *Windows 2000 Professional*. Is a stand alone or workstation[12] version.
- *Windows 2000 Server*. This version is intended for use as a server in departmental and office client server networks.[13] It supports up to four CPUs.
- *Windows 2000 Advanced Server*. For large centralized enterprise networking applications with high data processing requirements. It supports up to eight CPUs.
- *Windows 2000 Datacenter*. Intended for very large data warehouse applications it offers support for up to 32 CPUs.

### Windows XP/2003 versions

- Windows XP (2002) Home Edition
- Windows XP (2002) Professional
- Windows XP (2002) Tablet PC Edition
- Windows XP (2002) Media Center Edition
- Windows XP (2002) 64-Bit Edition
- Windows 2003 Web Server
- Windows 2003 Standard Server
- Windows 2003 Enterprise Server

[12] A workstation is one of a group of computers used in a network system.
[13] Networks are covered in Unit 4 of this book.

- Windows 2003 Enterprise Server 64-Bit
- Windows 2003 Datacenter Server

From a networking standpoint Windows 2000/XP Server is quite different to the earlier Windows NT4 Server edition. Even someone qualified in the use of NT4 Server will find they need to revise their knowledge base. The Active Directory in Windows 2000/XP user mode uses the Domain Name System (DNS). Therefore a good knowledge of TCP/IP is essential for any network administrator planning to use Windows 2000/XP Server. This is a subject in its own right and will not be discussed in detail on this book.

Windows 2000 and XP can operate as a workstation in a client/ server network or in a peer to peer network. It can also quite happily integrate into an existing Windows 95/98/ME/XP peer to peer network based on NETBEUI using file and printer sharing.

Windows XP has been in use now in for several months and it has proved a major success. Reports from all over the world suggest that it is indeed a very stable and reliable OS. With its user friendly environment, scalability and multitude of new features, many users regard it as the best OS from the Microsoft stable.

The few disadvantages reported seem to stem from users installing it on old hardware or attempting to add old hardware to existing Windows 2000/XP systems. This is not the fault of the OS, in fact it's a blessing, as it only willingly accepts certified drivers that operate correctly with HAL, thus maintaining system integrity. This 'driver certification' policy encourages device manufacturers to write stable, well supported drivers.

# Part 3 Systems support

# Unit 6 Routine maintenance and troubleshooting

*Analyse the symptoms to isolate the problem*

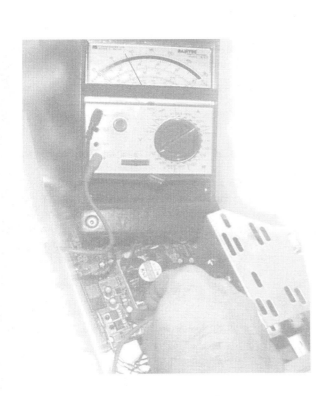

# Troubleshooting

To the uninitiated, troubleshooting today's PC systems must seem a bit like trying to find a lost pearl on a pea-pebble beach. You can probably see why if you have ever taken a peek at one of those massive 3 in. thick books on PC troubleshooting. You begin to think that the only way to diagnose a problem is to know every pebble, right? Wrong!

With the right knowledge and skills to logically analyse the situation, diagnosing and curing problems is based much less on remembering every fault condition and more on using your own skill and judgement.

To attempt serious trouble shooting it is important to have a good working knowledge of PC technology (*You are actively achieving this right now!*). You also need to know the correct techniques to diagnose problems and put them right. Troubleshooting can be summarized in three steps as shown below:

| Steps | What to do | What you must know to do it |
|-------|-----------|------------------------------|
| 1 | Analyse the symptoms and isolate the problem | What symptoms to check and how to check them methodically to isolate the problem |
| 2 | Rectify the problem | How to repair, replace, refit, reinstall, and upgrade the system, to fix the problem |
| 3 | Reduce the chance of the same problem re-occurring | When and where to look for underlying causes or potential causes of the problem. Know how to help prevent related faults from occurring on repaired systems (Figure 6.1) |

## Analyse the symptoms and isolate the problem

To become proficient in the first of the three troubleshooting steps (analysing the symptoms), you need:

(i) practical experience in the use of diagnostics methods and equipment;
(ii) a good working knowledge of the PC system hardware and operating systems (OSs).

If you have read up to this point in the book you are some way to achieving the latter.

Some parts of the system including, the BIOS, CMOS setup, plug and play (PnP), the POST and the Registry, have a direct role to play in making sure the system is configured correctly, they also help in diagnosing and sometimes rectifying problems once they occur. These important elements of the PC system were purposely only covered briefly in previous units of the book, in this unit you will expand your knowledge of them further.

### Rectify the problem

To be able to rectify problems quickly and efficiently requires:

(i) a good working knowledge of the PC system;
(ii) practical experience in handling system parts;

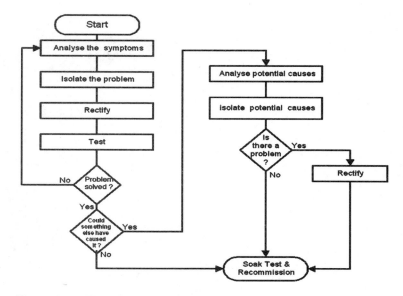

**Figure 6.1** *The troubleshooting process*

(iii)  a working knowledge of the OS;
(iv)  experience in using the right tools and equipment.

## Reduce the chance of the same problem occurring

This is the area that separates the novice from the professional. Knowing when and where to look for underlying causes of a particular fault and not just at the fault itself is the pièce de résistance of good troubleshooting. To develop this skill fully, requires a lot of practice. However you can apply this kind of reasoning to quite basic faults.

For example: if a CPU fails, do not just replace it and re-commission the system. First ask yourself, 'what could have caused it to fail?' 'Are the voltages and clock frequency settings correct?' 'Do they measure correctly?' 'Is the fan working properly or does it stick sometimes?' You should check these factors and correct them if necessary. It is also good practice to leave the system switched on and working for at least a day and then recheck it. This is known as 'Soak Testing' in the trade.

We will now look in more detail at the POST, the ROM BIOS, CMOS setup, PnP, and the registry, to increase your knowledge of the PC further, before we investigate good diagnostics techniques ...

## POST

Every time a PC is powered-up or reset, it performs the power on self-test or 'POST' as it is called in the fraternity. The POST is a set of machine code routines stored in the BIOS ROM. These routines carry out tests on the major parts of the system hardware to confirm their integrity. It is important to realize that the POST is not part of the BIOS itself, but it is closely linked to it. A POST routine may

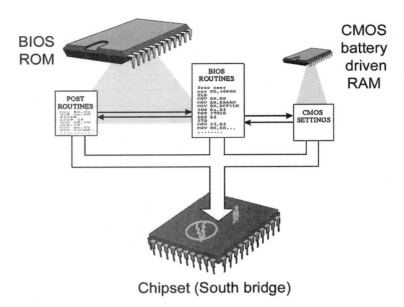

**Figure 6.2**    *The relationship between the POST, BIOS and CMOS setup*

call a specific BIOS routine to test the underlying hardware or it may do it directly. For the POST to work, the PSU, CPU, ROM and the first few kilobytes of RAM must be working. If one of these is faulty, then the POST will not function. This stands to reason, as these components are needed to execute the instructions making up the POST program. It must also have a small area of memory available as a 'scratch pad.'

The POST is inextricably tied up with the BIOS – they are both stored in the same ROM chip – so each BIOS ROM manufacturer will have their own specific POST error checking and coding system (Figure 6.2).

As we stated in an earlier unit, the POST is a suite of machine code routines stored in the ROM that initialize and configure the PCs hardware when the PC is powered up or Reset. Here are some of tasks performed by the POST:

- Tests the RAM.
- Performs an inventory of the hardware devices installed in the system.
- Configures the chipset and expansion buses including PCI, ISA, AGP and AMR slots if fitted.
- Configures the hard disk/s and floppy disk/s.
- Configures the keyboard, graphics card and the I/O ports including COM/LPT1/USB and PS2.
- Configures the CD-ROM drives and sound card.
- Initializes the PnP hardware.
- Initializes the power management hardware.
- Prompts on-screen, for the CMOS setup key or key combination.
- Refers to the Windows registry if present, for configuration data.
- Searches for a boot sector on the current bootable device (i.e. to install the OS). This routine is known as the *Boot loader*.

*Note*: These are not necessarily executed in this order.

## Post error checking cards

On a standard PC, prior to testing each section of the system hardware, the POST sends a unique two digit hexadecimal code called a checkpoint code to port 80H (ISA slot). The code is stored on the port until the next section is tested. If a fault is encountered, the POST attempts to send an audible beep code or an on-screen error message. Sometimes a fault is so serious that the BIOS routine hangs before it has send out an error signal. In this situation port 80H still holds the checkpoint code of the last item tested, so the faulty section can still be isolated.

To make full use of the error codes sent to port 80H, special POST diagnostics cards are available. These plug into the ISA and PCI bus slots of the faulty system. If a problem is encountered during the POST, the error code is conveniently shown in hexadecimal on the card's seven segment display. This can then be checked against the manufacturer's list of error codes.

You can obtain a no-nonsense POST error checking card for about £30.00 from several PC suppliers. They are available in ISA and PCI versions. Both versions do the same job.

'Post Probe,' by Micro 2000, is a more elaborate diagnostics card that basically carries out the same function with some clever additional features. It has two 7 segment LED arrays to display the contents of port 80H as well as a switch to select other ports used by more obscure BIOS manufacturers. It also sports many other features such as a built in logic probe and led indicators to give you an instant indication of the working state of the Clock, DMA, Reset and DC supply lines. These latter features can be quite useful (Figures 6.3 and 6.4).

> Some motherboards actually incorporate a form of POST diagnostics using an array of LEDs and a small look-up chart that sticks on the inside of the system unit. Taking this even further, a few motherboard manufacturers are incorporating speech chips into the motherboard to actually tell you when faults occur during the POST.

## Repairing motherboards

In the 1980s when XT and AT motherboards were covered in scores of fairly standard chips, the POST check facility was a real boon for locating faulty components. For example, when a faulty IC was

**Figure 6.3** *The Micro 2000 Post Probe diagnostics card*

**Figure 6.4** *A standard ISA POST diagnostics card*

located, it was easily replaced in the workshop. These components were readily available and relatively inexpensive. Today however the situation is quite different. Apart from the CPU and RAM, the whole of the active hardware on the motherboard is contained in one or two large VLSI chips, known as the 'chipset.' Unfortunately, if there is a fault on one of these chips, you will probably have to replace the whole motherboard. Even if you are a dab hand at desoldering these intricate devices, the total cost of replacing the chip is usually much greater than the cost of a new motherboard.

## Errors on the rest of the system

Apart from locating motherboard faults, the POST is also useful for detecting faults on the RAM, disk drives, keyboard and graphics adapter. Again using the POST check facility any faulty items can be quickly isolated and replaced if necessary.

## Using POST error codes

As mentioned earlier, POST errors that occur before a basic video display is established, are indicated by beep codes. Most BIOS ROMs send some sort of beep signal to the loudspeaker port. They vary from single beeps, to multiple, Morse-Code like, long and short beeps. The Phoenix BIOS beep codes are notable in that they relate directly to the hexadecimal error code displayed on port 80H.

To decode the Phoenix error beep codes, you can either look them up in a beep code chart or you can read them as a hexadecimal error code (see Appendix I). To do the latter, you must be familiar with the binary and hexadecimal numbering system.

For the interested reader, each group of beeps represents the two hexadecimal characters of the POST error code in 8-bit binary format. The 8-bit error code is broken down into four groups of 2-bits. As 0 cannot be distinguished very easily as no beep, one is added to each 2-bit group. So one beep = 00, two beeps = 01, three beeps = 10 and four beeps = 11.

For example: error code 02Ch is 00 10 11 00 in binary and is represented by 1-3-4-1 beeps

*Note*: 'h' signifies a hex number. The dash '–' on the beeps signifies the quiet passage between beeps.

---

### To use the Phoenix beep code

Write down the number of beeps and subtract 1 from each group, for example 1-3-4-1 becomes
   0-2-3-0

Convert each group into the corresponding 2-bit binary number
   00  10  11  00

Place the 2-bit groups together to form two 4-bit binary numbers
   0010  1100

Convert each 4-bit number to Hex
   2Ch

Look up the Hex number in the error code list
   For example, in Phoenix BIOS 4.0, 2Ch signifies RAM failure

---

*Note*: A list of POST error codes for AMI, Award, Phoenix, and the original IBM BIOS ROMs are included in Appendix I.*

# ROM BIOS

The BIOS itself is another group of machine code routines stored in the BIOS ROM. These mini-programs known as *service routines*, perform the important run-time communication link between the OS and the Hardware. In the hierarchy discussed briefly in Unit 2, they form lowest program layer. The next layer up is the OS followed by Applications (Figure 6.5).

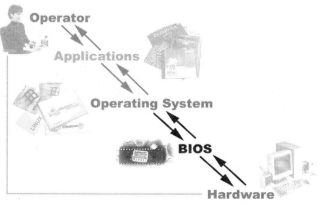

* Appendix I can be accessed via the publisher's website at http://books.elsevier.com/ companions/0750660740

**Figure 6.5** *Where the BIOS fits in the PC's communication hierarchy*

As the hardware and software technology changes more rapidly than the expected life of the motherboard, most BIOS manufacturers provide a facility to upgrade the contents of the BIOS to the latest version, to accommodate any radical changes. This is done using a special Flash ROM upgrade program supplied by the manufacturer. If you are having problems with the BIOS not recognizing a new piece of hardware, for example a large capacity hard drive, then it may prove fruitful to upgrade the BIOS. You should be able to download the latest BIOS ROM upgrade from your motherboard manufacturer's web site. Only do this if it is really necessary and refer to the BIOS manufacturers specific instructions on how to go about the upgrade.

## BIOS service routines

The service routines as we have mentioned briefly, are machine code programs stored in ROM. These important routines allow precise control of all the hardware devices in the system via software interrupt requests from the OS. Incidentally, these routines are generically referred to as *firmware* rather than *software* because the program code is locked electronically into a chip rather than onto magnetic disk.

The BIOS is effectively a collection of crucial routines that interact directly with the PC's input output devices, these include the disk drives, display, serial and parallel port, keyboard, mouse and other motherboard devices.

The OS software then calls the various BIOS routines to effect a communication link to the hardware, as shown in the hierarchy above.

The BIOS routines are accessed via *software interrupts*. A software interrupt is similar to a hardware interrupt in that it is serviced by temporarily interrupting the current program. The difference is that a software interrupt is activated directly in a program using an INT instruction, whereas a hardware interrupt involves an electrical connection to the interrupt controller circuit.

The interaction between advanced OSs such as Windows 95/98/2000/XP and the BIOS/hardware is a very complex one involving several intermediate software layers. To understand the Windows architecture in detail is useful but beyond the remit of current support technician exams (Figure 6.6). For more information

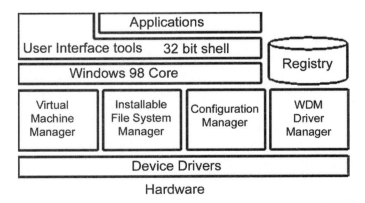

**Figure 6.6**   *Block diagram showing the components of the Windows 98 architecture*

*on the architecture of Windows 95/98/2000/XP you can view sections from the Microsoft Technet site: For example: Windows 98 info http://www.microsoft.com/technet/Win98/Reskit/Part6/wrkc28.asp.*

In the rapidly changing world of PC technology it is essential that the BIOS manufacturers keep pace with hardware and software updates and innovations. For the end user, with an older machine often the only way to get a new device to function is to change the motherboard which includes new hardware facilities and a BIOS to match.

### PnP compatible BIOS

Only a few years ago, to add an adapter card to your system required the use of jumpers and a good working knowledge of IRQs and DMAs. The major downfall of the manual jumper system is that you need to know all the IRQs and DMAs already in use, to avoid resource conflicts. This is still the case with legacy systems and the support technician will often come across these from time to time.

PnP automatically configures the PC hardware and peripherals without the need to manually set jumpers, you just plug in the new device and go – well that is the theory. In practice, sometimes things do not always go as you expect but on the whole PnP on Windows 98/ME/2000 and XP with a modern BIOS to match, works extremely well.

To work properly, PnP must be supported both in hardware and in software. This includes the BIOS, the device itself and the OS. In Windows 9*, PnP is controlled by a component known as the 'Configuration Manager' this utilizes the device drivers, the BIOS and several other Windows components to ensure that each PnP device is configured correctly and without confliction.

As PnP is now used almost exclusively on new adapter cards and peripherals, knowledge of the PnP system is essential for the budding PC technician. We will come back to PnP later.

## CMOS setup

The old IBM XT system used a dip switch arrangement on the motherboard to set basic system parameters, including the *graphics card type*, the *number of floppy disk drives* and the *memory size*. This system was simple and served its purpose reasonably well.

When IBM introduced the AT system, they decided to use a solid state memory device to store system information. Effectively they replaced the mechanical switches with memory cells. The memory device used was a battery powered CMOS SRAM chip. A closed or open switch, is easily represented in RAM by storing a 1 or 0 respectively. Being solid state, this system was more reliable and capable of holding far more information than the XT's mechanical switch system. Using only 64 or sometimes 128 bytes of SRAM, the equivalent of 512 or 1024 switches could be implemented.

Taking the idea one step further, IBM decided to use a real time clock (RTC) chip with a small amount of CMOS memory already built in. This eliminating the need for a separate RAM device. The chip originally used by IBM is the Motorola MC146818, RTC. So for the first time, the PC had its own built in time and date system and a more powerful, easy to use, setup facility. Another nice outcome of the system was that you no longer need to open the case to change parameters.

To allow the contents of the CMOS memory to be changed, early versions of the AT were supplied with a special setup program on floppy disk. Later the CMOS setup program was incorporated on the BIOS ROM.

Today most AT and ATX PCs including notebooks, etc., use the same system but the type and amount of information that is recorded, has changed considerably. To correctly accommodate the year 2000 rollover and the new millennium dates, most motherboards manufactured since early 1996 have a new RTC chip and BIOS supporting these changes.

To access CMOS setup, a prompt usually appears on-screen for a few seconds, just after the POST has completed. This informs the operator to press a certain key or combination of keys, to invoke the Setup routine. The keys used to do this vary from one BIOS ROM manufacturer to another. Some of the common ones are listed below:

| BIOS or PC manufacturer | Key or key combination |
| --- | --- |
| AMI | Del |
| Award | Del alternatively Ctrl Alt Esc |
| Gateway 2000 | F1 |
| Phoenix | Ctrl Alt Esc or Ctrl Alt S |
| Compaq | F10 |
| Hewlett Packard | DEL |

It is important to appreciate that the CMOS RAM, the CMOS setup program, the BIOS and the POST are separate items, however they are quite closely linked. What often tends to confuse people is the words used to describe them.

For example:

The words 'BIOS ROM' or 'ROM BIOS' are used to describe the system ROM. The confusion arises because the system ROM does not just hold the BIOS routines, it also contains the POST routines, the OS Boot Loader program and the CMOS setup program.

The phrase 'CMOS setup' refers to the special program stored on ROM that allows the system setup parameters stored in the CMOS RAM device to be edited. After the POST has executed, it checks to see if the operator executed the right key sequence to invoke the CMOS setup program. If the user has entered the setup program; any changes made can be saved to the battery powered CMOS RAM. This 'remembers' the settings while the PC is switched off.

The role of the battery powered CMOS RAM is to store the system setup information and the time and date. At startup, the POST reads the contents of the CMOS RAM to configure the various motherboard hardware devices. The battery keeps the CMOS RAM active and also keeps the RTC ticking to maintain the correct time and date.

CMOS is an acronym of *complementary metal oxide semiconductor* and refers to a particular circuit technology based on complimentary pairs of MOSFETs. It is notable for its extremely low power consumption when operated at low data rates, as it is with the RTC.

The contents of CMOS RAM for a typical AT computer are shown below. We have included this to give you some idea of what is stored in this tiny memory device. Please do not try to delve to deeply into its contents as there are no assessments on this in the A+ or C&G exams.

The first 14 bytes are dedicated to the RTC functions and consist of 10 read/write data registers and four status registers, two of which are read/write and two of which are read only.

## Typical contents of CMOS RAM for an AT machine

Actual contents of some memory locations vary widely between different BIOS manufacturers.

| Offset (address) | Description | |
| --- | --- | --- |
| 00h | Seconds | |
| 01h | Second alarm | |
| 02h | Minutes | |
| 03h | Minute alarm | |
| 04h | Hours | |
| 05h | Hour alarm | |
| 06h | Day of week | (01-07 Sunday = 1) |
| 07h | Date of month | (01-31) |
| 08h | Month | (01-12) |
| 09h | Year | (00-99) |

0Ah Status Register A

| | Bit 7 | Time update processing flag (i.e. read only) |
| --- | --- | --- |
| | Bit 6,5,4 | Setting for 22 stage frequency divider (32.768 kHz default) |
| | Bit 3-0 | Interrupt rate selection |
| | | 0000b – none |
| | | 0011b – 122 μs (minimum) |
| | | 1111b – 500 μs |
| | | 0110b – 976.562 μs (default) |

0Bh Status Register B (read/write)

| | Bit 7 | 1 enables cycle update, 0 disables |
| --- | --- | --- |
| | Bit 6 | 1 enables periodic interrupt |
| | Bit 5 | 1 enables alarm interrupt |
| | Bit 4 | 1 enables update-ended interrupt |
| | Bit 3 | 1 enables square wave output |
| | Bit 2 | Data mode – 0: BCD, 1: Binary |
| | Bit 1 | 24/12 h selection – 1 enables 24 h mode |
| | Bit 0 | Daylight savings enable/disable – 1 enables, 0 disables |

0Ch Status Register C (read only flags)

0Dh Status Register D (read only)

| | Bit 7 | Valid RAM | 1 indicates battery power good, 0 battery dead |
| --- | --- | --- | --- |

0Eh (PS/2) Diagnostic status byte

| | Bit 7 | When set (1) indicates clock has lost power |
| --- | --- | --- |
| | Bit 6 | 1 indicates incorrect CMOS RAM checksum |

| | | |
|---|---|---|
| | Bit 5 | 1 indicates that CMOS RAM configuration is incorrect |
| | Bit 4 | 1 indicates CMOS RAM size error |
| | Bit 3 | 1 indicates that the controller or hard disk drive failed initialization |
| | Bit 2 | 1 indicates that time status is invalid |
| | Bit 1 | 1 indicates installed adaptors do not match configuration |
| | Bit 0 | 1 indicates a time-out while reading adaptor ID |
| 0Fh | Shutdown status byte | |
| | 00h-03h | perform power-on reset |
| | 04h | INT 19h reboot |
| | 05h | flush keyboard and jump via 40h:67h |
| | 06h-07h | reserved |
| | 08h | used by POST during protected-mode RAM test |
| | 09h | used for INT 15/87h (block move) support |
| | 0Ah | jump via 40h:67h |
| | 0Bh-FFh | perform power-on reset |

10h Floppy drive type

| | |
|---|---|
| Bits 7-4 | first floppy disk drive type (A: drive) |
| 0h | No Drive |
| 01h | 360 kB 5 1/4 Drive |
| 02h | 1.2 MB 5 1/4 Drive |
| 03h | 720 kB 3 1/2 Drive |
| 04h | 1.44 MB 3 1/2 Drive |
| 05h | 2.88 MB 3 1/2 Drive ???) |
| Bits 3-0 | second floppy disk drive type (bit settings same as above) |

11h Used by various BIOS manufacturers for different purposes

12h Hard disk data

| | |
|---|---|
| Bits 7-4 | First hard disk drive |
| 00 | No drive |
| 01-0Eh | Hard drive type 1-14 |
| 0Fh | Hard disk type 16-255 (see actual type @ 1Ah) |
| Bits 3-0 | Second hard disk drive type (see actual type @1Ah) |

13h Used by various BIOS manufacturers for different purposes

14h Equipment byte

| | |
|---|---|
| Bits 7-6 | Number of floppy drives (system must have at least one) |
| 00 | 1 Drive |
| 01 | 2 Drives |
| 10 | 3 Drives |
| 11 | 4 Drives |
| Bits 5-4 monitor type | |
| 00 | Not CGA or MDA (i.e. EGA and VGA) |
| 01 | 40 × 25 CGA |
| 10 | 80 × 25 CGA |

11 MDA (monochrome)

Bit 3 Display enabled (1 = On)

Bit 2 Keyboard enabled (1 = On)

Bit 1 Math coprocessor installed (1 = On)
Bit 0 Floppy drive installed (1 = On)
15h Base memory in kB, low byte
16h Base memory in kB, high byte
17h Extended memory in kB, low byte
18h Extended memory in kB, high byte
19h Hard Disk C: (10h-FFh) First extended hard drive type 16d-255d
1Ah Second extended hard disk drive type (see 19h above)
1Bh to 2Dh Used by various BIOS manufacturers for different purposes
2Eh      Standard CMOS check sum (Most significant byte)
2Fh      Standard CMOS check sum (Least significant byte)
30h      Extended memory in kB, Low Byte
31h      Extended memory in kB, High Byte
32h      Century byte (BCD value for the century, i.e. 20)
33h information flag   (Bit 7 BIOS size 64kB/128kB)
34h-5Dh used by various BIOS manufacturers for different purposes

## CMOS setup parameters

When you enter CMOS setup after building a new PC, or when you are troubleshooting the system, it is important to check and set some of the more important parameters manually rather than relying totally on the default settings.

### Standard CMOS setup

To enter CMOS setup, power up the PC and wait for the POST to finish. This is usually a few seconds after the on-screen RAM check process has completed. You should see a prompt telling you to press a key or key sequence to enter CMOS setup (Figure 6.7).

On entering CMOS setup you are presented with a contents page. From here you can choose other setup pages. The simplest of these is 'Standard CMOS setup'. In the Award setup utility shown here, it is

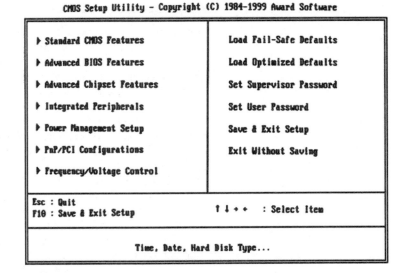

Figure 6.7

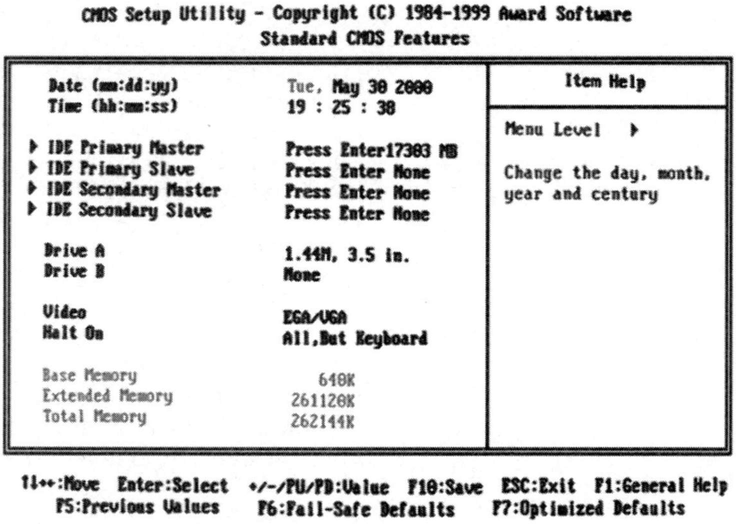

```
          CMOS Setup Utility - Copyright (C) 1984-1999 Award Software
                           Standard CMOS Features
 ┌─────────────────────────────────────────────────┬─────────────────────┐
 │   Date (mm:dd:yy)          Tue, May 30 2000      │      Item Help      │
 │   Time (hh:mm:ss)          19 : 25 : 38          │                     │
 │                                                  │  Menu Level   ▶     │
 │  ▶ IDE Primary Master      Press Enter17303 MB   │                     │
 │  ▶ IDE Primary Slave       Press Enter None      │  Change the day, month, │
 │  ▶ IDE Secondary Master    Press Enter None      │  year and century   │
 │  ▶ IDE Secondary Slave     Press Enter None      │                     │
 │                                                  │                     │
 │    Drive A                 1.44M, 3.5 in.        │                     │
 │    Drive B                 None                  │                     │
 │                                                  │                     │
 │    Video                   EGA/VGA               │                     │
 │    Halt On                 All,But Keyboard      │                     │
 │                                                  │                     │
 │    Base Memory                     640K          │                     │
 │    Extended Memory              261120K          │                     │
 │    Total Memory                 262144K          │                     │
 └─────────────────────────────────────────────────┴─────────────────────┘
   ↑↓→←:Move  Enter:Select  +/-/PU/PD:Value  F10:Save   ESC:Exit  F1:General Help
           F5:Previous Values    F6:Fail-Safe Defaults   F7:Optimized Defaults
```

**Figure 6.8**

entitled 'Standard CMOS features.' Highlight this feature using the mouse cursor or arrow keys, and click the left mouse button or press ENTER. The Standard CMOS setup screen allows you to edit the time and date, IDE hard disk drives, the floppy drives and the video mode.

Modern IDE hard disk drives automatically inform the BIOS of the various drive parameters so there is normally no need to alter these settings. Also do not try to enter your IDE CD-ROM/s in the list or SCSI hard disks.

The video mode setting is a remnant from the early IBM AT. This should always set to EGA/VGA to set a basic graphics standard for the Startup and DOS command mode screens. Once Windows starts to boot, the proper driver for your graphics card will be activated (Figure 6.8).

### Advanced CMOS setup

When you select Advanced CMOS setup (Advanced BIOS features), you will see many more editable features.

Some of the important ones are listed as follows (Figure 6.9):

**Virus warning** If you already have a virus checking program installed in Windows then it is best to disable this feature. It is not really a virus checking utility but a warning if a boot sector write is being attempted.

**CPU internal and external cache** Make sure all caches are enabled, particularly the CPU level 1 cache (internal) and level 2 cache (external).

**Drive boot sequence** This facility sets the order for the first, second and third drive to be searched, for a boot sector. This could temporarily be set by an installation technician, for example, to force the bootloader to boot from CD-ROM, for example, while installing a new OS from CD. For normal operator use it is best to set the primary hard disk drive C: as the first boot device and the A: drive as the second boot device. Setting A: as the first boot device will slow down the startup time slightly.

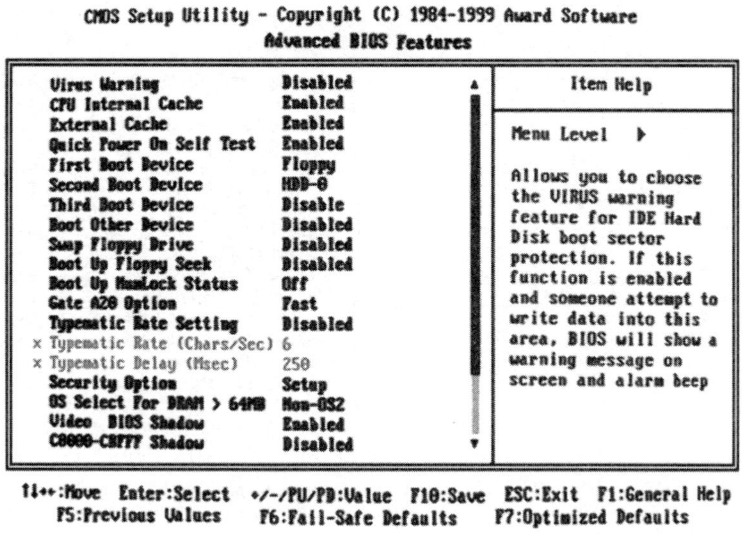

CMOS Setup Utility - Copyright (C) 1984-1999 Award Software
Advanced BIOS Features

| | | |
|---|---|---|
| Virus Warning | Disabled | Item Help |
| CPU Internal Cache | Enabled | |
| External Cache | Enabled | Menu Level ▶ |
| Quick Power On Self Test | Enabled | |
| First Boot Device | Floppy | Allows you to choose |
| Second Boot Device | HDD-0 | the VIRUS warning |
| Third Boot Device | Disable | feature for IDE Hard |
| Boot Other Device | Disabled | Disk boot sector |
| Swap Floppy Drive | Disabled | protection. If this |
| Boot Up Floppy Seek | Disabled | function is enabled |
| Boot Up NumLock Status | Off | and someone attempt to |
| Gate A20 Option | Fast | write data into this |
| Typematic Rate Setting | Disabled | area, BIOS will show a |
| x Typematic Rate (Chars/Sec) | 6 | warning message on |
| x Typematic Delay (Msec) | 250 | screen and alarm beep |
| Security Option | Setup | |
| OS Select For DRAM > 64MB | Non-OS2 | |
| Video BIOS Shadow | Enabled | |
| C8000-CBFFF Shadow | Disabled | |

↑↓→←:Move  Enter:Select  +/-/PU/PD:Value  F10:Save  ESC:Exit  F1:General Help
F5:Previous Values    F6:Fail-Safe Defaults    F7:Optimized Defaults

**Figure 6.9**

**Boot up floppy seek**   Best left as *disabled* otherwise it will take longer for the system to boot at power up.

**Gate A20 option**   Best left as *fast*. The CPU address lines A0 to A19 address the first 1 MB of memory (the address range of the old 8086/8). Gate A20 refers to the way the system addresses memory above 1 MB (extended memory). When set to fast, the motherboard chipset controls Gate A20. When set to Normal, the keyboard controller controls Gate A20 as in the original AT design. So setting this to fast will improve the system speed slightly when operating in Windows.

**Security option**   Allows you to choose to add password protection, to enter setup or to boot the PC. Passwords generally are not recommended as people inevitably forget them or loose contact with the person who originated the password. You may come across a PC where the manufacturers default password has been implemented. The AMI default password is 'AMI.' Award's default password is 'Award_SW' or 'BIOSTAR'

**Video BIOS shadow**   When enabled it copies the contents of the graphics card ROM to RAM. This can speed up video access times especially with older graphics cards incorporating an EPROM rather than a Flash ROM. Most modern high performance graphics cards especially AGP types, do not need this facility – in fact they might even run more slowly. Refer to the card manufacturers instructions to make sure. *Shadowing* refers to the process of copying the contents of a slower memory device like an EPROM to faster Main Memory. For a further speed increase, the RAM used as a shadow can be cached via the Level 2 SRAM cache.

## Advanced chipset features

**Video Ram cache**   When enabled it allows the contents of the graphic card's video RAM to be cached. Modern graphics cards use VRAM WRAM, SGRAM, or SDRAM and therefore will probably not benefit greatly from caching even when the system is used for games playing.

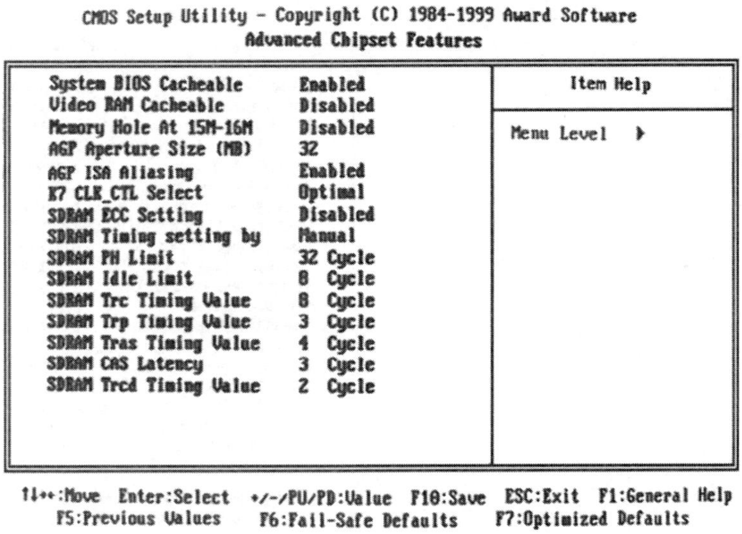

```
        CMOS Setup Utility - Copyright (C) 1984-1999 Award Software
                        Advanced Chipset Features
 ┌──────────────────────────────────────────────┬──────────────────┐
 │  System BIOS Cacheable       Enabled          │     Item Help    │
 │  Video RAM Cacheable         Disabled         ├──────────────────┤
 │  Memory Hole At 15M-16M      Disabled         │                  │
 │  AGP Aperture Size (MB)      32               │  Menu Level   ▶  │
 │  AGP ISA Aliasing            Enabled          │                  │
 │  X7 CLK_CTL Select           Optimal          │                  │
 │  SDRAM ECC Setting           Disabled         │                  │
 │  SDRAM Timing setting by     Manual           │                  │
 │  SDRAM PH Limit              32 Cycle         │                  │
 │  SDRAM Idle Limit            8  Cycle         │                  │
 │  SDRAM Trc Timing Value      8  Cycle         │                  │
 │  SDRAM Trp Timing Value      3  Cycle         │                  │
 │  SDRAM Tras Timing Value     4  Cycle         │                  │
 │  SDRAM CAS Latency           3  Cycle         │                  │
 │  SDRAM Trcd Timing Value     2  Cycle         │                  │
 │                                               │                  │
 │                                               │                  │
 └──────────────────────────────────────────────┴──────────────────┘
   ↑↓←→:Move  Enter:Select  +/-/PU/PD:Value  F10:Save  ESC:Exit  F1:General Help
        F5:Previous Values     F6:Fail-Safe Defaults    F7:Optimized Defaults
```

**Figure 6.10**

**BIOS ROM shadow**   When enabled it copies the contents of the system ROM to RAM. This can speed up some operations especially when the ROM is an EPROM rather than a modern Flash ROM (Figure 6.10).

**Memory hole at 15–16 M**   Best left *disabled*. This can be used to shadow ISA adapter card ROMs.

**AGP aperture size**   If you have an Accelerated Graphics Port (AGP card) then set this size to the Windows 9* default of 64 MB. The AGP is a clocked at 66 MHz – twice that of the generic PCI clock frequency. It also transfers data on the rising and falling edges of the clock thus achieving an incredibly fast data transfer rate as required in modern 3D textured scrolling games such as 'Tomb Raider' and 'Quake.'

**SDRAM timing**   Unless you have the full specification data for your installed RAM devices – you can usually find this from the manufacturers web site – choose the system ROM manufacturer's default settings. You could also try reducing the timings gradually and then testing the system for reliable operation but this is far to hit or miss and time consuming. The difference in performance is often unnoticeable anyway.

### Integrated peripherals

**IDE read/write prefetch**   Set this to enabled for faster drive accesses.

**IDE PIO settings**   The four IDE programmed input/output (PIO) settings shown opposite, can be set manually to PIO modes (0–4) for each of the four IDE drives supported. Modes 4 offers the highest performance. When set to Auto, the system automatically chooses the best mode for each device (Figure 6.11).

**IDE UDMA settings**   UDMA (Ultra DMA 33 or 66) is a direct memory access (DMA) data transfer protocol that utilizes ATA2/3 commands to transfer data at a maximum burst rate of 33 or 66 MB/s. It is best to leave this setting on auto. This will allow the

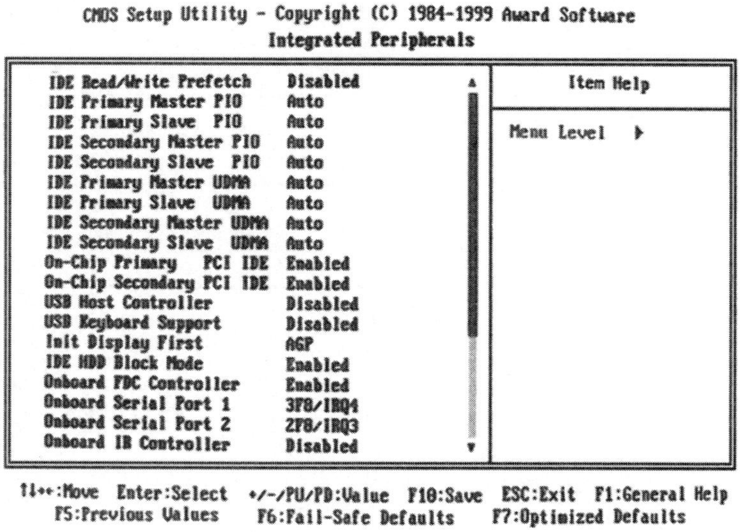

```
CMOS Setup Utility - Copyright (C) 1984-1999 Award Software
                    Integrated Peripherals
```

```
┌─────────────────────────────────────────────────┬──────────────────┐
│ IDE Read/Write Prefetch    Disabled          ▲  │    Item Help     │
│ IDE Primary Master PIO     Auto                 ├──────────────────┤
│ IDE Primary Slave  PIO     Auto                 │                  │
│ IDE Secondary Master PIO   Auto                 │ Menu Level    ▶  │
│ IDE Secondary Slave  PIO   Auto                 │                  │
│ IDE Primary Master UDMA    Auto                 │                  │
│ IDE Primary Slave  UDMA    Auto                 │                  │
│ IDE Secondary Master UDMA  Auto                 │                  │
│ IDE Secondary Slave  UDMA  Auto                 │                  │
│ On-Chip Primary   PCI IDE  Enabled              │                  │
│ On-Chip Secondary PCI IDE  Enabled              │                  │
│ USB Host Controller        Disabled             │                  │
│ USB Keyboard Support       Disabled             │                  │
│ Init Display First         AGP                  │                  │
│ IDE HDD Block Mode         Enabled              │                  │
│ Onboard FDC Controller     Enabled              │                  │
│ Onboard Serial Port 1      3F8/IRQ4             │                  │
│ Onboard Serial Port 2      2F8/IRQ3             │                  │
│ Onboard IR Controller      Disabled          ▼  │                  │
└─────────────────────────────────────────────────┴──────────────────┘
```

```
↑↓→←:Move  Enter:Select  +/-/PU/PD:Value  F10:Save  ESC:Exit  F1:General Help
    F5:Previous Values     F6:Fail-Safe Defaults   F7:Optimized Defaults
```

**Figure 6.11**

system to automatically select the optimal data transfer rate for the IDE devices installed.

**On chip primary/secondary IDE**  Leave as enabled unless you want to install an IDE adapter card in place of the built in interface.

**USB host controller**  Select enabled if the motherboard has a Universal Serial Bus (USB) controller and you want to install USB devices.

**USB keyboard support**  Select enabled if you have installed a Universal Serial Bus (USB) keyboard.

**Init display first**  If an AGP card is installed rather than an ISA or PCI graphics card, then select AGP so that this type of display is initialized first, that is set AGP as the primary display.

**IDE HDD block mode**  Most ATA2 & ATA3 IDE hard disk drives support block mode, so select enabled to allow the system to automatically detect the optimal number of block read/writes per sector for the drive.

**On board FDC controller**  Normally set to enabled to use the floppy disk controller (FDC) installed on the motherboard. If an FDC adapter card is being installed or the system has no floppy drive, select disabled. You might want to do this to overcome a fault on the built in FDC.

**On board IR controller**  Select disabled unless an infra red port is available on the motherboard and is being used.

**On board serial port 1 & 2**  This is best left on Auto, but you can manually select a logical COM port name, IRQ and matching address for the first and second serial ports, or they can be disabled. Each serial port must have a different port address and IRQ selected from the following:

| Port address | IRQ |
|--------------|------|
| 3E8 | IRQ4 |
| 3F8 | IRQ4 |
| 2F8 | IRQ3 |
| 2E8 | IRQ3 |

**On board parallel port**   This is best left on Auto, but you can alter the logical LPT port address and corresponding interrupts or disable the port if necessary. The parallel port address and IRQ can usually be selected from the following:

| Port address | IRQ |
| --- | --- |
| 278 | IRQ5 |
| 378 | IRQ7 |
| 3B7 | IRQ7 |

**Parallel port mode**   Modern parallel port interfaces now conform to the recent (1994) IEEE 1284 standard. This standard defines five operation modes, including the standard or compatibility mode. The other modes are *Nibble, Byte, Enhanced Capabilities Port (ECP)* and *Enhanced Parallel Port (EPP)*. Unless the installation instructions of the attached parallel peripheral device (usually a printer), states otherwise, it is wise to select Normal, Compatible, or SPP mode.

### Power management

**ACPI function** (*Advanced Configuration Power Interface*) Selecting enabled enables the ACPI function.
**Power management**   This option allows you to disable power management or set it to minimum power saving or maximum power saving using the *Doze, Standby and Suspend modes* (Figure 6.12).
**Doze**   After the user selected period of system inactivity, the CPU clock runs at slower speed while all other devices still operate at full speed. When the system enters Doze mode, you can select using the *Throttle Duty Cycle* field, the percent of time that the CPU clock runs.
**Standby**   After the selected period of system inactivity, the fixed disk drive and the video shut off while all other devices still operate at full speed.
**Suspend**   After the user selected period of system inactivity, all devices except the CPU shut off.

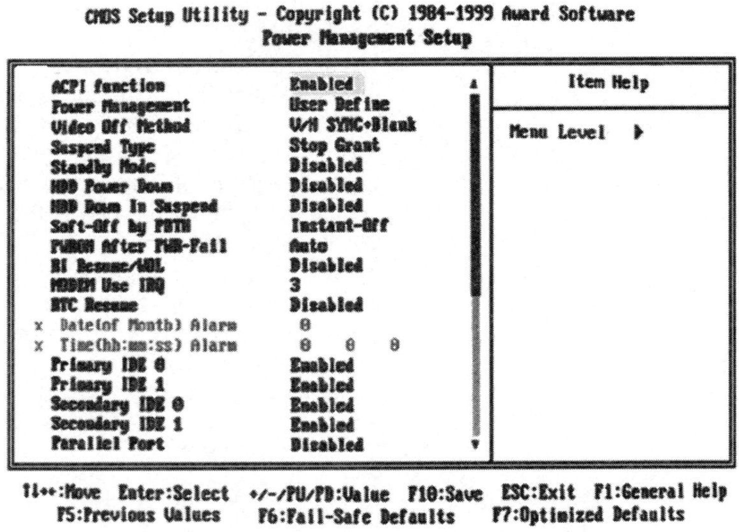

CMOS Setup Utility - Copyright (C) 1984-1999 Award Software
Power Management Setup

| | | Item Help |
| --- | --- | --- |
| ACPI function | Enabled | |
| Power Management | User Define | |
| Video Off Method | V/H SYNC+Blank | Menu Level ▶ |
| Suspend Type | Stop Grant | |
| Standby Mode | Disabled | |
| HDD Power Down | Disabled | |
| HDD Down In Suspend | Disabled | |
| Soft-Off by PWRN | Instant-Off | |
| PWRON After PWR-Fail | Auto | |
| RI Resume/WOL | Disabled | |
| MODEM Use IRQ | 3 | |
| RTC Resume | Disabled | |
| x Date(of Month) Alarm | 0 | |
| x Time(hh:mm:ss) Alarm | 0  0  0 | |
| Primary IDE 0 | Enabled | |
| Primary IDE 1 | Enabled | |
| Secondary IDE 0 | Enabled | |
| Secondary IDE 1 | Enabled | |
| Parallel Port | Disabled | |

↑↓←→:Move  Enter:Select  +/-/PU/PD:Value  F10:Save  ESC:Exit  F1:General Help
F5:Previous Values    F6:Fail-Safe Defaults    F7:Optimized Defaults

**Figure 6.12**

**Hard drive power down**   After the selected period of drive inactivity, the hard disk drive will power down while all other devices remain active.

The power management settings you choose will depend on individual circumstances but some scale of power management is advisable. Considering the number of computers in daily use in the UK alone, we should collectively show more concern about power conservation. The reduction in wear and tear on the system components by allowing them to power down when inactive for a set duration, is a worthwhile gain let alone the saving in the electricity bill.

### BIOS PnP setting

**PnP OS installed**   If running a PnP compatible OS such as Windows 9* and Windows 2000 then select yes.
**PCI/VGA palette snoop**   Only enable this feature if you have an MPEG video device such as a video capture card, connected to your video card's feature connector. It allows the device to look ahead and read the colour palette currently used by the graphics card. Refer to the card manufacturer's instructions (Figure 6.13).

### CMOS setup default settings

When you are troubleshooting or when you suspect that the CMOS settings are causing instability problems, try reverting to the CMOS setup default settings. Often the default settings are the best compromise between an over-tuned and temperamental system and a poorly set system that under-performs.

### Backing up CMOS RAM

If you have taken the trouble to optimize CMOS setup it makes a lot of sense to keep a record of the contents of CMOS setup either by writing them down on a sheet of paper or by saving the settings onto

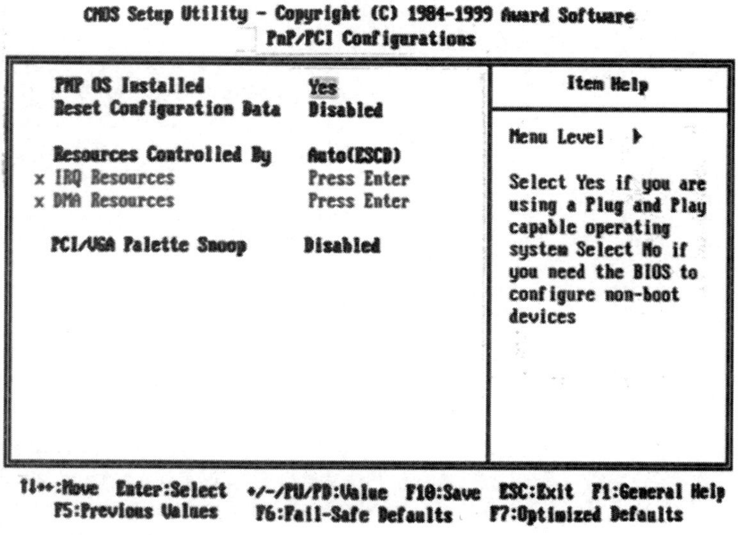

```
      CMOS Setup Utility - Copyright (C) 1984-1999 Award Software
                      PnP/PCI Configurations
┌──────────────────────────────────────────┬─────────────────────┐
│  PNP OS Installed          Yes            │     Item Help       │
│  Reset Configuration Data  Disabled       │                     │
│                                           ├─────────────────────┤
│  Resources Controlled By   Auto(ESCD)     │ Menu Level  ▶       │
│ x IRQ Resources            Press Enter    │                     │
│ x DMA Resources            Press Enter    │ Select Yes if you are│
│                                           │ using a Plug and Play│
│  PCI/VGA Palette Snoop     Disabled       │ capable operating   │
│                                           │ system Select No if │
│                                           │ you need the BIOS to│
│                                           │ configure non-boot  │
│                                           │ devices             │
│                                           │                     │
│                                           │                     │
│                                           │                     │
└──────────────────────────────────────────┴─────────────────────┘
  ↑↓→←:Move  Enter:Select  +/-/PU/PD:Value  F10:Save  ESC:Exit  F1:General Help
       F5:Previous Values    F6:Fail-Safe Defaults   F7:Optimized Defaults
```

**Figure 6.13**

a floppy disk using a suitable utility program. The record in whatever form, should be kept safe along with the PCs other bits and pieces. There are several freeware/shareware programs available, such as *CMOS_RAM2*, *CMOS.COM* and *BIOS* (BIOS1310.zip) that allow you to save the contents of CMOS setup to disk.

### Problem with passwords

As a PC trouble shooter you are bound to come across CMOS setups that have been password protected. Often the password has been lost or forgotten. On AMI and Award CMOS setup, the password is needed to enter the Setup menus. Other CMOS setup routines only need the password when you try to edit the settings. So at least you can note down the existing settings.

The freeware program 'BIOS' mentioned above, can in most cases read AMI and Award passwords. If this does not work, then you will have to disconnect the CMOS RAM battery for a few seconds to delete CMOS RAM. Then you can restore the battery and enter CMOS setup. It will now contain the default settings so you will have to manually set the time/date/and any other parameters you are sure of.

## PnP

PnP is not just confined to a particular bus such as PCI, it will work on a variety of technologies including ISA, VESA, PCI, USB, PCMCIA, SCSI, IDE, AMR, AGP, COM and LPT ports, as long as the manufacturer has produced a compatible PnP driver. Systems manufactured since 1995 are most likely to have PnP a compliant BIOS and chipset.

For a device to be PnP ready it must conform to three requirements:

- It must be capable of being uniquely identified.
- It must state its function and the resources it requires.
- It must be software configurable.

Once all the PnP devices are installed the PnP system must do the following:

- Allocate the required resources to each device every time the system initializes without creating a resource conflict.
- Load the device drivers needed to support each PnP device fitted.
- Notify any changes to the installed set of PnP devices. (When a new device is fitted it must try to identify it and install the appropriate drivers. Also when a device is removed, it must attempt to remove all trace of the device and its drivers.)

You know from an earlier unit, that there are only 16 Interrupts and 8 DMA channels available on a PC AT/ATX system. Of these, two are lost, as IRQ2 is vectored to IRQ9 and DMA 4 is used to link two DMA controllers chips, to provide more channels. So you may be wondering how PnP manages to assign so many individual devices to so few channels. One factor is the fact that the normal DMA channels are not used very much nowadays. This is because DMA is slower than CPU controlled data transfers.

This seems rather contradictory, as you learnt in Unit 1 that DMA was devised to relinquish the already overworked CPU from performing data transfers from a fast I/O device to memory. DMA forces the CPU to let go of the data and address bus, so that control could momentarily be handed over to the faster peripheral device transferring the data. This worked fine when CPUs clocked at 4.77 MHz and the ISA bus clocked at 8 MHz. Today the situation has changed dramatically. CPU speeds are now hundreds of times what they were then. So DMA is now a lot slower at performing block data transfers than the CPU itself. Due to this, DMA is only used where a relatively slow transfer rate is tolerable, that is the Floppy disk and Sound card for example. You can see this in the examples below.

## Self-assessment practical (SAP 6.1)

You can see the DMAs actually used on your Windows system.

| To do this in Windows 95/98/ME | To do this in Windows XP |
| --- | --- |
| **\| Start \| Settings \| Control Panel \| System \|** Device Manager \| Properties \| View Resources \| Finally select the Direct memory access (DMA) radio button\| | **\| Start \| Control Panel \| System \|** Device Manager \| View \| Resources by Type \| Direct memory access (DMA) (Figure 6.14) |

As you can see in in figure 6.14 below, the devices actually using DMA are slow peripherals where speed is not an issue.

That fine as far as the DMA channels are concerned but the short-age of IRQ channels poses a more serious problem because most devices need one. The other problem is that any legacy ISA devices fitted in the system demand their own unique IRQ. This is due to the design of the ISA bus system. Each ISA slot is not separately identi-fied from another slot. So the IRQ problem cannot be worked around logically by saying for example, *that the device in Slot 1 is using IRQ7 so ignore for the moment the other device using IRQ7 in Slot 2.*

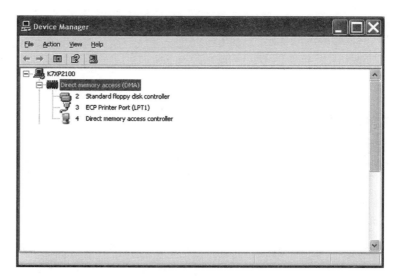

**Figure 6.14** *Typical list of devices using normal DMA modes*

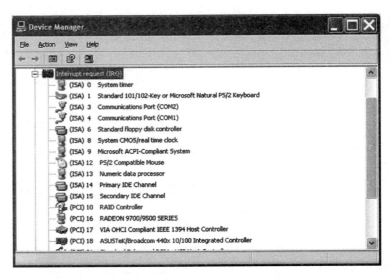

**Figure 6.15** *IRQ allocation of a typical Windows XP system*

The good news is that most modern expansion bus systems do allow this sort of thing. The PCI bus is one of them. On the PCI bus and other advanced buses there is more control over the allocation of IRQs. For example the PCI bus can use a technique called *Interrupt Request Steering*. This allows several devices to use the same IRQ.

## Self-assessment practical (SAP 6.2)

Again, you can see this for yourself in your Windows 95/98 system.

| To do this in Windows 95/98/ME | To do this in Windows XP |
| --- | --- |
| \| Start \| Settings \| Control Panel \| System \| Device Manager \| Properties \| View Resources \|. Finally select the Interrupt Request (IRQ) radio button | \| Start \| Control Panel \| System \| \| Device Manager \| View \| Resources by Type \| Direct memory access (DMA) (Figure 6.15) |

As you can see in our example many system devices require interrupts.

## Installing devices in Windows

### Installing legacy devices

ISA legacy cards are cards that need to be manually configured prior to being inserted in an ISA slot. Usually the available IRQ, DMA and I/O address settings on these cards are configured via jumpers (U-links). Some cards provide a DIP switch instead of jumpers. Since these legacy devices do not communicate their settings to the system, the PnP system cannot detect them. So you have to manually inform Windows that they exist using the *Add New Hardware* wizard. To add a legacy ISA card carry out the following steps:

1. Power down the system and remove the covers. If this is the only ISA card in the system, set the jumpers to the default settings

    recommended by the manufacturer. If there are other ISA cards fitted, choose settings that are not already in use.
2. Insert the card into a spare ISA slot.
3. Reboot the system.
4. Click *Start*, highlight *Settings* and click *Control Panel*, then double click the *Add New Hardware* Icon.
5. In the *Add New Hardware* wizard, click *Next* twice to allow for device detection and then if your device is not listed, as it probably will not be, click the '*No, the device isn't in the list,*' radio button.
6. A new Window appears informing you that, '*Windows can now search for hardware that is not PnP compatible or you can search for hardware from a list.*' Click *Yes* for '*Do you want Windows to search for your new hardware?*' Windows should then detect the device and prompt you to install new drivers.
7. If you have an installation disk provided by the manufacturer, click the '*Have Disk*' button and insert the disk in the appropriate drive. Now choose the drive and the appropriate directory as stated in the device manufacturer's instructions.

If Windows does not successfully detect the new hardware, you will have to do so manually. In Step 6 above, when asked '*Do you want Windows to search for your new hardware?*' choose '*No I want to select the hardware from a list.*' Now choose the relevant hardware from the list.

If you have trouble with IRQs and there are several adapter cards plugged into the ISA bus then one rather drastic solution would be to upgrade some or all of the ISA devices to a PCI version. Quite a few motherboard manufacturers are now ditching the ISA slot altogether in favour of more modern solutions. They now sport several PCI slots an Audio Modem Riser Slot and an AGP slot. This is a trend that will probably continue until ISA becomes obsolescent.

### Installing PnP devices

In the ideal world, all PnP devices including ISA, PCI, AGP, AMR, USB and PMCIA cards and other add-ons should be detected, configured and have their respective drivers assigned, completely automatically. Amazingly in a lot of cases this is exactly what does happen. We will call such an installation *fully automatic*.

In fully automatic installation the device drivers are already recorded in the Windows OS. (Indeed this is one of the major advantages of Windows 2000/XP, in that the bulk of the driver software is built in, allowing device manufacturers to concentrate more on the application software accompanying the device and less on the drivers.)

At present unfortunately, the majority of devices do not fall in the fully automatic category. They do need separate drivers and these are normally supplied on CD-ROM or 3.5 in. floppy disk. We will call this form of PnP installation *semi-automatic*.

To install a PnP device you would naturally follow the manufacturer's instructions so here we will just generalize. The sequence of events during the installation will depend on the version of Windows you have.

An automatic installation under Windows 95 Version 4.00.950 B might proceed as follows:

- Switch of the PC and wearing an earthed wrist band if it is a bare circuit board, install the device in the system, connecting any cables as necessary.
- After connecting the device, restart the computer. Windows 95 boots and starts the 'Update Device Driver Wizard' and automatically selects the driver for you.
- If you have a more up-to-date driver than the one Windows has chosen, make sure that the manufacturer's driver disk is inserted in the appropriate drive and confirm by clicking *Next*. When the current driver has been found, confirm once again by clicking *Next*.
- Click OK to acknowledge the '*Insert disk*' dialog box. The '*Copy files*' Window is now displayed.
- Click *Browse* to change to the drive the driver disk is in and double click OK to initiate the copy procedure.

A semi-automatic installation under Windows 95 Version 4.00.950 might proceed thus:

- Switch of the PC and wearing an earthed wrist band if it is a bare circuit board, install the device in the system, connecting any cables as necessary.
- After connecting the device, restart the computer. Windows 95 boots and opens with the '*New Hardware Found*' dialog box.
- Select '*Driver from disk provided by hardware manufacturer*' and click OK.
- Make sure that the manufacturer's driver disk is inserted in the appropriate drive and click '*Browse*.' The '*Open*' Window is displayed.
- Change to the appropriate drive path (e.g. D:\) and click '*OK*.'

A typical installation under Windows 98/ME/XP, might proceed as follows:

- Switch of the PC and wearing an earthed wrist band if it is a bare circuit board, install the device in the system, connecting any cables as necessary.
- After connecting the device, restart the computer. Windows boots and the '*Add New Hardware Wizard*' dialog box appears. Click '*Next*.'
- Windows now provides you with two options for searching for drivers. 'Search for the best driver for your device' or 'Display a list of all the drivers in a specific location so you can select the driver you want.' We will assume here that you have an updated driver disk on CD-ROM, so select the first option and click '*Next*.' (Windows 2000 and XP expects all drivers to be MS certified.)
- In the dialog box, activate the CD-ROM drive option and deactivate the other options. Insert the manufacturer's CD-ROM in the drive (e.g. D:\) and click '*Next*.'
- When the driver has been found, confirm by clicking '*Next*' in order to initiate installation.

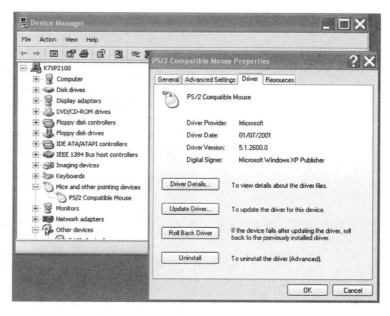

**Figure 6.16** *Using Device Manager to update a driver*

### Adding an updated driver after a device has been installed

If you want to add an updated driver after a device has already been installed again you should follow the manufacturer's instructions, but generally it would probably involve the following process:

- Click *Start*, highlight *Settings* then click *Control Panel*.
- Double click on the *System icon*.
- Now Select the *Device Manager* tab and choose the device you want to add the new driver for, from the list, and double click it.
- A new Window appears. Select the Drivers tab and click '*Update Driver*'.
- The '*Update Device Driver Wizard*' is displayed and installation then proceeds as before (Figure 6.16).

## Windows 95/98 registry

The registry is a database of the systems configuration data. It holds hardware and software configuration information in two files known as System.dat and User.dat. When things go awry within the Windows 95/98 software environment – especially after adding or deleting a new driver or application – knowledge of the registry and use of its tools, can often quickly get things up and running again.

As the names imply, System.dat holds *configuration data* pertaining to the PC system and User.dat holds *user specific* information.

Any legacy 16-bit MS-DOS and Windows 3.1 applications do not use the registry to store configuration data instead they use INI files. However configuration data for applications written to run in Windows 95/98 and all hardware devices – both modern PnP and legacy hardware – are stored in the registry.

When any new hardware and/or drivers are installed, the registry is automatically updated. The registry is vital to the operation of the system. So before you install a new driver or device, it is wise to back up the existing registry settings, just in case something later goes awry. To do this in Windows 95/98, open the registry by clicking 'Start,' then 'Run,' and type 'Regedit' in the text box and click OK. In Regedit, select the icon 'My Computer' top left of the list, click the registry menu item, and then click 'Export Registry File.' Save the file to a floppy disk using a name of your choice. If after installing the card or driver, the existing registry becomes damaged, the problem can be fixed by simply restoring the old working registry from your floppy. After the drivers have been installed, Windows will proceed to install as normal (Figure 6.17).

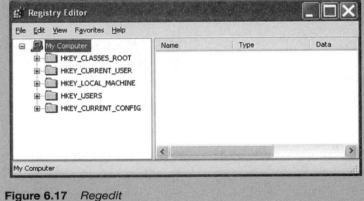

**Figure 6.17**   *Regedit*

In Windows 98/ME/2000/XP regular backups of the registry are performed automatically each day. In Windows 98/ME you can restore a previous version of the registry in an emergency thus:

1. Restart Windows in MS-DOS Mode.
2. At the DOS command prompt type, scanreg/restore. Then press enter to invoke 'Registry Checker'.
   You can now select a known good version of the registry (usually the most recent before the problem occurred).
3. Press enter and the registry is restored.
4. Press enter again to restart the computer.

The registry is a complex system that should only be edited by a person with a fair degree of knowledge about its structure and how data is represented. It is a subject in its own right.

# Diagnosing faults

Right at the beginning of this unit we mentioned three pointers to successful troubleshooting. These next few pages concentrate on these best practices.

PC compatibles are highly reliable pieces of electronic equipment, with few moving parts. In fact they are so reliable, that we happily use them for important day-to-day applications. However when a PC breaks down, it is inconvenient and often wastes a lot of the operators time as the system is gradually restored back to its original state.

It would be desirable if all repairs could be carried out quickly and reliably with the minimum of fuss and inconvenience to the customer. All too often however this is not the case, and the customer is often kept waiting with a machine down for several days.

If a hard disk fails catastrophically then often several Gigabytes worth of system and applications software must be reinstalled back onto the hard disk. It is often quite a time consuming task to get it all back exactly as it was before the failure – it can take days. One way to prevent this situation occurring is to persuade the customer to make a full backup of the hard disk at least once a month and regular backups of important data at least once a day. Then if a hard disk failure occurs a new hard disk can be fitted and backed up in a couple of hours, instead of taking days.

There are swings and roundabouts in all servicing jobs, happily some problems can be sorted out in a few minutes.

A few common instances of simple PC failure are:

(a) Failure to power up, although it worked fine the last time it was used.
(b) After the customer disturbed a previously working system, for example after inserting a card or a cable, or switching on a peripheral.
(c) After installing a new piece of software.

### Rule 1 of fault analysis: Always look for the obvious first

(a) If the power indicator on the base-unit or monitor, is not lit, and there is no noise from the fan and hard disk, check to see if the system is switched on at the supply socket or that the power cables are pushed home fully. It sounds obvious, but it is surprising the number of times people are caught out on this one. If necessary, plug in an appliance, such as a table lamp, to confirm that power is reaching the mains socket.

If the base-unit has still not powered up, try another mains lead or check the fuse and cable for continuity. If the base-unit still refuses to power up, you will have to perform a more rigorous checking procedure. See *Problem Isolation Chart* 1 (PIC 1)

(b) If someone has just added hardware to the system, reboot the computer and see if the system returns to normal. If it does not, then remove the newly added item, and then reboot. If the system now returns to normal, at least you now that the rest of the system is working OK. The new item itself may be faulty or it could be causing a device confliction error, this happens when two devices are using the same *Interrupt*, *DMA* or *Port Address Settings*.

(c) If you have just installed a new application program or a software driver, it could be conflicting with an existing program. The obvious thing to do is remove it, reboot the system, and see if the system returns to normal. Unfortunately, it is not always straight forward to remove all the program fragments once they have been installed, as they often alter the DOS configuration files, *config.sys* and *autoexec.bat* and in the case of Windows applications the ini and registry files. The best way to prevent this sort of problem in DOS and Windows 3.1, is to make a backup copy of autoexec.bat, config.sys, win.ini and system.ini, prior to installing new software. In the event of a problem after loading a new application program, you can then simply replace the suspect files with the original backed up versions and then reboot the system. For Windows 95/98 and Windows NT there are several commercial programs available that 'undelete' application programs for you, these save you the time and hassle of doing it manually.

Using proprietary uninstall software is a far safer option, but even these can delete wanted system files, so take care.

As mentioned previously, Windows 95/98 and Windows NT has a clever system known as the registry to keep a record of all the installed software and hardware items in the system as well as configuration data. It is effectively a complete database of the system.

A person highly conversant with the registry – and there are not many of you out there – can edit the data files using the registry editor. This is called Regedit.exe in Windows 95 and Windows 98 and Regedt32.exe in Windows NT and Windows 2000. In Windows 98 you also have SCANREG a useful utility that checks the validity of the registry and allows you to restore previous versions using SCANREG/restore.

### Rule 2: Don't compound a problem by adding too many unknowns

A classic example: Imagine for the moment that you are this PC buff who has decided to add an old second hand hard disk and controller card to an old AT system (the card is fitted with a hard disk and floppy disk connector).

*You have just added the hard disk and controller card and now the system will not recognize the floppy drive or the hard disk. Unfortunately you have no information on the hard disk or the adapter card, they were sold to you at a bargain price at an auction. You think the problem is the controller card, so you try another old card you have lying around in your spares box but again you have no instructions. The system still does not recognize the drives. You think, 'good! it can't be the controller card, it must be the hard disk.' You try another hard disk and still the problem persists. You then think, 'maybe, the card is configured incorrectly on the 'U' links or the hard disk settings are wrong in CMOS setup?'. ... you just don't know. You mess around changing jumpers here and there and several hours pass and still you have got nowhere ....*

The message here is: do not add hardware unless you have full installation and configuration instructions. Also, ensure that you have the correct drivers for the OS. In most cases, it is not worth your time and effort trying to guess settings or trying to get the wrong drivers to work.

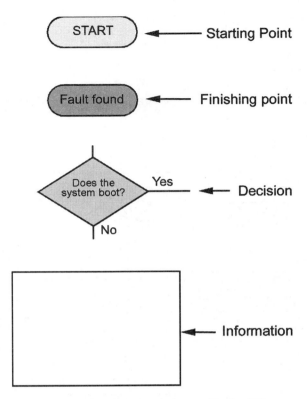

**Figure 6.18** *A key to the chart symbols used in the PICs*

### Rule 3: Think logically

When you are learning the technique of fault diagnostics, use a PIC. This will help you think in a logical, step-by-step manner. Eventually you will automatically think about faults this way.

The series of PICs that follow, will help you diagnose simple board level faults on any PC system.

Before you carry out replacements that entail the removal of fly leads and ribbons, always switch off the computer before hand and make a note of the position and polarity of cable and wire connectors. It is all too easy to forget where a crucial lead should go or how it should be fitted. Also remember to wear an antistatic wrist band before you insert or remove ICs, adapter cards and drives (Figure 6.18).

## Software diagnostics tools

Software diagnostic tools, including those built into the OS offer a good range of error checks. These diagnostics routines are fine as long as the essential elements of the PC, including the CPU, ROM, and part of RAM, are working properly. Under more serious fault conditions however, software diagnosis is useless, for example, diagnostics routines cannot be executed if the CPU or the lower part of RAM is faulty. In such instances, only hardware diagnosis or complete unit replacement is effective.

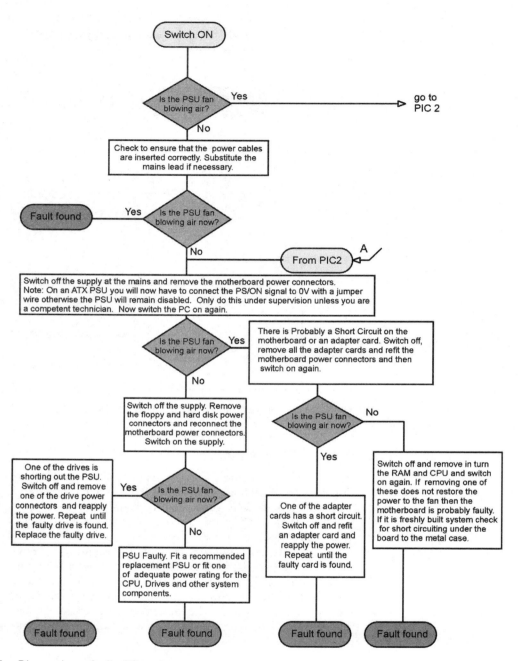

**PIC 1**   *Diagnosing a faulty PC system*

# Hardware diagnostics tools

Hardware diagnosis involves checking the electronics of the system. The minimum equipment required is a multimeter for checking the DC supply voltages and for checking the continuity of wires and cables. A POST diagnostics adapter card can also sometimes prove useful.

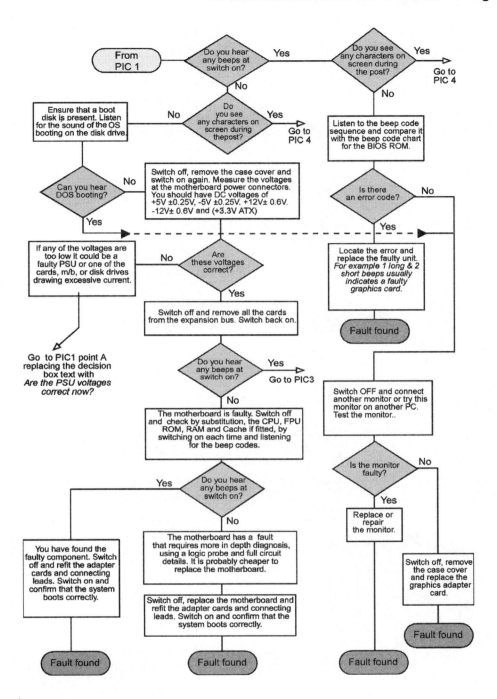

**PIC 2**

Some technicians use a logic probe. This can help to diagnose basic logic faults but you do need to be proficient at digital electronics to use one properly. A good logic probe is useful for checking the presence of digital signals up to 100 MHz. Therefore they are not suitable for checking the fast signals at the North bridge end of the motherboard but circuits around the slower south bridge end can be checked. For example a logic probe could be employed to check the handshake lines on a faulty parallel port connection.

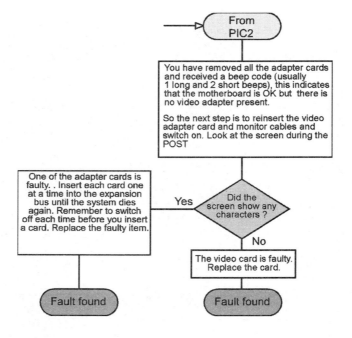

**PIC 3**

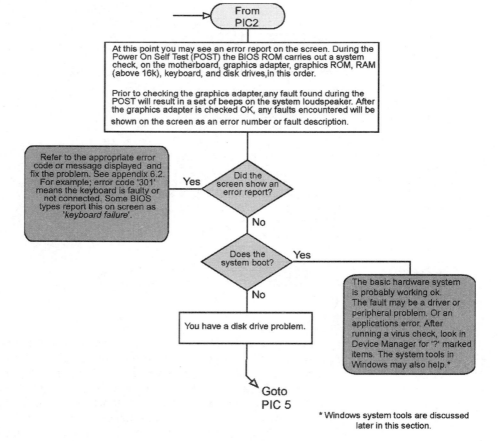

\* Windows system tools are discussed
later in this section.

**PIC 4**

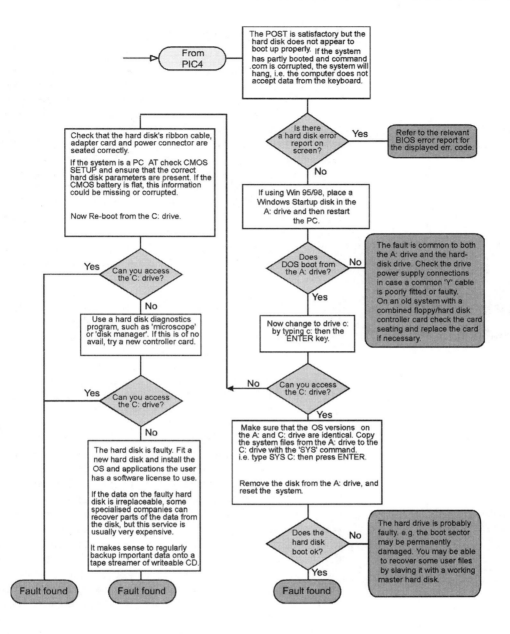

**PIC 5**

Usually logic probes obtain power from the host PC via two leads that clip onto the DC 5 V supply rails. The needle shaped probe tip is then placed at convenient points in the circuit under test, for example an IC pin. The logic state of the signal whether pulsing, static or floating, is then indicated on a row of LEDs. Normally the device is only used for component level fault finding in the workshop, where full circuit details are available.

Nowadays it is increasingly more economical to fault find down to board level and replace the complete unit rather than replace individual electronic parts. This trend unfortunately does not bode well for the experienced electronics technician.

# Substitution method

It is important to carry out basic routine maintenance work prior to digging deeper for fault symptoms. For example a fault floppy drive or CD-ROM drive could just be due to the head in need of cleaning.

After confirming that common elements are not the cause of a fault: often the quickest and most effective way to diagnose hardware faults is to check, by substitution, the major parts of the system, including the CPU, adapter cards, RAM and disk drives.

When a faulty component on the motherboard is suspected, the main pluggable ICs including the CPU, and RAM could be checked by substitution. When you are not sure if the motherboard is faulty or not, a quick test is to remove all the adapter cards including the graphics card and listen for the BIOS beep code informing you that the graphics adapter is missing or faulty. If the beep codes are present, then the CPU, BIOS and part of the RAM are probably working ok.

# Finding faults by visual inspection

Here we will examine some real examples that fall under Rule 1 (Always look for the obvious first (Figure 6.19)).

It is surprising the number of times a seemingly nasty fault turns out to be something simple like a loose, card, connector, cable or IC. So after checking the external cable connections, remove the cover and take a good look inside the base-unit. Use a PIC by all means, but do not get bogged down with serious fault finding until some of the more simple causes have been investigated.

For disk drive problems, check the ribbon connections, particularly those that plug directly onto the motherboard. The plugs are often located in inaccessible places and tend to become dislodged, particularly during the initial build stage or after adding a new card.

Check all pluggable devices, even adapter cards have been known to pop out of their edge connectors. Another common fault source is badly seated DIL ICs. In time they can creep out of their holders. This was particularly a problem with the older dual in line ram chips. Modern SIMM and DIMM devices have a retaining clip to prevent this from happening but it is still worth reseating them when RAM faults occur. Wear an antistatic wrist band, and push down any raised cards or socketed ICs and ensure they are a snug fit in their holders.

If faulty RAM is suspected on a modern motherboard, remove the SIMMs/DIMMs and check the contacts for signs of corrosion. Even with gold plated SIMM and DIMM boards, a form of corrosion can sometimes occur when tin plated SIMM/DIMM holders are used. Carefully wiping the contacts with a linen cloth impregnated with isopropyl alcohol can often cure the fault.

# Peripheral faults

When a peripheral device appears faulty, first ascertain if the problem is with the peripheral itself or with the system.

### If the equipment does not power up

Make sure the mains lead is pushed home properly then try the mains lead on another piece of equipment. If a fuse has blown, check

Using a logic probe to check for the presence of a clock pulse on a motherboard.

An IC clip in use providing easy access to the pins on a 14 pin Dual In Line (DIL) IC.

**Figure 6.19** *Hardware diagnostic methods are not used so much on PCs nowadays*

to see if the correct rating and type of fuse is fitted. The supply voltage, power consumption and fuse rating is usually specified on the rear of the equipment. If the rating is too low, then replace the fuse with the correct rating and power up the equipment and retest it. If the blown fuse is of the correct type and rating, then the chances are that something else has failed and blown the fuse. Use your common sense. If a fuse of the correct rating has blown catastrophically, that is the wire has vaporized completely inside the glass, then a power supply component has probably failed and short circuited. In this case do not try another fuse as it will almost certainly blow the fuse again and possibly cause further damage. Never insert a fuse greater than the recommended current rating or insert a slow blow fuse where a quick blow fuse is specified, as doing so could cause a fire under certain fault conditions. At the very least, it could damage the equipment even further.

### If the equipment powers up but malfunctions

After making sure the connectors on the PC end and Computer end are fitted correctly, carry out some basic house keeping checks, for example if it is a printer, make sure the consumables have not run out (e.g. an empty toner or ink cartridge).

Now power up the PC and check the driver software. Try reinstalling it. If it is a Windows 95/98 PnP device, go into *Device Manager* and make sure the device is recognized and configured properly.

If after these basic checks it appears that the equipment itself is faulty, if possible, confirm this by attaching an identical unit.

If the equipment is still under warranty it is imperative not to invalidate this by attempting an internal repair. Repack the equipment in the original box if possible and tighten down any transit screws. Obtain a returns number from the supplier. Write a returns note stating clearly the returns number and the address where the replacement goods are to be sent and write a brief description of the fault. Enclose this in an envelope and place it inside the box. Stick a suitable sized label on the outside of the box containing the suppliers address and returns number. Tape the box down securely and arrange for a courier to collect and deliver the goods on an insured delivery service.

Most modern peripherals require specialist knowledge and specific replacement parts, so it is unwise to dabble around inside them. This type of servicing is best left to a service organization recommended by the manufacturer. These companies have access to specialist components and equipment that are difficult or sometimes impossible to obtain from normal sources.

The following pages contain a short troubleshooting guide for some typical problems in a modern Windows 95/98/2000 system. This is not intended to be a full treatise on the subject, it is to give you some idea of the types of problems that are encountered and typical cures.

# Specific troubleshooting guide (Windows 95/98/2000 systems)

# Floppy drives

### New system, just built

| | |
|---|---|
| Symptom 1.1 | The floppy disk drive light comes on at power up and stays on. |
| Cure | Refit ribbon cable correctly. Pin 1 of the motherboard and drive connector should go to the to leader stripe of the ribbon cable. |
| Symptom 1.2 | A: drive is not recognized. |
| Diagnosis/cure | Make sure that the A: drive in standard CMOS setup is set to 1.44 MB 3.5″. If ok, reseat the drive's ribbon and power cables. |

### Previously working system

| | |
|---|---|
| Symptom 1.3 | The A: drive used to work fine but now it will not read or write to a floppy disk. |

| | |
|---|---|
| Remedial work that should be done on a regular basis | If the drive is more than a few weeks old it may benefit from a good internal cleaning. Remove the dust and fluff from inside the drive using a brush and vacuum cleaner. The heads should also be cleaned with a proprietary cleaning disk daubed with a small drop of isopropyl alcohol. |
| Diagnosis/cure | Check the drive using a brand new preformatted floppy disk. If fault still persists, click on *Control Panel* and then double click *System*. Click on the *Device Manager* tab and select *Standard FDC* properties then click the *Resources* tab. Check to see if there is a DMA or an IRQ conflict. If there is a conflict, find out what device is using the same settings and try an unused DMA or IRQ. |

Run the *Add New Hardware Wizard* from control panel it might automatically clear the device conflict.

Even though it is not freshly built system, someone may have altered the settings in CMOS setup or disturbed the cables, so check that the A: drive in CMOS setup is set to 1.44 MB 3.5″. If ok, check the seating of the drive's ribbon and power cables.

At this point if the fault persists, quickly test if the drive is at fault by trying a new floppy disk drive. You could also try a new ribbon cable.

For most mechanical problems it is usually cheaper to replace the drive rather than spend too much time attempting a repair.

# Hard disk drives

## New system, just built

| | |
|---|---|
| Symptom 2.1 | The hard drive is completely dead (i.e. you cannot hear it spin up when the system-unit is switched on. |
| Suggested cause and cure | This is could be due to a badly fitting power connector. If the drive is powered via a 'Y' connector, make sure the intermediate connection is pushed home fully. We have known incidents where the Y connector joints do not make a satisfactory connection – due to poorly crimped terminals. If you find this, either replace the Y connector or remove the terminals and neatly re-crimp or solder them to the wires. See illustration below. To remove a terminal: |
| Useful tip | 1. Slide a thin metal tube over the terminal from the front of the connector to flatten the retaining barbs (e.g. you could make a useful tool from a piece of an old broken telescopic |

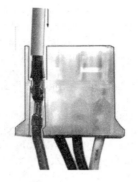

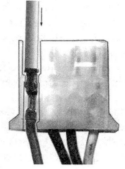

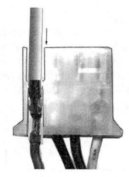

**Figure 6.20** *Removing a faulty terminal*

aerial). Then pull out the terminal from the rear of the shell (Figure 6.20).

2. Re-crimp or solder the terminal.

3. Reshape the barbs if necessary and then push the terminal back into the shell.

Symptom 2.2

On an ATA drive the drive spins up at switch on but it is not recognized by the system, that is the drive has not been detected by the BIOS and listed in CMOS setup.

Suggested cause

Check the ribbon cable: make sure the connectors are pushed home fully into their respective sockets and it is the right way round, that is leader stripe to pin 1 at both ends. If the drive is the master or slave with another drive on the same ribbon cable. Recheck to make sure the drive jumpers are set correctly. If everything checks out ok the drive is probably faulty.

Symptom 2.3

The ATA drive spins up then slows down and stops or the drive spins up and down erratically.

Suggested cause and cure

Make sure the power cable is fitted home firmly and that the terminals are not poorly crimped – waggling the power lead gently should prove if this is the case.

If it is a single drive make sure the jumper is set to *master* or *normal*. If two drives are to be used on the same ribbon make sure the drives are set as master and slave. If cable select mode is being used make sure someone has not swapped the two drives around with the bootable drive in the slave position.

Symptom 2.4

The ATA hard drive parameters listed in CMOS setup are different to the parameters of the installed drive. So the drive cannot be correctly partitioned or formatted.

Suggested cause

Go into CMOS setup and make sure the hard disk utility is set to AUTO DETECT the hard drives. Modern BIOS routines can read the drives parameters directly from the hard disk so always set the hard disk utility to auto detect. Some CMOS setup programs permanently set the BIOS to auto detect so this problem will not occur.

Symptom 2.5

The ATA hard drive spins up but is not detected when the Windows setup CD-ROM is inserted. During Windows 95/98/2000/XP installation, a new hard disk will automatically be detected, partitioned and formatted after prompting the user for information. So if this is not happening make sure the hard drive has been detected by the BIOS (i.e. see if the parameters are listed in CMOS setup). If they are not listed, check the cable connections.

Make sure the CD-ROM drive is set to master or normal and is inserted into the secondary IDE connector on the mother-board. The IDE hard disk should be using the primary IDE connector.

Symptom 2.6

You have installed a SCSI hard drive in a new system. The drive spins up ok but is not detected when the Windows 95/98/2000 Installation CD-ROM is inserted.

Notes and suggested cause

*Note*: SCSI drives are not listed in CMOS setup. To use a SCSI drive go into CMOS setup and set the hard drive parameters to None (no drive installed).

Most systems fitted with a SCSI drive need a separate SCSI controller card. This contains an onboard ROM with special SCSI BIOS extension routines. When installing a SCSI drive, follow the SCSI card and drive manu-facturers instructions to the letter.

In particular, ensure that a suitable ID is selected for the card and drive and that termi-nating resistors are fitted as instructed. Usually the controller card is set to ID 7 and the hard drive to ID 0. If another SCSI peripheral is fitted make sure the selected IDs do not clash. Many SCSI hard drives do not have on board termi-nating resistors so it is usually necessary to insert an in-line terminator – again refer to the drive manufacturer's installation instructions.

## Previously working system

Symptom 2.7

Hard drive (ATA type) will not boot. The drive has been detected by the BIOS – the correct parameters are listed in CMOS setup. After the POST has completed you see an on-screen prompt asking you to insert a bootable disk or an error report such as '*Boot Sector Error*', or an error number appears on-screen, for example error 1780/81 or 1790/91. Other common error reports are '*Invalid System Disk*', '*Incorrect MS-DOS Version*', and '*Missing or Corrupted Command.com*'

Suggested causes

*Cause 1*: Run a virus checking program from floppy to check if a virus has caused the prob-lem. There are numerous viruses around that can corrupt the boot sector but fortunately for most people, this is still quite rare. To prevent virus problems, install a resident anti-virus utility.

*Cause 2*: On a fairly well used drive, the boot sector could be damaged slightly, for example the hidden system files may just need reinstal-ling. To check if this is the case in a Windows 95/98 system:

Reinstalling the hidden boot files in Windows 9*

Boot off your Startup disk and at the command prompt type C:\. If you can access the C: drive and the contents look intact, type

SYS C: and then press ENTER. After a few seconds of disk activity, the following files are copied to your hard disk:

- IO.SYS
- MSDOS.SYS
- COMMAND.COM

After a successful installation the 'System Transferred' message appears.

If the SYS C: command does not work, make sure there are no anti-virus programs resident. Also make sure the BIOS boot sector virus checker is turned off. Then try again.

**Reinstalling the hidden boot files in Windows 2000**

In Windows 2000 a similar problem can occur with the boot files. To fix the problem you need the actual *emergency repair disk* (ERD) created when Windows 2000 was originally installed. You cannot use any ERD, it must be one created from the installed Windows 2000. However there is a way to create a generic boot disk from a working Windows 2000 system. To do this carry out the following:

Format a floppy disk using Windows 2000 explorer (i.e. right click the floppy drive letter then click Format. (*Note*: You cannot use a preformatted floppy disk as it will be DOS compatible not Windows 2000 compatible.) Now copy the following hidden boot files from the root directory to your floppy disk:

- NTLDR
- NTDETECT.COM
- BOOT.INI

(*Note*: To see the hidden files, click *View* then *Options* and click Show All Files)

When you boot off this disk it will automatically use the files to repair the corrupted boot files on the hard drive. (*Note*: You will not be able to boot to the command prompt with this disk.)

**Symptom 2.8**

The drive (IDE type) spins up at power up and it is detected by the BIOS – the correct parameters are listed in CMOS setup. After the POST has completed you see hard disk errors. Also after booting from a Windows 9* Startup disk, the hard disk cannot be accessed by typing C:\ then enter. Instead you get an invalid drive message.

**Probable cause**

The drive is faulty.

**Action**

However first make sure the IDE connectors at each end of the ribbon cable are pushed home fully. You could quickly prove if the drive is faulty after checking the cables by disconnecting the hard disk and connecting another working drive with an OS already installed. After ensuring that the BIOS has correctly detected the drive's parameters, you should be able to access the test drive satisfactorily. If this is the case then the original hard drive is faulty.

If the drive is fairly new and of decent capacity, it is worth trying to resurrect it. You could try re-partitioning the drive using FDISK from the Startup disk.

If that fails, as a last resort, try a low level format using the drive manufacturer's recommended low level format utility.

If the low level format proceeds without a hitch, try reinstalling the OS onto the drive. *Note*: Any user data on the hard disk will certainly be lost after performing the above operations. If there is any irreplaceable material on the disk that the user has not backed up, you might consider letting a data recovery company look at the disk. These companies specialize in recovering at least some of the user data from a failed hard disk. The service is expensive but often money is not an issue in these cases.

|  |  |
|---|---|
| Prevention | To prevent the loss of valuable data always ensure that the user is aware of the need to make daily backups onto tape or recordable CD. |
| Symptom 2.9 | The drive (IDE type) spins up but it is not detected by the BIOS, that is incorrect parameters are listed in CMOS setup and drive errors are reported during the POST. |
| Probable cause | The drive is faulty. |
| Action | After checking the drive connections, go into CMOS setup and invoke the hard disk auto detect and see if the correct parameters are detected. If this does not work, try to manually enter the correct parameters as specified by the manufacturer. |

## Working system

|  |  |
|---|---|
| Symptom 2.10 | The performance of the drive has deteriorated over several weeks. Files take longer to load and save. |
| Probable cause | Disk is fragmented. After a lot of use and as the disk capacity is used up, common files that should ideally be closely spaced on the same portion of the disk surface – so that the head only needs to travel a small distance – are written as fragments all over the disk's surface. This is the consequence of the disk drive's inherent |

efficiency. Files are written in clusters to any part of the disk surface where there is a gap. As files are saved and deleted over time, more and more gaps appear over the disk surface. When a file is subsequently saved, it is divided up into fragments and saved in some of the gaps.

Solution

Defragment the disk. In Windows 95/98 the Disk Defragmenter is found in the *System Tools* folder. (Click Start, Programs, Accessories, System Tools, Disk Defragmenter.) In Windows 2000, the disk defragmenter is built into the Computer Management console in Administrative Tools.

The quickest way to invoke Disk Defragmenter in Windows 95/98 and Windows 2000, is to right click the volume you want to defragment in My Computer. Then select *Properties*, click *Tools* and click the *Defragment Now* button.

Defragmenting gathers fragments that belong together, like organizing files in a cabinet. This process can often take several hours and is therefore best performed when the machine is not in use. Overnight is probably the best time to perform the operation.

In Windows 98 you can use the Windows Maintenance Wizard to perform the task automatically on a regular basis. To run it, click Start, Programs, Accessories, System Tools, Maintenance Wizard.

Symptom 2.11

Bad Sectors are reported in Windows, for example when installing a new program or when running ScanDisk or another hard disk utility.

Solution

Run ScanDisk (click Start, Programs, Accessories, System Tools, ScanDisk) or right click the volume you want to defragment in My Computer. Then select *Properties*, click *Tools* and click the *Check Now* button.

Warning ScanDisk can cause the system to crash especially if cross-linked system files are deleted. So make sure this option is set to either *Make copies* or *Ignore* rather than Delete. You can do this by selecting Advanced from the ScanDisk menu. See the Advanced Options check window screen below. This shows the safest options selected (Figure 6.21).

- *Cross-linked files*: These are two separate files that erroneously occupy the same hard disk cluster. The options available are *Delete*, *Make Copies* and *Ignore*. It is best to play safe and set this to *Make Copies* or *Ignore*.

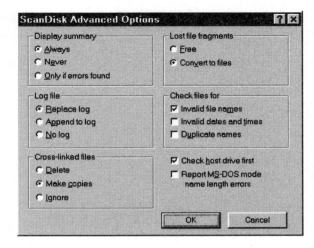

**Figure 6.21** *Setting options in ScanDisk*

- *Lost file fragments*: These can occur if the system crashes or if the computer is powered down without first closing open files. Converting them to files is probably the best option as sometimes you can recover lost work when the system crashes in the middle of an application.
- *Check host drive first*: If you have compressed the hard disk using DriveSpace (not recommended), then this option tells scandisk to automatically check the whole of the drive.
- The other advanced options are reasonably self-explanatory.

# CD-ROM drives

### New system, just built

Symptom 3.1      The CD-ROM drive is completely dead. When you press the disc eject button, nothing happens.

Suggested cause and cure      This is probably the power connector not fitted or not pushed home sufficiently. If you are using a 'Y' connector to give you more drive power lines, make sure the intermediate connection is pushed home fully. We have known incidents where the Y connector joints do not make a satisfactory connection – due to poorly crimped terminals. If you find this, either replace the Y connector or remove the terminals and re-crimp or solder. See illustration in Symptom 2.1.

Symptom 3.2      The tray opens and accepts a CD but it cannot read it. This applies to music CDs and data CDs.

Suggested causes
and cure

Make sure the drive ribbon cable is inserted correctly. Check the slave/master jumper setting: If the CD-ROM is connected to the same ribbon cable as the hard disk on the primary IDE port, make sure it is set to Slave. If possible, it is best to place the CD-ROM as the master on the secondary IDE port.

If you have connected the CD-ROM to the secondary IDE port, go into CMOS setup and make sure that the secondary IDE port has been enabled.

Windows 95 onwards automatically detects an IDE CD-ROM drive and installs protected mode drivers. To check if this is so: open *My Computer* and see if the drive icon is present. If you only have one CD-ROM and one hard disk partition it will listed as the D: drive. If the hard disk is partitioned into several logical drives the CD-ROM should be the next letter after the last logical drive.

If it is not present in My Computer go into Device Manager and click on CD-ROM (i.e. click on *Setup*, *Settings*, *Control Panel*, double click *System*, click on the *Device Manager* tab, click on *CD-ROM* and then click on the named CD-ROM device.) If there are any problems a Yellow exclamation mark will be displayed against the 'CD ROM' entry in the device manager Window. Click on the General Tab and note the Device Status. It should say *This device is working properly*. If there is a device confliction or driver problem, run the hardware Installation Wizard and see if this cures the problem. Also try changing the drive letter to an unused one.

## Newly built or working system

Symptom 3.3

CD-ROM seems slow when reading a data disk (Windows 95/98/2000).

Suggested things to
check

Make sure the CD-ROM cache is set to a maximum size. To do this select System Properties and click on the File System button at the bottom of the box (i.e. click on *setup, Settings, Control Panel*, double click *System*, click on the *File System* button then click the CD-ROM tab) (Figure 6.22).

If the drive is quite old and it takes an inordinate amount of time to read any disc then the lens on the CD optical block may need cleaning. To do this, use a proprietary CD-ROM lens cleaning disc. This a normal CD with a minute fine bristled brush attached to the read surface of the disc. After applying a spot of isopropyl alcohol to the brush, the disk is inserted in the drive and played as

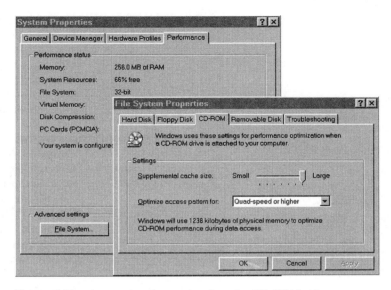

**Figure 6.22** *Increasing the cache size of a CD-ROM drive*

normal. Usually an audible message is recorded onto the disk to inform you when the cleaning process is over.

**Symptom 3.4**
When a music CD is inserted nothing is heard in the speakers.

**Suggested thing to check**
Make sure an audio lead is inserted from the audio connector on the back of the CD drive to the correct audio input connector on the sound card. Also ensure that the speaker system attached to the sound card is connected and working satisfactorily.

# Motherboards

### New system, just built

**Symptom 4.1**
The system is dead on power up. The monitor screen is blank. Even the hard disk does not spin up.

**Suggested things to check**
Check that the RAM, CPU and adapter cards, have been fitted correctly. Refit them if necessary. Also check the PSU connector and jumper settings and make sure they are correct.

Pull out the motherboard power connector/s and see if the hard disk spins up now. If it does, you have a short circuit on the motherboard. This may be a metal pillar or post in the wrong position underneath the motherboard.

**Symptom 4.2**
The system is dead on power up. The monitor screen is blank. However there is some activity – you can hear the hard disk spin up and the CPU fan is running.

Suggested things to check

Check that the RAM, CPU and adapter cards have been fitted correctly. Refit them if necessary. Also check the CPU and RAM jumper settings and make sure they are correct. Undo the motherboard fixing screws and make sure that there are no metal objects fouling the motherboard underneath.

### Previously working system

Symptom 4.3

The system is dead on power up. The monitor screen is blank. However there is some activity – you can hear the hard disk spin up and the CPU fan is running.

Suggested things to check

Place a POST diagnostics card in a spare expansion bus slot (ISA or PCI depending on the type of POST diagnostics card in use). Switch on and see if there is any activity on the seven segment display. If the display does not change at all from initial switch on, then the CPU, RAM or Motherboard is probably faulty. Remove all adapter cards except the POST diagnostics card as see if there is activity now.

Switch off the system, remove the RAM devices and try a substitute RAM device. Even a small capacity RAM device from the spares box will do just to test the system. (If you are using SIMMs remember to use them in pairs.) If the system works now, find out which RAM device is faulty and replace it with the same or similar device. Alternatively, take this opportunity to upgrade the RAM to a larger capacity.

Suggested things to check

If the RAM is not the problem, next try the CPU. Before doing this make sure the CPU fan is working properly: if the fan has failed or sticks occasionally, this will almost certainly have caused the original CPU to overheat and subsequently fail. It is also wise, if you are reasonable competent at using a multimeter, to measure the CPU Core and I/O voltage with a multimeter.

This will confirm that the on board voltage regulators are producing the correct voltage as set by the jumpers/dip switches. This also helps to ensure that the same fault will not occur again in a few weeks or months time. For example: if the voltages are too high, due to a faulty regulator on the motherboard for example, the replacement CPU will almost certainly fail again, sooner or later.

Taking the time and effort to check out causal factors are what separates the professional from the novice. If something has failed, always ask yourself: 'What could have

An increasingly popular trend is for motherboard manufacturer's to incorporate a voltage monitoring system into the motherboard along with Window's device drivers. If an over- or under-voltage situation occurs a pop-up window is displayed listing the state of the various supplies. The voltage states can also be viewed whilst carrying out general maintenance.

**Figure 6.23** *Checking the CPU core voltage*

caused the failure?' and check it out. The failed item is often only the end result of a more obscure failure elsewhere in the system (Figure 6.23).

If the CPU is ok then the motherboard itself is probably the cause of the problem. For example the chipset may have developed a fault. Today it is cheaper to replace the motherboard rather than try to desolder and replace the on board ICs. For one thing you need very expensive surface mount soldering and desoldering equipment and for another you need a supply of replacement ICs. These individually are often difficult to obtain and more expensive than a new motherboard.

| | |
|---|---|
| Symptom 4.4 | After the POST, 'CMOS Checksum Bad', 'CMOS Display Type Wrong' or a similar message appears on-screen. |
| Probable cause | The CMOS setup RAM has lost power due to a flat battery |
| Action: | Replace the battery. This is usually a 3 V lithium cell located on the motherboard. Under normal circumstances the battery should last for 10 years or more but sometimes a faulty cell can shorten this period considerably. |

After replacing the battery and powering up the system you should see a message asking you to press a key to run CMOS setup. You will now have to set some parameters back to what they were before the problem.

Other problems
associated with
the motherboard

As long as the CPU, memory and the expansion buses are functioning, other motherboard problems are probably best diagnosed using a POST diagnostics card. If the floppy disk is also functioning then software diagnostics can be also be used. The latter method is ideal for diagnosing serial and parallel I/O problems and for carrying out a more detailed analysis of the motherboard hardware.

# Graphics adapters

### New system, just built

Symptom 5.1

No display. The screen is blank after powering up the system and monitor. You may also hear a series of beeps usually one long and two short beeps.

*Note*: When you first install a graphics card and power up the system the BIOS treats the card like a generic VGA device so that a basic display system can be quickly established. Later, after installing the OS, the full power of the card can be realized using the manufacturer's recommended drivers.

Probable cause

The graphics adapter is probably not fitted in its slot correctly. Switch off the system-unit and monitor and leave their mains cables in place to earth the equipment. Wait a few seconds, then remove monitor connector from the graphics card and remove the card from the slot. Now refit the card by pushing it down firmly into the edge connector and refit the card retaining screw.

If locating and tightening down the retaining screw puts tension on the card tending to pull it out of the edge connector: remove the card and carefully bend the metal retaining strip slightly to counter act this.

In extreme cases, the tolerance is such that the card retaining screw will not locate without this tendency to pull the card out of the slot. To correct this, remove all the adapter cards and the signal leads from the rear of the case and try relocating the motherboard slightly by loosening the fixing screws and moving the motherboard towards or away from the rear of the case until the cards locate properly, without undue strain.

*Note*: Remember to always wear an earthed wrist band when fitting adapter cards, etc.

## Previously working system

| | |
|---|---|
| Symptom 5.2 | There is no display on the monitor, the screen remains blank. |
| Diagnosis | Listen for a beep code error after switching on the system. For example the Award BIOS issues one long and two short beeps to signify a graphics card error (see the beep code error charts in appendix). If you hear the relevant beeps, then the graphics adapter is probably faulty or the card is making bad contact with the motherboard edge connector (see Symptom 5.1). |
| | If there are no beep codes this does not necessarily mean the card is ok. The best way to diagnose the fault in this case is to temporarily attach another monitor. |
| | If this monitor works then the fault is probably the monitor. |
| | If the screen remains blank then the graphics card is the culprit. You could prove this by temporarily installing a similar graphics card but do check first to make sure the problem is not simply the seating of the card. |
| Symptom 5.3 | There is a display on the monitor during the POST but the characters on the screen are illegible and some are flashing on and off. Alternatively you may see vertical stripes. The system is working apart from this and after a while, on a Windows system, you may see what looks like the Desktop through the maze of flashing characters. |
| Probable cause | This is a fault on the graphics card. It is probably a faulty RAM device or a faulty character generator in the graphics engine chip. If the card's memory devices are fitted in sockets you could try reseating them or even replacing them one by one. |
| | This symptom can also occur on a system when someone has added a new graphics adapter with the old one still in place, for example on some systems the graphics circuitry is built into the motherboard. This must be disabled (usually via a jumper) to allow the new card to function properly. |
| Symptom 5.4 | You have just added a modem card and there is now a confliction with the graphics card. |
| Probable cause | If your modem is configured to use COM4 and you are using a graphics card with the S3 graphics engine. Some of these older chipsets decode the x2E8 address range incorrectly with the I/O address mapping performed in Windows 9*. These chipsets include the 801, 805 and 928 versions. |
| Symptom 5.5 | On a Windows 95/98/2000 system, during the POST and Windows initialization, the display |

is fine then suddenly the screen bursts into a series of horizontal lines as the desktop is loaded.

Probable cause

Make sure the correct driver software is in use for your graphics card and monitor.

To see if the correct driver is installed, click *Start*, *Settings*, *Control Panel* and double click *System*. Select the *Device Manager* tab and double click the *Display adapters* icon.

In Device manager, double click on the *Monitors* icon to check whether the monitor driver is working satisfactorily. If the monitor is not a PnP device you must choose the correct monitor type from a list in Display Properties (i.e. click on *Setup*, *Settings*, *Control Panel*, double click *Display*, then select the *Settings* tab). Click on *Advanced* to see the Monitor and Graphics Adapter details. If your monitor is not listed choose a generic device.

If this symptom has always been a problem since Windows was installed, then the refresh and synchronizing rates of the monitor may be incompatible and a new monitor may be required. Try setting the display resolution to a lower setting and see if that restores a stable display.

Symptom 5.6

When viewing high quality full colour graphics images (e.g. photographs), the colours look dull and dithered. Not vibrant as the images used to appear.

Probable cause

Someone has probably changed the display setting to 256 colours or less, perhaps to play a game that needs a low colour setting. Go into Display Properties and in the settings Window change *Colours* to 16-bit, 24-bit, or 32-bit – depending on the capability of your graphics card (i.e. click on *Setup*, *Settings*, *Control Panel*, double click *Display*, then select the *Settings* tab).

# Monitors

## Previously working system

Symptom 5.7

No display. The graphics adapter card is working perfectly as another monitor has been tried, and this worked fine.

Diagnosis

If the monitor power light does not illuminate when the monitor is switched on, test the fuse in the monitor's mains lead. If this is blown then make sure the fuse rating is correct for the monitor.

For example the rating on the back of the monitor will probably be stated in a similar way to this. *Rating AC*: 110–240 V; 2.5–1.5 A; 60–50 Hz.

This shows that the maximum current is 1.5 A when operated on the UK 240 V 50 Hz AC mains and 2.5 A when operated on the American 110 V 60 Hz AC mains.

Plug fuses are rated at 1, 2, 3, 5, 7, 10 and 13 A in the UK. So for the above monitor, in the UK, the fuse in the mains plug should be rated a 2 A device.

If the fuse in the mains plug is a 1 A device and has blown, then this may well be the cause of the fault. Replace the fuse and retest the monitor.

If the fuse in the plug is a 5, 7, 10 or 13 A device and it has blown, then something more serious has caused it to blow. If this is the case, do not replace the fuse. The monitor will need to be serviced by a competent electronics technician/engineer experienced in monitor fault finding and repair.

|  |  |
|---|---|
| Symptom 5.8 | The colour cast on the screen sometimes suddenly changes to too much red, green or blue and then changes back to normal again spasmodically. |
| Suggested cause | This may be a fractured wire or joint on the monitor signal lead. To prove this, gently sway the cable near the monitor end and then at the graphics card end. If the fault occurs during this operation, then the lead is faulty. |

If the lead is detachable, replace the lead. If you are competent at soldering, cut off the faulty end, a few centimetres beyond the fracture and solder a new 15-way VGA connector to the end.

## Sound cards

### New system, just built

|  |  |
|---|---|
| Symptom 6.1 | No sound from the speakers. |
| Suggested cause 1 | If there is an external loudspeaker amplifier, make sure it is powered up. Check to make sure the sub-miniature jack plug on the end of the loudspeaker/amplifier signal lead is pushed into the correct socket on the sound card (Figure 6.24). |
| Suggested cause 2 | Generally, if you are using an external stereo amplifier then you should connect the jack to the *line out* socket if there is one fitted, otherwise use the loudspeaker socket. |
| Suggested cause 3 | As with all adapter card faults, check the seating of the card. |
| Symptom 6.2 | In Windows 95/98/2000 make sure the device has been detected by PnP and that the device |

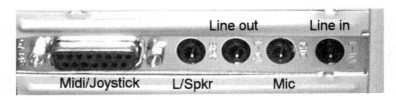

**Figure 6.24** *The rear panel of a typical sound card*

drivers are working properly (i.e. click *Start*, *Settings*, *Control Panel* and double click *System*. Select the *Device Manager* tab and double click the *Sound Video* and *game Controllers* icon, double click the relevant sound card, icon then select the *General* tab to check the device status. If there is a device confliction, try re-running the Hardware Installation Wizard.

Probable cause Windows 95/98/2000 sounds are working ok but no sound is heard when a music CD-ROM is played even though the CD player box appears.

The audio cable from the CD drive to the sound card is either missing, or inserted in the wrong audio connector on the sound card.

### Previously working system

Symptom 6.3 WAV files do not work in Windows 95/98/2000. Audio CDs work fine however. (WAV files are normal Windows sound files with the '.wav' extension, for example button clicks, file opening and closing sounds, etc. are in the wav format.)

Probable cause The fact that other sound files work ok suggests that the problem may be that the WAV volume control slider has been turned down to minimum in Windows.

To check this click *Start*, *Programs*, *Accessories*, click on *Multimedia* and click *Volume Control*. You will see several volume control sliders for different sound sources. Ensure that the 'Wave' volume control is turned up sufficiently.

*Note*: If you cannot see the 'Volume Control' and Wave slider you may have to install these options from your Windows CD-ROM.

Symptom 6.4 The sound system once worked fine but now there is no sound.

Diagnosis First check that the loudspeaker/amplifier system has powered up, and test to see if it is working ok. A quick way to do this with an amplified pair of speakers is to pull out the loudspeaker jack lead and touch the tip of the jack plug with a screw driver and listen for

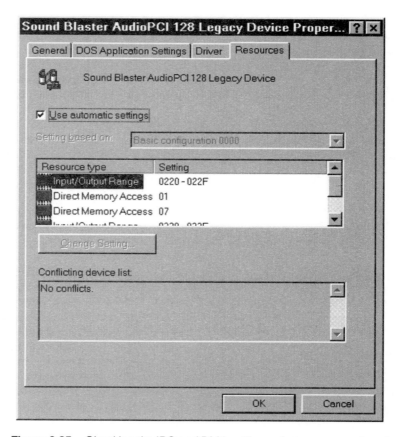

**Figure 6.25** *Checking the IRQ and DMA settings of a legacy sound card*

hum and clicks from the loudspeaker. If there is no sound it could be the amplifier at fault or the loudspeakers but it is unlikely that both speakers of a stereo pair have failed.

If this is ok, check that the sound card is still set up correctly in device manager, that is click *Start*, *Settings*, *Control Panel* and double click *System*. Select the *Device Manager* tab and double click the *Sound Video* and game *Controllers* icon, double click the relevant sound card icon then select the *Resources* tab to check the IRQ, DMA and port address settings. If there is a device confliction, try re-running the Hardware Installation Wizard.

If this is ok the sound card itself could be faulty, however try reseating it first (Figure 6.25).

# Modems

### New system, just built

Symptom 7.1        Modem fails to work, that is, not recognized by the OS.

Suggested checks

*Internal modem*:    First make sure the device is configured correctly. Internal modems have a unique IRQ and port address. Read the manufacturers instructions. To check the settings: click *Start*, *Settings*, *Control Panel* and double click *System*. Select the *Device Manager* tab and double click the *Modem* icon, double click the relevant modem card icon, then select the *Resources* tab to check the IRQ and port address settings. The most common resource confliction is with older graphics cards. Graphics cards with the S3 engine are known to cause a confliction with modems using COM 4. If the latter problem occurs the best option is to change the graphics card for a newer type.

If this is ok, next check the driver. Click *Start*, *Settings*, *Control Panel* and double click the *Modems* icon and select the *General* tab. Examine the modems properties Window. Your modem card should be listed. Now click the diagnostics tab and see if Windows has detected the modem on a port (not always but usually the port is set as COM 4). If you click the *Driver* button, your driver should be listed.

If the device is not listed, then reinstall the modem driver or run the *Add New Hardware Wizard* from control panel. If it does not automatically find your modem, check the '*No, I want to select a new driver from a list*' option. Then choose *modem* and follow the on-screen instructions to detect and install your modem and driver.

If none of the above applies, unplug the cable from the rear of the modem card and remove the modem card retaining screw and remove the card from the slot and reseat it firmly back into place, making sure there is no undue tension on the board when the retaining screw is refitted. Retest the modem. *External modem*:    Check the cabling from the com port on the rear of the system-unit to the modem. Ensure that the modem powered up.

Make sure the modem driver has been installed correctly, that is click *Start*, *Settings*, *Control Panel* and double click the *Modems* icon and select the *General* tab. Examine the modems properties Window. Your modem card should be listed. Now click the diagnostics tab and see if Windows has detected a modem port (usually this is set to COM 2, if you have a serial mouse or COM 1 if a PS/2 mouse is used. Make sure that the modem is actually connected to the port listed.

### Previously working system

| | |
|---|---|
| Symptom 7.2 | Modem has stopped working. |
| Suggested checks | *Internal modem*:    First make sure that external connection to the telephone socket is fitted correctly. Test the modem by running the *diagnostics* facility in the *modems* folder in Control Panel, that is, click *Start*, *Settings*, *Control Panel* and double click the *Modems* icon and select the *Diagnostics* tab. Check to see if your modem has been assigned to a port. If it has, highlight your modem port and click the *More Info* button. This sets sends a series of AT commands to the modem and lists the responses (Figure 6.26). |

If your modem is not listed against a port, then reinstall the modem driver or run the *Add New Hardware Wizard* from control panel. If it does not automatically find your modem, check the '*No, I want to select a new driver from a list*' option. Then choose *Modem* and follow the on-screen instructions to detect and install your modem and driver. If you still have no joy, remove the case cover and try reseating the modem card. If this fails to cure the problem, consider replacing the modem.

*Internal and external modems*:    If the modem is detected correctly in Windows, the problem could be in the application program using the modem. For example if the modem is used mainly for Internet access, check that the

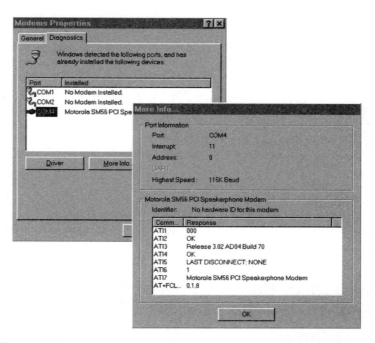

**Figure 6.26**   *Diagnosing your modem using the* More Info *facility*

**Figure 6.27**   *Accessing your ISP's phone number*

Internet Service Provider's telephone number is still present. Someone may have deleted or partially installed another ISP program. To check this, click *Start*, *Settings*, *Control Panel* and double click the *Internet Options* icon and select the *Connections* tab. Then click on the *Settings* button. In the Internet settings Window, click on the *Advanced* button to see your ISP phone number (Figure 6.27).

If your ISP telephone number is incorrect or not present either re-enter it in the boxes or re-run the Internet connection wizard by clicking on the setup button in the Internet Properties Window.

Hint

In Windows 98 and Windows 2000 to help you diagnose modem problems click *Start*, *Settings*, *Control Panel* and double click the *Modems* icon and select the *diagnostics* tab. Now click on the help button for a useful troubleshooting guide.

## Windows 98 specific problems

### New system just built

Symptom 8.1

Windows 98 setup fails during hardware detection and hangs.

Note about Windows 98 setup

During hardware detection in Windows setup, if a device causes the installation to fail, a hidden log file called *detcrash.log* is created in the root directory. When the system is rebooted, setup resumes and uses this file to bypass the item that caused the failure.

Several log files are created during setup. *detlog.txt* is a list of all the devices found during the hardware detection phase. *netlog.txt* lists all the detected network components and *setuplog.txt* is a hidden file created in the root directory that records each step of the whole installation.

Suggested things to check

Reboot the system and see if the installation proceeds smoothly. If safe recovery does not solve the problem, view detlog.txt and setuplog.txt in a text editor and try to isolate the cause of the problem. Look at the end of the list at the last items to be detected.

If this fails to isolate a cause, check to see if the BIOS ROM anti-virus program is active. If it is, this will cause the system to hang as it will prevent writes to the boot sector of the hard disk. During Windows setup the boot sector must be written to as the system files, etc. are installed. Go into CMOS setup and disable the anti-virus program. If you wish you can reactivate it after Windows 98 has been installed correctly.

If power management is activated and set to send the system to sleep after a chosen duration of time, the installation process could fail. Turn off power management in CMOS setup temporarily until the installation has completed.

Symptom 8.2

Windows starts to load normally and then it hangs.

Suggested things to do

Start Windows in Safe Mode* to see if the system still hangs. If it boots up fine in Safe Mode, a Device Driver is probably incorrectly configured. To fix the problem go into Device Manager and determine which device is incorrectly configured. In Windows 98 an exclamation mark (!) appears next to a suspect driver. Install the latest device driver.

If the system still does not boot properly or seems erratic and crashes unexpectedly, then run CMOS setup and set the system to a default, non-optimized setting. If this still fails to cure the problem check the CPU and RAM settings. Refit the RAM if necessary.

*To run Safe Mode: press Ctrl while Windows is booting. This takes you into the Startup menu. For information on the Startup menu facility see below.

# The Startup menu

The Windows 98 Startup menu allows you to boot the system in a number of different ways. For example, if Windows fails to start properly and hangs, as in this example, you can choose to boot in safe mode. To start Windows using the Startup menu, hold down the

Ctrl key during the Windows boot process and the following menu appears:

Windows 98 Startup menu
1.   Normal mode
2.   Logged mode
3.   Safe mode
4.   Step-by-step confirmation mode
5.   Command-prompt-only mode
6.   Safe mode command-prompt-only mode
7.   Previous version of MS-DOS

### Normal mode

This is the normal operation mode of Windows 98. Select this mode if you boot to the Startup menu but then decide to complete the boot process normally.

### Logged mode

When you select Logged mode, the entire boot process is logged to a file called bootlog.txt in the root directory. You can view bootlog in a text editor to determine where a boot failure occurred. Booting in this mode is the same as normal mode except that it produces a log of the startup process.

### Safe mode

In this extremely useful mode, many of the advanced features of Windows are disabled, including the following:

- Config.sys and autoexec.bat
- The registry
- [Boot] and [386Enh] parts of system.ini
- 'Load =' and 'Run =' parts of win.ini
- The startup program group
- All device drivers except the keyboard, mouse and VGA.

### Step-by-step confirmation mode

This is another useful utility. This one allows you to single step through the various stages of the boot process. It allows you to specify whether a particular stage should or should not be completed. Using this utility you can isolate the parts of the startup process that are causing problems. It is also very useful to see if problems occur as each line of Config.sys and Autoexec.bat is executed.

### Command-prompt-only mode

This mode is similar to the MS-DOS command prompt. Only config.sys, autoexec.bat, command.com and the registry are activated in this mode. This mode is useful if you need to troubleshoot MS-DOS applications.

### Safe mode command-prompt-only mode

In this mode the more powerful features of Windows are disabled as in Safe Mode except that command.com is executed to allow command processing. Himem.sys and the Windows interface are not loaded. This mode is very useful if the system fails to boot in safe mode.

***Previous version of MS-DOS***
If Windows 98 was an upgrade from Windows 3.1, for example it allows the system to be booted from the previously installed versions of MS-DOS.

# Windows 98 specific problems

### Working system

| | |
|---|---|
| Symptom 8.3 | After running a certain application you get a general protection fault (GPF) or the application hangs. |
| *Notes*: | A GPF is usually caused by an application attempting to violate the integrity of the system. For example the most common cause of a GPF is an application trying to utilize the memory space assigned to another application. A GPF error message is shown on-screen when the OS closes the offending application. |
| Suggested course of action | If the symptom occurs on a regular basis. Check the application manufacturer's web site for information on known problems and patches/updates. You could also run Dr. Watson. Then start the suspect application and make the problem occur again. As soon as the fault occurs Dr. Watson generates a snap shot of the system. You will then hopefully be able to analyse what caused the problem. |
| Dr. Watson | Dr. Watson is a clever little utility that runs in the background and logs information about the system at the precise time an application crashes. When this occurs, Dr. Watson logs the system information to a file in the \Windows\Dr. Watson directory called Watsonxx.wlg |
| Starting Dr. Watson | To run Dr. Watson, click on the Start menu and highlight, *Programs*, *Accessories*, *System Tools*. Click on *System Information*. Now click the Tools button and from the pulldown menu choose Dr. Watson. |
| | A Dr. Watson icon will appear in the system tray at the right of the screen. Click on the Dr. Watson icon, and it proceeds to generate a log of the current state of the system. The Dr. Watson Window then appears on-screen. You have two basic views, standard view and advanced view. The image below is in advanced view (Figure 6.28). |
| Symptom 8.4 | After installing a new application some existing applications are reporting problems. |

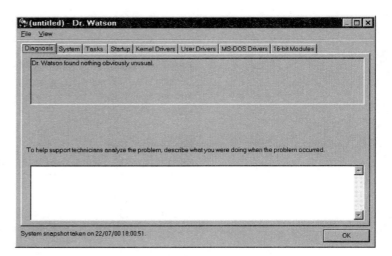

**Figure 6.28** *Dr. Watson used to report the state of affairs prior to a system crash*

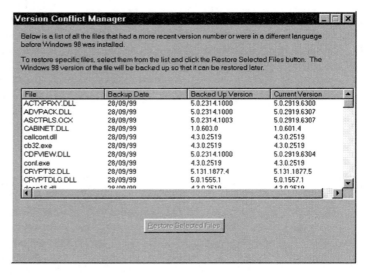

**Figure 6.29** *Using Version Conflict Manager to restore a driver file*

| | |
|---|---|
| Suggested course of action | It is possible that the new applications has inserted a newer version of a driver or other system file. Run *Version Conflict Manager* to check if some old application files have been replaced. Version Conflict Manager lets you restore a backed-up version of the file that was present before the application was installed. |
| Version Conflict Manager | Click on the Start menu and highlight, *Programs*, *Accessories*, *System Tools*. Click on *System Information*. From the Microsoft System Information Utility *Tools* menu, select *Version Conflict Manager* (Figure 6.29). |

To restore a particular file or files, select them from the list and click on the *Restore Selected Files* button.

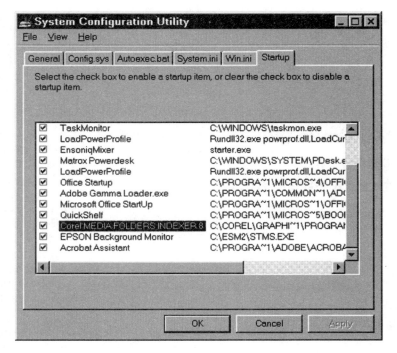

**Figure 6.30** *System Configuration being used to isolate a troublesome program from startup*

| | |
|---|---|
| Symptom 8.5 | After Windows 98 starts the floppy disk drive keeps being accessed intermittently by an unknown program. |
| Suggested course of action | To isolate the program causing this phenomenon, you could open the *Close Program Window* by pressing Ctrl Alt Del and turn off suspect programs one by one. A better way is to use the *System Configuration Utility*. This is described below (Figure 6.30): |
| System configuration utility | The system configuration utility (Msconfig. exe) allows you to turn on and turn off programs and files to isolate problems in the system, that is, you can modify the system configuration through a process of elimination using the check boxes. Configuration utility can also create a backup copy of your system files before you begin a troubleshooting session. It is important to do this so that modifications made during your troubleshooting session can be reversed if necessary. |

To start system configuration utility click on *Start*, then highlight *Programs*, *Accessories*, *System Tools*, and then click *System Information*. In System Information, click the *Tools button*, and then click *System Configuration Utility*.

Back up your current system configuration before you begin troubleshooting. To this, click the *General* tab, click *Create Backup*, and then click *OK*.

To isolate the program causing the above symptom, click on the Startup tab and disable a program you suspect by clicking the relevant check box. Now exit and restart Windows. Repeat until the culprit has been isolated.

## Windows 98 general troubleshooting

General fault finding in Windows 98

Apart from the useful set of tools in the Windows 98 'System Information folder,' Windows 98 is equipped with a set of 15 troubleshooters (Figure 6.31).

To access these useful hypertext files, click on *Start* then *Help* and select the *Contents* tab. In the contents Window click on *Troubleshooting* then the *Windows 98 Troubleshooters* icon. To run a troubleshooter, click on the question mark against the topic you need help on.

These hypertext files can often provide a useful initial focus on a problem.

## Self-assessment practical (SAP 6.3)

To become proficient at locating and fixing faults on a PC system, the one thing most budding PC support technicians need more than anything at this point is, 'hands-on experience.' The best way of getting this is to have a go at diagnosing and repairing real faulty PC systems. However these are not always that easy to obtain.

**Figure 6.31** *Windows 98 built in troubleshooting guides*

The following self-assessment practical based on nine system faults is no substitute for real hand on experience but it should help you build your confidence.

In this practical we have provided several fault symptoms and each one could be caused by one or more faulty parts or incorrectly configured drivers. We want you to suggest just one item that could cause the symptom listed and suggest a cure. In a real fault finding exercise there would naturally be real symptoms and faults. Here you will have to use your own skill, judgement and imagination.

Fill in the diagnosis and cure for each of the following symptoms. Use the PICs to help you. The first fault example has been done for you, to show you how to go about it.

**(Fault 1) Symptoms** The customer states that everything works ok except that the colour ink-jet printer prints streaky black text and images.

The PC is an ATX system with Windows 98.

**Diagnosis** Print a page with black and coloured text and images and see if the same problem occurs, that is do a print self-test.

(a) Examine the image for tell tale signs of an exhausted ink cartridge, for example lines of missing ink.

(b) Check for smears and smudges in the image.

**Possible cure** If (a) above, check the black ink cartridge to see if it is indeed nearly empty and replace it with a new one if necessary.

If (b) above, the ink head is probably clogged or some dirt and fluff has been picked up near or around the ink head. Clean the head and surrounding area as instructed in the printer user manual.

**Faulty part**
**Customer**
**interaction if**
**necessary**
*No faulty part*, just routine maintenance and every day printer house keeping. If the problem was (a) above, mention politely to the customer about the need to replace the cartridge when it is empty and show them how to detect the warning signs on the printer status lights and on-screen reports. Also show them how to replace the cartridge, if they are unsure how to do it.

**(Fault 2) Symptoms** The owner of the PC says that the machine suddenly stopped working.

You switch on the PC: the PSU fan is blowing air but there is no on-screen activity and there is no noise from any of the drives, the system seems to be dead, apart from the blowing fan.

**Diagnosis** *Note*: There are several things that can cause this, just go through a diagnosis that results in just one of them being diagnosed as the cause.

**Cure**

**Faulty part**
**Customer**
**interaction if**
**necessary**

**(Fault 3) Symptoms**  In an AT system, the PSU fan is not blowing air, the system is dead. You turn on the monitor and the power indicator does not come on either.

**Diagnosis**

**Cure**

**Faulty part**
**Customer**
**interaction if**
**necessary**

**(Fault 4) Symptoms**  Immediately following the POST you see a message saying CMOS checksum error and a prompt asking you to enter CMOS setup.

**Diagnosis**

**Cure**

**Faulty part**
**Customer**
**interaction if**
**necessary**

**(Fault 5) Symptoms**  On a previously working Windows 98 system: after the POST the message 'No OS detected. Please insert a bootable disk' is displayed.

**Diagnosis**

**Cure**

**Faulty part**
**Customer**
**interaction if**
**necessary**

**(Fault 6) Symptoms** No display. You can hear the hard disk spin up correctly and the Windows Startup sound from the loudspeaker system.

**Diagnosis**

**Cure**

**Faulty part**
**Customer**
**interaction if**
**necessary**

**(Fault 7) Symptoms** You have just installed a new 20 GB hard disk into an old system with a 1994 AMI BIOS. When you install Windows 98, the maximum hard disk size reported is just 2.1 GB. Typing FDISK Status from the command prompt confirms this, there is only one partition of 2.1 GB.

**Diagnosis**

**Cure**

**Faulty part**
**Customer**
**interaction if**
**necessary**

**(Fault 8) Symptoms** Customer reports that they can still send FAX messages from their modem but they can no longer connect to the Internet.

**Diagnosis**

**Cure**

**Faulty part**
**Customer**
**interaction if**
**necessary**

**(Fault 9) Symptoms** The customer has just installed a new PC sound system featuring large bass ported loudspeakers either side of the PC work desk. They have now reported a fault with their monitor. The screen display is patchy with weak desaturated areas and coloured patches near the screen edges.

**Diagnosis**

**Cure**

**Faulty part**
**Customer**
**interaction if**
**necessary**

**(Fault 10)** The IDE CD-ROM drive no longer works.
**Symptoms** The PC is a modern Windows 98 Multimedia system. This happened since a new CD-R/CD-RW IDE drive was installed in addition to this drive. Both CD-ROMs share the same ribbon cable on the secondary IDE port.

Pressing the disc eject button will not open the CD tray on the standard CD-ROM drive. However the CD tray does open on the CD writer drive.

Diagnosis

Cure

Faulty part
Customer
interaction if
necessary

# Self-assessment questions

**SAQ 6.1** Imagine you have a fault on a PC where the monitor screen is blank with no visible sign of a raster, but the rest of the system seems to be working correctly, that is the system boots. Have a go at producing your own PIC. Use substitution were necessary such as 'try another monitor'.

**SAQ 6.2** The fault on a PC is a blown chip on the sound card. This is shorting out the $+5$ V rail, so the system does not boot up, in fact it seems dead. The PSU fan is also rendered inoperative. Follow the PICs starting at PIC 1 and see if the chart leads you to the correct fault.

**SAQ 6.3** State the disadvantage of software diagnostics compared with hardware diagnostic methods.

**SAQ 6.4** Name the three best practices for effective troubleshooting.

**SAQ 6.5** Where are the POST routines stored?

**SAQ 6.6** How does the memory device storing the CMOS setup information retain its contents when the power is switched off and yet allow you to edit it at anytime?

**SAQ 6.7** What does a POST diagnostics card do?

**SAQ 6.8** Before the AT PC with its CMOS setup system, the XT system used a totally different way to configure the system hardware. What method was used?

**SAQ 6.9** What key do you normally hold down to enter AMI CMOS Setup?

**SAQ 6.10** What does interrupt Request Steering allow?

**SAQ 6.11** Name the two Windows 95/98 registry files that hold system configuration data and user specific information respectively.

**SAQ 6.12** In an emergency when Windows has failed catastrophically due to a recently installed item, what does the command line entry scanreg/restore do? (Use a one sentence answer)

**SAQ 6.13** When would you probably resort to typing *SYSC*: after booting from your Windows Startup disk?

**SAQ 6.14** What is the most common cause of a hard disk appearing to load and save files progressively more slowly?

**SAQ 6.15** What Windows system utility could you use to take a snap shot of the state of the system after a General Protection Fault?

**SAQ 6.16** What Windows 98 utility would you use to selectively disable programs from running at startup, to isolate an unknown troublesome application?

**SAQ 6.17** What useful facility does the Windows 98 utility, Version Conflict Manager offer?

# Unit 7 Health and safety

*Looking after yourself and others at work*

NO SMOKING

Wear mask

**DANGER**
**Electric**
**shock risk**

FIRST AID

Fire
Extinguisher

NO RUNNING

HEAD GEAR
MUST BE WORN

**DANGER**
**Toxic hazard**

EMERGENCY
EXIT

FIRE HOSE

Red & Black & White     Blue & white     Yellow & Black     Green & White     Red & White

# Computers and safe working practices

Computer systems are inherently safe when used for their intended purpose, however computer users are subjected to possible injuries if they fail to comply with safe working practices. This health and safety (HS) unit will look at the hazards and safe working practices for the following user categories:

- Operators
- Service personnel

## Operators

- Use of hardware
- Posture
- Monitors

### Use of hardware

The Health and Safety at Work (HSW) Act states that, '*employees must take reasonable care for the health and safety of themselves and others effected by what they do at work*'.

Computer operators should not remove the covers off equipment or replace cables and fuses unless they have the appropriate skills and authorization to carry out microcomputer systems servicing. Unfamiliarity with equipment is one of the major causes of accidents.

Food and drink must not be placed near computer equipment, as well as the risk of electric shock due to spillage, remnants of food and liquid can wreck equipment and increase the risk of infection.

### Posture

Incorrect posture, particularly with long term computer use, can result in repetitive strain injury (RSI) in certain parts of the body. In particular, prolonged use of a keyboard and mouse can cause painful spasms in the hands and arms. Also bad seating and equipment positioning can cause neck and back ache.

### Repetitive strain injury

RSI occurs from repeated physical movements damaging the muscles, nerves and tendons. Symptoms are tingling, tightness or numbness in the fingers, hands, wrist and arms. This leads to loss of strength and co-ordination in these limbs.

### Prevention

Correct posture and typing technique, correctly positioned furniture and equipment, are important factors in the prevention of RSI. Figure 7.1 shows a correctly set up system with the operator sitting comfortably. The height of the table and chair is arranged so that the operator's forearms and thighs are parallel to the floor, with the back straight. The hands and wrists are straight and parallel to the keyboard, and are free to move while typing. The operator can reach everything without stretching.

### Good typing technique

Do not rest your wrists on anything while typing and do not hammer the keys, press them lightly.

**Figure 7.1** *An operator demonstrating the correct typing posture*

### Take plenty of breaks

Ensure you take plenty of short breaks between typing and do regular exercises to stretch your limbs and relax your muscles.

### *Monitors*

Monitors based on the cathode ray oscilloscope (CRT) can cause short term health problems if the operator sits close to the screen for long period of time without rest. There are four principle hazards:

- High frequency EMR
- Electrostatic charge
- Low frequency EMR
- Glare and poor visibility

### High frequency EMR

Monitors produce visible electromagnetic radiation (EMR) (light) by bombarding the screen phosphor with a high energy electron beam. When atomic particles such as electrons are accelerated or decelerated, as they are in the CRT, EMR is produced. The principle radiation is visible light but other high frequency emissions are also present. For example, some colour CRTs radiate short wavelength EMR in the form of X-rays in the vicinity of the shadow mask. These are low energy X-rays which are almost totally absorbed by the metal and glass surrounding the mask. Modern monitors, often termed 'low radiation monitors', must conform to stringent EMR standards and produce little or no X-ray radiation outside the monitor case. Any radiation that does exist will be well below background radiation levels.

## Electrostatic charge

All monitors produce high electrostatic voltages across the face of the CRT. The positively charged surface attracts negative ions from the surrounding air and neutralizes them, creating an **ion depleted region** around the monitor. The lack of negative ions can cause short term health problems for some operators particularly those who sit close to the monitor for long period of time. Symptoms are either headache, vertigo, fatigue, skin rashes and sometimes even respiratory disorders. These symptoms can also be caused by ion depletion due to other factors such as, poor air conditioning, synthetic carpets, crowded offices and passive smoking.

You can feel the static charge on a monitor by gently resting the back of your hand on the screen. The hairs on the back of your hand will probably tickle and stand on end if the static electricity is high enough. After doing this test, discharge your hand by touching an earthed antistatic mat or strap.

## Low frequency EMR

Monitors radiate low frequency EMR at the fundamental vertical and horizontal scan frequencies between 50–75 Hz and 15–30 kHz respectively. The radiation emanates from the scan coils and line output transformer situated towards the rear of the monitor. The radiation causes radio frequency interference (RFI) so manufacturers are required to add metal screening around the offending components.

Low frequency EMR from monitors, overhead power cables and house-hold appliances is generally considered to be harmless but research into the effects of this form of radiation is still not complete.

## Glare and poor visibility

The most common health hazard directly related to monitor use is **eye strain**. The biggest cause is a badly sited work station relative to the prevalent light source, causing shadows and reflections across the screen. Poorly maintained or inferior quality monitors with soft focus, interlace flicker or low screen resolution will also drastically increase the likelihood of eyestrain, which can manifest itself as sore eyes, headache and fatigue.

Reflection and glare can be reduced quite easily by relocating the workstation or the light source or installing window blinds if stray sunlight is a problem. Manufacturers of monitor tubes now incorporate special screen coatings that filter out reflected light thereby reducing reflection and glare to a minimum. For older monitors without the special screen coating, several companies manufacture clip-on filter screens to reduce glare and reflection but some makes, particularly the ones made of a thin mesh, can reduce the resolution slightly. Polarizing filters seem to produce the best performance. Some makes of filter also incorporate a clear conductive coating that can be earthed to eliminate static electricity (Figure 7.2).

The use of high resolution LCD monitors in the workplace helps considerably in overcoming many of the drawbacks found with CRT monitors. However in some lighting conditions CRT type monitors can provide a more comfortable image. Importantly nowadays, cost permitting there is a choice and the operator hopefully has some say in what he or she feels comfortable with.

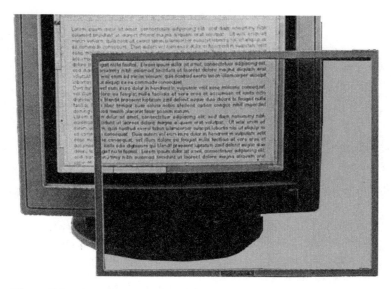

**Figure 7.2**   *An add-on anti-flare screen can sometimes improve visibility*

# Service personnel

- Use and handling of equipment
- Service of equipment

## *Use and handling of equipment*

The HSW Act states, *'employees must take reasonable care for the health and safety of themselves and others effected by what they do at work'*.

Be aware of general HS issues. Remember the major causes of accidents are usually human factors such as

- Unpreparedness
- Distraction
- Overconfidence
- Unfamiliarity
- Rushing

### Preparation

**Prepare yourself:**  First make sure you are prepared, illness, tiredness or the effects of drink and drugs can seriously effect your performance.

Use the correct handling technique when lifting heavy items like monitors and printers. When stooping down to lift such items, **bend your knees and keep your back straight** (Figure 7.3). The wrong way, is to do exactly the opposite of this. This is one of the biggest causes of back injury in the workplace and it is so easy to prevent.

**Prepare your equipment:**  Before using equipment make some simple but essential safety checks. For example, check for, proper earthing, correctly fitted mains plugs, undamaged mains sockets and plugs, and ensure that the mains voltage settings are correct. Make sure the equipment you use is in a safe condition for other people to use.

**Figure 7.3** *Bend your knees when lifting heavy items*

**Prepare the environment:** Do not use any mains equipment in wet or damp conditions. Isolate the power before adjusting rotary tools such as power drills. Think about other people using the environment, do not leave leads trailing across the floor, or leave tools and equipment in an untidy manner. Food and drink should always be kept away from the mains supply and equipment.

It is your duty to recognize dangerous equipment and work conditions and report them using your company's HS hazard reporting procedure. The details of the correct reporting procedure should be stated in your company's HS policy.

### Distraction

Keep your mind on the job in hand, particularly when doing something hazardous like using power tools or connecting mains cables. It is so easy to loose concentration during a crucial moment and that is when accidents usually happen.

### Overconfidence

Never be blasé in your work procedures, particularly when you are handling mains powered equipment and power tools. Be concerned about safe practices at all times. *All too often, an experienced, but poorly trained, person may take short cuts (e.g. remove guards and safety shields, refuse to wear safety gear, etc.) because they have done it before.*

### Unfamiliarity

When confronted with potentially dangerous equipment that you are not familiar with, do not touch it! Ask a qualified person to show you how to use it properly. If you need extensive training to use it, make sure you are trained before you use it.

### Rushing

Before carrying out a task, plan your time. Give yourself enough time to complete the job without rushing. *A rushed job often leads to short cuts and skimpy work. Tools can be left in equipment, covers may be not fully tightened down and safe practices may be overlooked.*

### *Service of equipment*

When servicing equipment, be aware and adhere to general HS guidelines, at all times. In addition, remember that any neglect, on your part, of HS issues, could directly effect other people that subsequently use the equipment.

It is important to ensure that any work carried out in the course of the repair does not impair the safety standards adopted by the manufacturer.

For example screening covers, and earth straps must be refitted as intended by the manufacturer and any electrical safety components removed during fault rectification must be replaced with the correct types and ratings.

To highlight the safety requirements of electrical equipment it is informative to take a brief look at the manufacturing regulations set by the European Commission (EC).

Equipment manufacturers must ensure that any equipment for use in the EEC meets or exceeds the various safety directives laid down by the EC. In the computer industry the important directives are the '1994 Electrical Equipment (Safety) Regulations', based on the EC 'Low Voltage Directive', the 1992 Electromagnetic Compatibility (EMC) Regulations and the 1992 'Telecommunications Terminal Equipment' (TTE) Directive.

## The EEC Low Voltage and CE Marking Directive

*Note*: This is not an authoritative interpretation of the regulations, but is intended to help the student develop a greater awareness of safety requirements in the computer industry.

In 1973, to ensure that all manufactured electrical goods supplied in the EEC conform to a common safety standard, the Council of Ministers adopted the European Commission Directive 73/23/EEC, known as the 'Low Voltage Directive'. In 1993, the 'CE Marking Directive' was adopted. In the UK both of these directives were adopted in the national 'Electrical Equipment (Safety) Regulations' of 1994. The UK regulations are effectively a modified version of the Low Voltage Directive, incorporating the new CE marking requirements.

The regulations effect manufacturers and all other suppliers of electrical goods including importers, wholesalers and retailers. A wholesaler or retailer in the course of their business has a statutory duty to ensure that the equipment they supply satisfies the safety requirement and bears the CE marking.

From 1st January 1997, all electrical goods placed on the market must comply with the 1994 regulations.

The important safety principles embodied in the directive are:

1. only electrical equipment which does not jeopardize the safety of people, domestic animals and property shall be placed on the market;
2. only electrical equipment which satisfies the CE marking requirements will be taken as complying with the 'Electrical Equipment (Safety) Regulations of 1994 and is thereby entitled to free circulation throughout the EEC.

From 1st January 1996 under the 1992 EMC Regulations, manufacturers must ensure that their products made or sold in the UK, including imported goods must be so constructed that they do not cause or are effected by, electromagnetic interference beyond a certain set standard.

**Figure 7.4** *The CE marking*

The implications of the regulations are as follows:

• That only electrical equipment which satisfies the CE marking requirements will be taken as complying with the 1992 EMC Regulations, and is thereby entitled to free circulation throughout the EEC (Figure 7.4).

## Use of cleaning agents and aerosols

Several chemical substances are used in the PC support industry for equipment cleaning and maintenance. Some can be harmful if mishandled. In accordance with the Chemical (Hazard Information and Packaging) Regulations, all chemical substances supplied to an individual or company must contain a safety data sheet identifying the substance, its composition, possible hazards, first aid measures, fire fighting measures, and recommended handling, storage and disposal instructions.

The law requires all employers using chemicals and substances that pose a possible health risk to comply with the 1999 Control of Substances Hazardous to Health Regulations (COSHH). There are seven guidelines to help employers and employees comply, which are as follows:

1. Assess the risks to health arising from hazardous substances used in or created by your workplace activities.
2. Decide what precautions are needed. You must not carry out work which could expose your employees to hazardous substances without first considering the risks and the necessary precautions, and what else you need to do to comply with COSHH.
3. Prevent or adequately control exposure. You must prevent your employees being exposed to hazardous substances. Where preventing exposure is not reasonably practicable, then you must adequately control it. The advice in this leaflet, and in the other guidance it refers to, will help you to make correct assessments and to put the appropriate controls into place.
4. Ensure that control measures are used and maintained properly and that safety procedures are followed.
5. Monitor the exposure of employees to hazardous substances, if necessary.
6. Carry out appropriate health surveillance where your assessment has shown this is necessary or where COSHH sets specific requirements.
7. Ensure employees are properly informed, trained and supervised.

# A brief introduction to the Health and Safety at Work Act of 1974

The HSW Act of 1974 is in four parts. The most important part of the Act for employees and employers is Part 1. The main objectives of Part 1 are as follows:

(a) securing the health, safety and welfare of people at work;
(b) protecting the HS of other people against risks arising out of work activities.
(c) controlling the storage and use of explosives, inflammable substances and other dangerous substances and preventing people from unlawfully having and using such substances;
(d) controlling the release of noxious or offensive substances into the atmosphere from premises to be prescribed by regulations.

# The Health and Safety Commission and the Health and Safety Executive

Two bodies were established by the act to promote its objectives and to ensure that its provisions are implemented. They are the Health and Safety Commission (HSC) and the Health and Safety Executive (HSE).

## Health and Safety Commission

The HSC's role is to promote the objectives of the act, encourage research and training and provide an advisory service. It also reports to the government any changes and/or proposals under the act.

## Health and Safety Executive

The HSE and its staff act for the commission and carry out the daily duties to allow the commission to carry out its role. One of its major tasks in the field is as the inspectorate of factories, farms, quarries, mines and railways. The HSE also provides the Employment Medical Advisory Service (EMAS), whose role includes advising the government, the HSC and the Training Enterprise and Education Directorate (TEED) on improving and safeguarding the HS of people at work, those seeking work and those training for work.

There is a HSE office in most cities.

# Duties under the Health and Safety at Work Act

## Duties of everyone involved in work activities

The act imposes duties on everyone involved in some way with work activities where potential HS hazards exist. This applies to employers, employees, the self-employed, manufacturers, designers, suppliers, importers, caretakers and also the general public. The duties are broadly generalized as taking responsibility for safety and adopting safe working practices. This requires that employers, employees and other people, are aware of their roles and responsibilities under the HSW Act.

## Duties of employers to their employees

The general duty imposed on all employers is to ensure, so far as is reasonably practicable, the health, safety and welfare of all their employees, at work. The employer's duties also extend to employees working in another person's premises. Which is particularly relevant to the service industries, such as electronics and electrical servicing.

Employers must, so far as is reasonably practicable, provide machinery, equipment and other plant that is safe and well maintained that does not increase the risks to HS. They must also ensure that, so far as is reasonably practicable, the layout of the workplace, and the method of working is planned to minimize the risk to HS. For example, a machine itself, and the way it is operated, must both be safe.

Employers must also ensure that the ways in which the materials are used, handled, stored and transported, are safe and without risks to HS. This duty extends to end products as well as the materials used in their manufacture. They must ensure that the ways in which end products are transported and stored are safe and without risks to HS of their employees.

Employers must provide HS training to all their employees to include instruction in safety and emergency procedures such as routine checking of equipment, fire drills and first-aid. Also special training may be needed for work involving a high degree of risk. Retraining must also be given when the equipment or work schedule changes. In essence, employers must ensure that all their employees are competent to carry out their jobs in a safe manner, that is, with the minimum of risk to themselves or others.

The work place itself must also conform to safe standards. The basic structure of the workplace must be of a sound structure, that lifts are properly maintained; staircases are well lit; and that passages and roadways are kept free from obstacles. The working environment must also conform to acceptable standards, for example, heating, lighting, ventilation and noise reduction. There must also be adequate arrangements for the welfare at work of their employees, such as seating, washing arrangements and toilets.

## Employers' Safety Policy

The HSW Act requires that every employer employing five or more employees, must prepare a written statement of their safety policy. An employer with less than five employees is not required to produce such a policy.

The purpose of the safety policy is to ensure that employers think carefully about the nature of the hazards at the workplace and the actions needed to reduce those hazards and make the workplace safe and healthy for their employees. The policy must be made available to all employees to make them fully aware of HS issues and practices. When revisions to the policy are necessary a new statement must also be brought to the notice of all employees.

## Safety Representatives and Safety Committees

Trade unions have the right to appoint safety representatives to represent the employees in consultation with their employer, in all HS matters. Two or more safety representatives can request in writing that the employer sets up a safety committee to ensure that HS practices are maintained in the workplace.

### Duties of self-employed

As well as having responsibilities towards other people, the self-employed must run their businesses in such a way as to ensure, so far as is reasonably practicable, that they do not risk their own HS.

### Duties of employees

All employees must take reasonable care for the HS of themselves and of other persons who may be affected by what they do or fail to do at work. Each employee must understand the hazards in the workplace and comply with the company safety policy and other statutory safety rules and regulations to ensure that nothing they do or fail to do, puts themselves or others at risk.

Employees who fail to comply with their statutory duties make themselves liable to be prosecuted in the courts.

### Duties of all people

The Act imposes one duty on all people, both people at work and members of the public, including children. This is not intentionally to interfere with or misuse anything that has been provided in the interests of health, safety or welfare, whether it has been provided for the protection of employees or other people. The purpose of the provision is clearly to protect things intended to ensure people's safety. The things covered by the provision include fire escapes and fire extinguishers, fencing, warning, protective clothing, guards on machinery and special containers for dangerous substances.

People who fail to comply with these statutory duties make themselves liable to be prosecuted in the courts.

## The European Safety Signs Directive

## Safety signs

The European Safety Signs Directive 92/58/EEC, standardizes safety signs throughout Europe to ensure that with the free movement of labour, there is no risk of safety signs being misinterpreted. In the UK it comes under the jurisdiction of the Health and Safety (Safety Signs and Signals) Regulations of 1996. These Regulations became Law in the UK on 1st April 1996 with a transitional up to 24th December 1998, to allow all fire and safety signs to be changed or added. All employers must use safety signs whenever there is a risk to HS that cannot reasonably be controlled by other means. Each sign must contain a pictogram to make it easily understood by people of all nationalities.

The EEC Directive puts signs into the following categories as shown in Figures 7.5 and 7.6.

**Figure 7.5** *Categories of signs from the 92/58 EEC Directive and a few examples*

Orange & Black

**Figure 7.6** *Example signs used to label hazardous substances*

- **Prohibition** – YOU MUST NOT DO (red and white background with black figures)
- **Mandatory** – YOU MUST CARRY OUT OR DO THE ACTION (blue background and white figures)
- **Warning** – CAUTION/HAZARD/DANGER (yellow background and black figures)
- **Safe Condition** – THE SAFE WAY/EMERGENCY EXIT (green background and white figures)
- **Fire Equipment** – INDICATES THE LOCATION OF FIRE EQUIPMENT (red background with white figures)

# Self-assessment questions

**SAQ 7.1** What is COSHH?

**SAQ 7.2** The Health & Safety at Work Act of 1974 is in four parts, which part is the most important for people at work?

**SAQ 7.3** When lifting heavy items up from a low position that can be handled by one person, you should bend your knees. Why?

**SAQ 7.4** Describe the duties you as an employee have to yourself and to others.

**SAQ 7.5** What action can be taken against an employee who fails to comply to the duties above?

**SAQ 7.6** State the one duty that ALL people, including children, must legally abide by.

**SAQ 7.7** Signs with a triangular surround are generally classed as _____ signs.

**SAQ 7.8** Mandatory signs (signs that must be obeyed), have a _____ background colour with _____ symbols or text.

**SAQ 7.9** Prohibition signs, have a _____ & _____ background colour with _____ symbols or text.

# Appendix A: Glossary

**A**   **Actuator**   The mechanism used on a magnetic disk drive to move and position the read/write head to any track of the disk. Most of today's hard-disk drives use a *speech-coil* actuator, older drives use a stepper motor.

**AGP** (accelerated graphics port)   An interface specification recently developed by Intel corporation. AGP is based on PCI, but is designed especially for the throughput demands of 3D graphics. Rather than using the PCI bus for graphics data, AGP uses a separate dedicated channel allowing the graphics controller to directly access the main system RAM. The AGP channel is 32-bit wide and runs at 66 MHz, but supports data transmission at both the rising and falling ends of the clock cycle for an effective speed of 133 MHz. This translates into a total bandwidth of 533 MB/s, as opposed to the PCI bandwidth of 133 MB/s.

**Allocation unit**   An allocation unit – formerly known as a cluster – is a group of one or more disk sectors treated by the operating system as the single smallest unit of disk space. The smaller the size of an allocation unit the more efficient the use of disk space. Allocation unit size is not fixed but determined by the size of the disk partition. In FAT 16 an allocation unit can contain a maximum of 64 sectors and as a sector is fixed at 512 bytes the maximum allocation unit size is 32 kB. Fat 32 has improved things somewhat by reducing the maximum allocation unit size even on large capacity drives.

**Alpha blending** (used on graphics cards)   Combines background information and the colour and transparency of an object to make it look realistic. For example, an image of a *frosted glass window* could be placed over a background scene. The transparency factor and colour of the window is combined with the colour of the background to form a natural looking lighting effect.

**Alpha channel** (used on graphics cards)   An extra-channel of information in addition to the red, green and blue channels, used to define a graphic object's transparency or edge characteristics.

**AMR** (audio/modem riser)   A slot specification announced by Intel in 1998 and developed by a group of leading audio/modem and PC manufacturers, to make it easier to add new built-in audio and modem functions to the motherboard. Rather than adding analogue circuitry to the motherboard, the functions will be implemented on a small riser printed circuit board (PCB) that plugs into the AMR slot.

**Anti-aliasing** The smoothing out of the jagged edges of a graphic object, particularly diagonals, by blending the edges of the object with the background.

**API** (application program interface) An API is a set of software tools and routines to help programmers interface their programs easily to the OS.

**Application program** Any program that carries out a useful task, such as a word processor, spreadsheet, database or desktop publisher, to name but a few.

**Archimedes screw** A rotating helical-shaped shaft devised by Archimedes, the Greek mathematician and inventor, to draw water from a reservoir. A modified form is used as a simple drilling device and as a mechanism for toy spinning tops.

The same principle is used as the head actuator on some floppy-disk drives. A shaft with a helical thread is passed through the head mounting block threaded to the same pitch as the shaft. By rotating the shaft with a small electric motor, the head can be positioned anywhere along the length of the shaft.

**Areal density** The amount of data that is stored on a disk per square inch. This is calculated by multiplying the number of bits per inch along the track, by the number of tracks per inch.

**ASCII** (American Standard Code for Information Interchange) Originally a 7-bit code that defines a set of 128 alphanumeric characters and control codes. See Appendix D.*

**AT** (advanced technology) The AT is an advanced version of the IBM PC-XT. It has a more powerful central processing unit (CPU) based on the 80286 or higher, a faster clock rate, a wider external data bus width and a greater addressing range beyond the 1 MB limit imposed by the 8086/8088. It also uses a 16-bit extension to the 8-bit PC bus.

**AT bus** The IBM PC-AT expansion bus, to which various adapter cards can be added. Later to be standardized and named the industry standard architecture (ISA) bus, it normally runs at 8 MHz, with a 16-bit wide data bus.

**ATA (**AT attachment) A very popular hard drive/CD drive I/O standard. The IDE drive and host adapter combination is known as the ATA. The salient features of the ATA system are high performance, high reliability at low cost.

ANSI CAM (*common access method committee*) supports the ATA specification and recommends performance enhancements on a regular basis: ATA-1, ATA-2, ATA-3, ATA-4, ATA-5, etc.

**Athlon** The first seventh generation X86 compatible CPU manufactured by AMD corporation. It used a PCB slot design mechanically similar to Intel's Slot 1 design used on the PII, PIII and Celeron, but with completely different pinouts, so the two CPUs were not interchangeable.

**Autoexec.bat** A set of device setup options and DOS commands stored in a file with a *bat* extension that are automatically invoked after the PC has booted and config.sys has been searched. This is used to configure various startup options such as the DOS prompt, date and time, keyboard layout, and to optionally run any DOS commands. For example, the user could set up autoexec.bat to automatically run a favourite DOS-based application program.

* Information on ASCII can be obtained from Appendix D on the Elsevier website at http://books.elsevier.com/companions/0750660740

**B** **Backup** Duplication of a whole disk or file for security.

**Bad sector** A sector on the hard drive or floppy diskette that will not reliably store data, often due to a surface defect.

**Bandwidth** Imagine the bandwidth as a gate to a field of sheep. The sheep representing data. The wider the gate, the greater the number of sheep that can pass through the gate in a given time interval. Wide bandwidth systems transfer more information in a given time interval, than narrow bandwidth systems.

**Bank switching** A method of expanding memory particularly on an 8086/8088-based XT system, beyond the 1 MB memory limit. By switching banks of memory in and out of a portion of the upper memory area (the 384 kB of memory above the 640 kB DOS limit). See expanded memory.

**Base memory** (conventional memory) The first 640 kB of memory available on a PC. On an MS-DOS-based machine this is the maximum memory available to application programs not making use of an extended/expanded memory manager.

**Batch file** A file containing a collection of DOS commands and named with a *bat* extension. MS-DOS executes these commands in sequence when the file name is entered at the DOS prompt.

**Baud rate** The number of *bit groups* transferred in one second. The baud rate is the bit rate in *bits per second (bps)* divided by the number of bits in each bit group (*n*).

**Beep code** A sequence of short and long beeps from the PC loudspeaker during the power on self-test (POST) that indicate certain errors. See Unit 7.

**BIOS** (basic input output system) The BIOS is a set of routines permanently stored in ROM that are able to interact directly with the hardware devices installed on the system. It effectively acts as an interface between the OS and the computer hardware. The OS can communicate with the BIOS using a series of interrupt routines.

**Bit** (from **bi**nary di**git**) The smallest possible unit of information in a digital computer system. A binary digit has only two states, '0' and '1'.

**Bitmap** An image stored pixel by pixel, with every pixel represented by individual locations in the memory device, be it in RAM or an auxiliary storage device. Bitmaps take up large amounts of memory or disk space. However, unlike vector-based graphics, a bitmap requires no mathematical calculations to regenerate the image. Images from a scanner or digital camera can only be sourced as a bitmap.

**Boot** The loading process that places the OS software into RAM, to bring the system to a user ready state. The boot process is initiated from a routine in the BIOS ROM, which in turn invokes a boot loader program stored on the boot sector of the disk.

**bps** (bits per second) The number of binary digits transferred in 1 s.

**Breakout box** A connection box connected in parallel with the cable under test that gives access to all the signal wires on both sides of the box. This allows signals to be monitored and connections to be reconfigured with ease. This piece of test equipment is used mainly on RS232 and network cables.

**Buffer** (i) A block of memory used to store data temporarily to speed up the apparent read and write speed of a slow peripheral device. The buffer behaves in a similar way to the fuel tank in a car. The car engine consumes the fuel from the tank at a steady rate even though the tank is only filled periodically. In a similar way the computer can quickly fill the buffer up with data and then carry on with something else while the peripheral empties the buffer. The trouble is this only works for small bursts of data. If continuous data transfer takes place, the buffer will soon fill to the brim and the computer will have to wait for it to empty before it can feed in more data, thus using up valuable processor time. (ii) An integrated circuit (IC) that increases the drive capability (current) of a signal.

**Bug** A defect in a program. Sometimes serious enough, that the program writers have to produce a patch (a bug fixing program) or sometimes so slight that the majority of users fail to notice the problem.

**Burst timing** In a Pentium-based PC, data is fetched from memory, four blocks at a time, where each block is the width of the data bus, that is, 64 bits (8 bytes). The first block always requires the maximum number of clock cycles compared to the subsequent three blocks. Depending on the type of RAM installed, the timings in clock cycles for the four blocks in a system with a 66 MHz system clock vary from say 5–5–5–5 for ordinary DRAM to as good as 5–1–1–1 for some types of SDRAM.

**Bus** A group of signal wires and printed circuit tracks with a common function, used to convey signals and data to and from the CPU to the rest of the system. A CPU usually has three buses: the data bus, address bus and control bus.

**Byte** A group of 8 bits. The storage capacity of both primary and secondary storage devices is measured in bytes. See megabyte (MB) and gigabyte (GB).

**C**

**Cache memory** A set of very fast RAM chips in addition to the main RAM, used to store prefetched program instructions. This saves valuable time as the CPU does not need to access the slower main RAM so often. *Primary cache or Level 1 cache* is a cache memory area built into the CPU. *Secondary cache or Level 2 cache* is a set of cache memory chips separate to the CPU, usually situated on the motherboard or on the CPU sub-assembly, as in the Pentium II.

**Capacity** The storage capacity of a hard disk is expressed in MB or GB, but the way it is defined is different from the normal binary definition. A kB is defined as $2^{10}$ bytes (1024) bytes. An MB is $2^{20}$ bytes (1024 $\times$ 1024). A GB is $2^{30}$ bytes (1024 $\times$ 1024 $\times$ 1024). This is the way RAM memory capacity is defined. However when it comes to drive capacities, some drive manufacturers define an MB as 1000 kB (1000 $\times$ 1024), and a GB as 1,000,000 kB (1,000,000 $\times$ 1024). For example, a 5 GB hard disk has a capacity of 5,000,000 $\times$ 1024 = 5,120,000,000 bytes, not $5 \times 2^{30} = 5,368,709,120$ bytes. This is a difference of 248,709,120 bytes!

**Cathode ray tube** (CRT) The large evacuated glass tube used as the display device in the majority of monitors to display text and graphics. Notebook PCs adopt a different technique using a liquid crystal display (LCD). A smaller version of the CRT is used as the display in oscilloscopes.

**CCD** (charge coupled device) A photosensitive device used in scanners and digital cameras that holds an image in the form of electrical charges on an array of minute metal oxide semiconductor field effect transistor (MOSFET) capacitors (cells). Each cell in a row is linked to its neighbour. The image is read out of the CCD by shifting the charge from each cell to its neighbour in a bucket brigade fashion. See Section 4 of the course on digital cameras.

**CD**    Acronym of compact disc.

**CD-I** (compact disc interactive)    A CD-ROM format specially designed for interactive multimedia. A CD-I player is a stand-alone consumer product that simply connects to a television set. It combines sound, graphics, animation and video that runs in real time and allows the user to interact via a keyboard or joypad.

**CD-R**    Refers to a CD that can be recorded once on a CD-R drive. The disc can then be read on any CD-ROM drive just like a normal CD-ROM.

**CD-ROM** (compact disc read only memory)    A CD-ROM is a CD used to store digital data suitable to be read in a CD-ROM drive attached to a computer.

**Central processing unit** (CPU)    In a PC system it is a single chip that decodes instructions, performs arithmetic and logical operations, and controls the system hardware. It contains four main elements: an execution unit (EU), a bus interface unit (BIU), internal registers and a set of data, address and control buses. Most modern CPU including the 80486 and above have an internal cache memory.

**Centronics interface**    A popular printer interface standard that uses an 8-bit parallel data bus and special control signals for handshaking and specific printer functions such as paper out and line feed. It was originally developed by the *Centronics* printer manufacturing company.

**CGA** (colour graphics adapter)    An old bit-mapped graphics system that can display four colours at a time in graphics mode and 16 colours in text mode.

**CISC** (complex instruction set computer)    This refers to CPUs that have a large instruction set with many elaborate commands. Compare this with reduced instruction set CPU (RISC).

**Clock**    A term used in electronics to describe a reference timing source usually based on a *quartz crystal* controlled square wave oscillator circuit. In a PC, the *system clock* provides a regular pulse waveform that controls the transfer of data throughout the system.

**Clusters**    The old name for allocation unit. See allocation unit.

**CMOS** (complementary metal oxide semiconductor)    A chip fabrication technology based on complementary pairs of P-channel and N-channel MOSFETs. When one device is on, the other device is off, so very little power is consumed. See transistor/transistor logic (*TTL*) and *MOSFET*.

**CMOS setup memory**    This refers to the special CMOS RAM chip on the motherboard. The chip is fabricated using CMOS, so it consumes only a tiny amount of power. This allows it to be operated continuously from an NICAD or lithium battery, even when the PC is switched off. The device is used to store essential system setup information such as the date, time, drive parameters, RAM type, and a whole host of other motherboard and peripheral parameters. The CMOS setup information being RAM based, can be altered to take account of system changes by invoking the BIOS CMOS setup routine during power up.

**Cold start**    Starting or rebooting the computer by turning it on at the power switch. In this case the full *POST* is executed. In a warm start, the Ctrl/Alt and Del keys are pressed in unison and the full POST is not executed. Continually performing a cold start, particularly after only a short delay, can stress the electronic components in the system.

**Command.com**  An essential OS file that interprets all DOS commands typed on the keyboard. It is automatically loaded when the PC is booted. If this file becomes corrupted, the system hangs and the prompt '*Bad or missing command interpreter*', appears on the screen.

**Composite video**  A single video signal that contains the line-by-line picture brightness information, synchronizing signals and colour signals.

**Computer**  A digital computer incorporates four main elements: a CPU, memory, an input unit and an output unit.

**Config.sys**  A file that is automatically looked for and executed, after the system has booted and prior to executing autoexec.bat. It is mainly used to set up device drivers in DOS.

**Constant angular velocity**  See constant linear velocity below.

**Constant linear velocity**  If a disk spins with a *constant angular velocity*, the surface at the outer edge of the disk passes the head at a much higher velocity than that at the inner portion of the disk. So data recorded near the outer edge is spread out more than data recorded near the inner hub. With constant linear velocity disk drives, the head to surface velocity is held constant. This means that as the head moves nearer to the outer edge, the rotational speed of the disk is reduced. This allows the longer outer tracks to hold more data than shorter inside tracks.

**Conventional memory**  See base memory.

**Convergence**  The process of aligning the red, green and blue guns of a monitor CRT, so that they overlap precisely to form a single image.

**Co-processor**  A processor which works alongside the CPU. In a PC, a maths co-processor is used in addition to the main CPU, to handle floating point arithmetic – sometimes called the floating point unit (FPU).

**CSEL** (cable select)  This allows ATA hard-disk drives to be automatically set as master and slave depending on their position on a special ribbon cable. In a single-drive system the drive must sit on the far end connector. To implement the auto-master/slave feature of this system, both drive jumpers must be set to CSEL.

On a modern 80-wire ribbon cable, the CSEL signal wire is connected on the end connector and this configures the drive connected there as 'master'. The CSEL signal on the intermediate connector is not connected (the terminal has been removed), so a drive connected here is configured as slave.

Some earlier implementations of CSEL on 40-wire ribbon cables may use a different arrangement.

**Cursor**  A character-sized blinking line or square on the screen that marks the spot where a character can be placed from the keyboard. The arrow keys move cursor *up*, *down*, *left* and *right*.

**D**  **DAC** (digital to analogue converter)  A device used to convert binary data to continuous analogue signals. A DAC is used on a graphics card to produce the analogue RGB signals for the monitor. See Unit 5.

**Daisywheel printer**  An old type of impact printer that uses the petals of a plastic 'daisy-shaped' wheel to impress characters onto paper via an inked ribbon. The print quality is similar to that of a typewriter.

**Data bus**　A set of wires and printed circuit tracks that link the data bus of a device to other devices in the system. The CPU data bus, for example, is a set of internal and external tracks that connect to the RAM, ROM and system data bus often indirectly through buffer or bridging chips.

**DDR-SDRAM**　Double Data Rate-Synchronous DRAM is currently the most popular type of DRAM used on today's PCs. Supplied in several different form factors and speeds. 184 pin dimm versions are usually supplied in 200, 266, 333 or 400 MHz versions labelled PC1600, PC2100, PC2700 and PC3200, respectively.

**Defragment**　After a hard-disk drive has been in use for several days and many files have been installed or removed, parts of files become scattered all over the disk, instead of in contiguous blocks. These fragmented files inevitably reduce the read/write performance of the drive. To defragment the files back into contiguous blocks the DOS *DEFRAG* command can be used. In Windows 95 a disk defragmenter is provided in the Accessories/System Tools folder.

**Degauss**　This is to remove stray magnetic fields. All colour monitors use an internal degaussing coil that operates for a few seconds when the monitor is switched on. The degauss cycle removes stray magnetic fields around the monitor screen to prevent purity errors – this manifests itself as coloured patches on an otherwise even hued display. This effect is often noticed on a multimedia PC, when the speakers are moved too near the monitor screen.

**Device driver**　A software routine that links a peripheral device to the OS. Usually provided by the original equipment manufacturer (OEM) and written using API tools.

**Diagnostic software**　Software which is used to diagnose faults in the system hardware or software.

**Digitize**　To convert an image or signal into digital form suitable for processing by a computer.

**DIL** (dual in-line)　A device package with connecting pins arranged in two parallel rows.

**DIMM** (dual in-line memory module)　This is an RAM package. DIMMs have 168 pins so they can support the full P54/P55 data bus width of 64 bits in one device, or 72 with parity.

**DIP/DIL switch**　A set of PCB mounted switches often used on motherboards, adapter cards and peripherals, to switch in or out various options. and does the same job as a set of jumpers.

**Direct 3D**　An API developed by Microsoft, allowing the programmer to access advanced 3D graphics effects on graphics cards fitted with a 3D accelerator chip.

**Direct X**　A set of APIs developed by Microsoft that simplifies the programmers task when writing hardware device drivers. Instead of writing complex low-level code directly to the hardware device, Direct X acts as an intermediate interface. Direct X is particularly useful when writing applications that use the advanced features of a graphics accelerator card or other multimedia device.

**Directory**　A magnetic or optical storage disk is like a filing cabinet. It holds folders containing groups of files. Each folder is called a directory. For example, all the external DOS files on a PC hard disk may be stored in a directory named DOS.

**Disc**　Optical storage medium in the form of a disc of plastic with pits and lands that diffract a laser beam in two different directions to form binary '0' and '1's.

**Disk**   A medium which enables data to be read and written magnetically from the surface of the disk.

**Disk operating system** (DOS)   An OS for computers based on the use of magnetic disks for auxiliary storage.

**DMA** (direct memory access)   This is a method of quickly transferring data between an intelligent peripheral and memory. The peripheral sends a DMA request to the DMA controller – part of the motherboard chipset. The DMA controller responds by sending a DMA hold request to the CPU. The CPU then completes the current instruction and places its address and data bus pins into the floating state – effectively disconnecting itself from the buses. It then sends a *hold acknowledge signal* (HLDA), to the DMA controller which in turn sends an acknowledge signal (DACK) to the I/O device to begin the data transfer. While the HOLD signal is high, the CPU is disconnected from the bus, and the peripheral has access to the buses. As soon as the hold line goes low again, the CPU regains control of the bus. In the PC, the DMA control system is quite elaborate, with DMA transfers taking place during intervals when the buses are idle. See *multi-word DMA*.

**Dot clock**   The graphics card displays pixels at a rate determined by a quartz crystal controlled clock known as the *dot clock*. The dot clock oscillates at a frequency of several Megahertz. In a VGA or SVGA card there are several different dot clock frequencies available for different resolutions and refresh rates. As an example, the standard VGA $640 \times 480 \times 60$ Hz display has a dot clock rate of 25.175 MHz. The dot clock frequency is often referred to as the *video bandwidth*.

**Dot-matrix printer**   A type of printer which prints characters using an array of pins fired from the print head. The resulting character is in the form of a matrix of dots. The print quality varies from poor (draft) to typewriter (letter) quality, depending on the number of dots used in the print head. A modern dot-matrix printer can exceed 300 dots per inch.

**Dot pitch**   The distance between the phosphor dots triads on the monitor screen. The smaller the dot pitch the more defined the image.

**DRAM** (dynamic RAM)   This type of RAM relies on a small capacitor to store each binary digit. To prevent loss of data, the capacitors must be topped up regularly in a process known as *refreshing*. In the PC-XT the refresh cycle is controlled by *DMA* 0. In a modern PC-AT it is taken care of by a refresh circuit built into the chipset.

**DURON**   AMD's first 453-pin CPU featuring a similar specification to the Athlon but built on a pin grid array socket (named Socket A) rather than the expensive to produce, Slot A design.

**E**

**EGA** (enhanced graphics adapter)   This is the first high-resolution graphics card to be used on the PC, now classed as obsolete. It superseded the earlier low-resolution CGA standard. An EGA card can be switched to both MDA and CGA modes. The EGA was capable of displaying 16 colours in text mode and graphics mode up to a resolution of $640 \times 350$ pixels.

**EISA** (extended industry standard architecture)   An extension of the ISA, now classed as obsolete. An EISA adapter card used an elaborate double-decker arrangement on the PCB edge connector, to allow downward compatibility to the ISA bus as well as providing a 32-bit wide extended bus.

**EMS** (expanded memory specification)   See expanded memory.

**EPROM** (erasable programmable read only memory)   See ROM.

**ESDI** (enhanced system device interface)   An improved hard-disk standard that enabled faster data transfer than the earlier MFM ST506 standard.

**Expanded memory**   A system enabling extra-RAM memory above the physical 1 MB addressable on XT systems. The 1 MB limit is due to the limited addressing range of the 8086/8088 CPU, with its 20 address pins, that is, $2^{20} = 1,048,576$ bytes $= 1$ MB. To overcome this memory limit, Lotus, Intel and Microsoft (LIM) introduced the EMS. The last version of this now obsolete system was LIM4.0. The system allowed XT- and AT-PCs fitted with an expanded memory card, and an expanded memory manager such as LIM4.0, to run applications requiring more memory than the 640 kB DOS limit.

This was achieved using a technique known as *bank switching*. In this technique, 16k blocks of memory from the expanded memory board are switched in and out of a portion of the upper memory area. Each 16k block is known as a *page* and four such pages make up a *page frame*. To use this facility the application program must be specially written to make use of the LIM memory manager.

**Expansion card (adapter card)**   A PCB with an edge connector on one edge that fits into an expansion slot on the motherboard. Such a board can add new features to the system, such as graphics or sound.

**Expansion slots**   Edge connectors fitted to the motherboard that accept expansion cards. The two main types used on a modern Pentium motherboard are 16-bit ISA slots and PCI slots.

**Extended DOS partition**   One of the two types of hard-disk partition allowed on an MS-DOS-based system. The extended DOS partition can hold any file types and can be split into a number of logical drives. The other partition is the *primary DOS partition* that contains the boot sector and the essential MS-DOS system files IO.sys, MS-DOS.sys and Command.com.

**Extended memory**   This is the area of memory above 1 MB that can be accessed by a computer using a 286 CPU or higher.

**F**   **FAT** (file allocation table)   Part of the DOS file system, stored on a portion of the hard disk, that contains information on the files stored on the disk, such as their location and size.

**Field replaceable unit**   A component of a PC system that may be easily identified as faulty and quickly replaced during field servicing, without the need to return the system to the workshop for repair.

**File**   A record, document or program that is stored on a disk with its own file name.

**Floppy disk**   A name used for removable magnetic storage disks made from a thin mylar (plastic) sheet. The main floppy disks used in today's PCs are high-density (HD) 3.5-in. disks. These are housed in solid plastic cases and offer a storage capacity of 1.44 MB. Earlier double-density (DD) 3.5-in. disks had a capacity of 720 kB.

**Floppy-disk controller** (FDC)   A part of the motherboard chipset or a chip on a disk/drive interface adapter card that controls the floppy-disk drive.

**Format**   A DOS command that prepares a magnetic disk – floppy of hard disk – for use by the OS. It wipes any existing data off the disk and places new track and sector information on the disk.

**Formatted capacity**    The amount of storage space remaining on the diskette after formatting. This is always less than the capacity before formatting.

**Frame buffer**    An area of video memory used to store a complete frame of an image while it is being updated.

**FTP** (file transfer protocol)    It is a protocol used to log onto a network, list directories and copy files.

**G**    **Gigabyte** (GB)    A GB is $2^{30} = 1,073,741,824$ bytes. This is the true value based on powers of two. Most hard-disk manufacturers use another definition based on a million kilobytes ($1,000,000 \times 1024 = 1,024,000,000$) or sometimes based on the decimal Giga, which is 1,000,000,000 bytes.

**Gouraud shading**    A form of shading which blends together colours based on the general direction of a light source, to form a realistic shading on the surfaces of an object. This gives the object a realistic illusion of depth.

**Graphics accelerator**    A CPU specially designed to manipulate images and carry out graphical transformations. Nowadays the term is used to describe a graphics adapter card with a graphics accelerator chip on board. Sometimes known as a *windows accelerator card*, as it speeds up graphics in a graphical user interface (GUI) environment.

**Graphics card**    A system expansion card used to drive a monitor (visual display unit). Modern cards incorporate elaborate ICs to produce a range of 2D and 3D graphics effects.

**Graphics engine**    The chipset on the graphics card that takes raw image instructions from the CPU and processes and writes image data to the frame buffer. Most modern graphics cards can apply many of the imaging techniques listed in this glossary.

**Green book**    The recommended standard for CD-interactive (CD-I).

**Ground** (Gnd)    An electrical current connection to the earth. The chassis of the PC case is normally grounded via the power supply unit (PSU).

**GUI** (graphical user interface)    Unlike the DOS command-based environment that relies on commands typed at the keyboard, a GUI provides an interactive mouse driven drag and drop system. There are several GUI driven environments available for the PC such as Windows 3.1, Windows 95, Windows NT and OS2.

**H**    **Half-height drive**    A drive unit with a height of 4.13 cm (1.625 in.) and width of 14.6 cm (5.75 in.). This is the standard drive height for CD-drives. This is half the height of the floppy drives and hard drives used on the early IBM PC-XT. Most 3.5-in. 1.44 MB floppy drives are only quarter height.

**Hard disk**    See hard drive.

**Hard drive**    A magnetic storage medium based on one or more permanently fixed metal disks coated in ferromagnetic compounds. The disks spin at high speed inside a hermetically sealed case. Due to fine mechanical tolerances and the ultra-clean interior, the read/write heads are able to float a few micrometers above the disk's surface without touching it. The contactless head allows very high head to disk speeds to be achieved with virtually no surface wear and hence offers a reliable long life. Some hard drives allow the disks to be removed and replaced for archival purposes. These are known as removable hard disks.

**Hardcopy**   A general name for paper output from a printer or plotter.

**Hardware**   The name given to the electronic and mechanical parts of a computer, as distinct from software.

**Head crash**   When the head of a hard drive touches the media while it is spinning at high speed, surface damage can result. If this happens loss of data may result. This is often caused when the base unit is suddenly jolted.

**Heat sink**   A piece of finned aluminium or copper physically attached to a chip to dissipate heat. Often a fan is also attached to aid cooling.

**Hercules graphics card**   A graphics card that offered a better resolution than the MDA card in text mode and $720 \times 348$ pixels in monochrome graphics mode.

**Hertz** (Hz)   The SI unit of frequency, named after Helmholtz Hertz. The number of cycles of a repetitive waveform, such as a clock waveform, that occur in 1 s.

**Hexadecimal** (hex)   A number system based on 16 digits 0–15. The first 10 characters are '0' to '9', the same as the decimal system. However, the numbers 10–15 are represented by the characters 'A' through to 'F', respectively, that is, A = 10, B = 11 and so on. The neat thing about hex is that only two characters are required to represent any 8-bit number. To show that hex is in use, a letter 'H' or 'h' is usually placed after the number, for example 255 is shown as FFH in hex. See the Appendix B.*

**High-level format**   After low-level formatting and partitioning a hard disk, it is necessary to high-level format each partition using the OS's FORMAT command. High-level formatting creates a file management structure on the disk.

To high-level format a drive under MS-DOS, the 'FORMAT' command is used.

**Horizontal scan rate**   The rate at which each line of the screen raster is drawn. In the default Windows VGA $640 \times 480$, 16 colour mode, the horizontal scan rate is 31.5 kHz. Higher rates are used on today's high-resolution modes.

**I/O**   Acronym for input/output.

**IBM PC compatible**   PCs that are downward compatible to the IBM PC/XT and AT computers.

**IDC** (insulation displacement connector)   A special connector with terminals spaced to match the pitch of the associated ribbon cable. Each terminal incorporates an insulation cutter. When the connec-tor is clamped onto the ribbon cable, each terminal cuts through the insulation and makes intimate contact with the underlying conductor.

**IDE** (integrated drive electronics)   A modern IDE hard-disk drive conforms to either the ATA2 or ATA3 standards. The word IDE refers to the fact that the electronics for the hard-disk controller and analogue/digital interface is integrated onto a built-in PCB, attached underneath the drive itself.

**Impact printer**   A printer which prints by using physical force between the printhead, inked ribbon and paper.

* Information on Hexadecimal can be obtained from Appendix B on the Elsevier website at http:/books.elsevier.com/ companions/0750660740

**Ink-jet printer**   A non-impact printer that squirts ink onto the paper to form characters, with no substantial physical contact (a non-impact printer).

**Input device**   A hardware device that produces data that can be *read* by the computer.

**Integrated circuit** (IC)   A complex electronic circuit fabricated onto a slice of semiconductor material – usually silicon.

**Interface**   Generally, the connection between one subsystem and another. Often some form of data conversion process is involved.

**Interlaced scan**   A method used to reduce the apparent screen flicker with low-cost, low-vertical refresh monitors. More expensive monitors use non-interlaced scan. See Unit 5.

**Internet**   A highly popular wide area network used by millions of people worldwide. It uses the transfer control protocol/Internet protocol (TCP/IP) protocols. Any user can hook up to the Internet by linking their modem to an Internet service provider (ISP). The ISP charges a nominal rate for using their services. A few of the popular ISPs in the UK are listed below:

| | | |
|---|---|---|
| AOL | www.uk.aol.com | 0800 2791234 |
| BT Internet | www.btinternet.com | 0800 800001 |
| CompuServe | www.compuserve.com | 0870 6000 800 |
| Demon Internet | www.demon.net | 0800 0279200 |
| Force 9 (Plusnet) | www.force9.net | 01909 531923 |
| Freeserve | www.freeserve.net | 0839 517517 |
| Freezone | www.freezone.co.uk | 0845 117 0366 |
| Global Inter't | www.global.net.uk | 0870 909 8100 |
| MCI | www.global.mci.com/uk | 0800 279 4585 |
| MSN | www.uk.msn.com | 0345 002000 |
| Netcom | www. netcom.net.uk | 01344 395600 |
| NetNames | www.netnames.com | 0800 269049 |
| Softnet | www.softnet.co.uk | 0845 080100 |
| Virgin Net | www.virgin.net | 0500 558800 |
| Virtual Inter't | www.vi.net | 0800 597 5225 |

**IRQ**   Acronym for **I**nterrupt **ReQ**uest.

**ISA** (industry standard architecture)   This is the old AT 16-bit standard bus which runs at 8 MHz. Many of today's low-bandwidth adapter cards (e.g. sound cards) use the ISA bus. The PCI bus however is now the default bus for high-bandwidth applications such as network cards and graphics cards. Although for the latter the AGP bus is now the most popular.

**ISDN** (integrated services digital network)   This is an international telecommunications standard that enables a communication link to carry digital data simultaneously with voice and video information, at a higher-data rate than a normal telephone line.

**ISP** (Internet service provider)   See Internet.

**J**   **Jumpers**   Electrical shorting links that allow adapter cards and the motherboard to be customized.

**Joystick**   An input device in the form of a hand grip that sends north, south, east, west positional information to a computer, depending on the precise position of the hand grip relative to a centrally defined position. Used mainly with games and flight simulator programs. An analogue joystick sends binary data to the input port relating directly to the sticks position. A digital joystick on the other hand sends a simple up, down, left and right signal only.

**K**

**kB or K** (kilobyte)   1 kB is $2^{10} = 1024$ bytes.
**Keyboard**   The main device for manually entering data to a computer.
**Keyboard keys**   See Appendix D.*
**Keyboard scan codes**   See Appendix D.*

**L**

**Laser printer**   A printer that uses a laser to form an optical image on a photosensitive drum. The image on the drum is then coated in toner before being transferred to the paper.
**Latency**   In disk drives, the average time in milliseconds that it takes for a particular sector to appear in front of the head after the head has reached the correct cylinder. It works out as the time the disk takes to do half a revolution.
**LBA** (logical block addressing)   This is a method of uniformly addressing disk sectors, for example LBA0 is sector 1 of head 0, cylinder 0 and LBA1 is sector 2 of head 0, cylinder 0 and so on in a linear fashion. Many AT BIOS programs prior to 1996 address the hard disk using the cylinder, head and sector number (CHS).
**LCD** (liquid crystal display)   This display device is often used on digital watches and other portable equipment including notebook PCs. This display uses a special *nematic* liquid crystal sandwiched between two pieces of glass, that changes its polarity to light when a voltage is applied across its two faces, sealed between two pieces of polarized glass.
**LED** (light emitting diode)   This is a semiconductor diode used for indicators and displays. When the positive end of a current limited DC supply is connected to the anode terminal and the negative end to the cathode terminal, the device emits light of a narrow wavelength.
**LED printer**   A printer that uses an array of LEDs to form an optical image on a photosensitive drum. The image on the drum is then coated in toner before being transferred to the paper.
**Level 1/Level 2 cache**   See cache.
**LIM/EMS standard**   A old DOS expanded memory standard, that enables software to work with expanded memory above the normal 640 kB DOS memory limit. See expanded memory.
**Logical drive**   The extended DOS partition can be subdivided into smaller portions and treated as a normal drive. Designated using letters from D onwards. The primary DOS partition is normally designated as the *C:* drive.
**Low-level format**   Low-level formatting is a *physical* process where tracks are laid down on the disk's surface and divided into sectors. During this process, a header and trailer list is recorded on each sector along with a block of dummy data. Inter-track and sector gap positions are also recorded.

* Information on keyboard can be obtained from Appendix D on the Elsevier website at http://books.elsevier.com/companions/0750660740

**M**

**Mainframe computer**   A mainframe computer is a large, very fast, multi-user computer containing an array of central processors and a vast data storage capacity. Most international companies and universities, for example, have mainframe computers, serving large numbers of people.

**Maths Co-processor**   A CPU specially designed to handle mathematical operations. It frees the main CPU to carry out normal processing operations while it handles floating point and other maths intensive operations.

**MCGA**   Acronym of multi-colour graphics array.

**MDA** (monochrome display adapter)   The early IBM text-only display adapter card.

**MDC** (mobile daughter card)   See AMR. A similar specification to the AMR slot for mobile PCs (laptops).

**Megabyte** (MB)   1,024 kB ($2^{20}$ bytes). Often abbreviated to Meg or MB. Sometimes interpreted as 1000 kB by some drive manufacturers.

**Megahertz** (MHz)   A frequency of 1 million cycles per second.

**Metal-in-gap head** (MIG head)   The MIG head is an improvement on the ferrite head, where metal alloys of aluminium, iron and silicon are deposited in the magnetic gap of the head. This substantially increases the magnetic field strength in the gap, thus allowing higher bit densities to be achieved.

**MFM** (modified frequency modulation)   A method used to store data on a magnetic disk and Used on floppy disks and older types of hard disk.

**Micro-computer (PC)**   A PC, is a stand-alone microcomputer designed for use by one person at a time. The most popular PC in the world today is the IBM PC compatible, nowadays simply referred to as the PC. Other non-IBM compatible PCs such as the Apple Mac are also very popular, particularly in the USA, however the extensive range of application software and inexpensive upgrade options available for PC make it the first choice for home and office users alike. In this course we will focus our attention on the PC and in particular on the latest generations of this very popular computer.

**Microprocessor**   The CPU used on a microcomputer. The microprocessor is so called, because the full CPU complete with registers, ALU and control unit is fabricated on one single IC. In a mini- or mainframe computer, the CPU can extend over several ICs or even several circuit boards.

**Minicomputer**   A minicomputer is like a scaled down mainframe, supporting many users within a single organization like a large family business or college. They are inherently fast machines with a large storage capacity.

**MIP mapping**   A texturing technique used on advanced graphics which adds more detail to an object as it approaches the observer and loses detail as it recedes from the observer.

**MIPS** (million instructions per second)   The average number of machine instructions that a computer can execute in 1 s is usually measured in so many MIPS.

**MO technology**   A technology that combines the best properties of both magnetic and optical recording/playback techniques to achieve reliable, fast-access, high-density, long-term data storage. To write data to an MO disk, both a laser beam and a magnetic recording head are used to change the light polarizing property of the disk's surface.

During the read operation, a low-intensity laser is focused onto the surface of the spinning disk and the polarity of the reflected light is detected and converted to a stream of '0' and '1's. See Section 1.

**Modem**   An abbreviation for **MO**dulate/**DEM**odulate. The modulation process of a modem converts the RS232 signals from the computer into a form that can be sent over a telephone line (tones). The demodulation is the converse of this. The signal on the telephone line is demodulated and converted back to an RS232 signal for the computer.

**MOSFET** (metal oxide semiconductor field effect transistor)   An important semiconductor device used as an electronic amplifier and switch. It can be packaged as a discrete device or as the tiny building brick of a much larger IC. It can be fabricated to an incredibly tiny scale, with several million on a silicon chip not larger or thicker than a finger nail. There are two main types of MOSFET, known as P-Channel and N-channel, depending on the type of silicon used for the channel region. See CMOS.

**Motherboard (system board)**   The main PCB inside the base unit containing the CPU, chipset, RAM, ROM and expansion slots.

**MPC**   Acronym of multimedia PC.

**MPC Level 1**   This specifies that a multimedia PC should have the following minimum specification: CPU 80386SX CPU; 2 MB of RAM; $640 \times 480 \times 256$ colour VGA graphics adapter; 8-bit sound blaster compatible sound card + MIDI; single-speed CD-ROM drive and a 30 MB hard disk.

**MPC Level 2**   This has a minimum specification as follows: CPU 80486SX 25 MHz; 4 MB of RAM; $640 \times 480 \times 65535$ colour SVGA graphics adapter; 16-bit stereo-sound blaster compatible sound card + MIDI; dual-speed CD-ROM drive and a 160 MB hard disk.

**MPC Level 3**   This has the following minimum specification: CPU Pentium 75 MHz; 8 MB of RAM; $640 \times 480 \times 65535$ colour SVGA graphics adapter + MPEG1; 16-bit stereo-sound blaster compatible sound card; quad-speed CD-ROM drive and a 500 MB hard disk.

**MR head** (magnetoresistive head)   Pioneered by IBM in 1991. The MR head has a separate read and write head layered together. The write head is a conventional thin film inductive head (TFI) and the read head is based on an alloy film of nickel–iron that exhibits a change in resistance in the presence of a magnetic field. The major advantages of this type of head, compared to conventional heads, are higher-production yields and greater sensitivity, allowing high-areal densities beyond 1 Gbit/in.$^2$.

**MS-DOS**   The standard PC DOS developed by Microsoft Corporation.

**Multimedia PC Marketing Council**   A group founded in 1991 by Microsoft, AT&T, CompuAdd, Creative Labs, Fujitsu, Media vision, NEC Technologies, Olivetti, Philips, Tandy, Video seven/Headland technology and Zenith Data Systems that defined a minimum specification for multimedia personal computing. The MPC Working Group now call this specification MPC Level 1.

**Multimedia PC Working Group**   A group active in the development of multimedia PC specifications. The executive members are Creative Labs, Disney Interactive, Dell, Fujitsu Personal Systems, Gateway 2000, Horizons Technology, IBM, Intel, NEC Technologies, Philips Consumer Electronics, Quicksilver Software, Sigma Designs, Software Publishers Association and Zenith Data Systems.

**Multimedia system**   A system that uses more than one type of media simultaneously such as sound and music, text, graphics and video. Many PCs are multimedia systems in that they are capable of providing all of these media types. For a microcomputer to qualify as a multimedia machine it must reach certain recommendations laid down by the Multimedia PC Marketing Council – a group of prominent PC software and hardware manufacturers.

**Multi-word DMA mode 1 and 2**   This is an alternative protocol to programmed I/O (PIO). ATA-2 defines the data rate for DMA transfer between the drive and the computer interface as 13.3 MB/s for mode 1 and 16.6 MB/s for mode 2.

**N**   **Nematic crystal**   A substance that exists in a state between that of a true solid and that of a true liquid. For example, the nematic substance used in an LCD can change its light polarizing properties on the application of a voltage. When a voltage is applied, the molecules line up in the direction of the field and consequently polarize the light passing through the material. See LCD.

**NLQ** (near letter quality)   Usually used to describe the print quality of a dot-matrix printer in some operating modes. Some 9-pin dot-matrix printers produce NLQ by printing a line, then reprinting it again after slightly shifting the position of the paper and/or head. This considerably reduces the overall print speed.

**Non-interlaced display**   Most high-quality monitors are designed to produce a high-vertical refresh rate without having to resort to interlacing two frames of the raster to form a complete picture. Each horizontal line of the raster is formed one after the other in a sequential manner. Non-interlaced displays do not suffer from the interline flicker (interlace flicker) that cheaper interlaced displays suffer from.

**O**   **OEM** (original equipment manufacturer)   Many manufacturers produce versions of their products in large quantities for other companies who either use their name on them or use them as components for their systems. OEM products often make it to the retail sales arena where they are sold at lower prices.

**OPC** (organo-photo-conductor)   The photosensitive drum or sheet used in photocopiers and laser/LED printers. It stores an electrostatic analogue of the image scanned by a laser or LED array, prior to transferring it to a paper sheet.

**Open Architecture (in engineering)**   A system whose design specifications are available to the public. The advantage of such a system is that other manufacturers can produce add-on products for the system. The IBM PC being a classic example.

**Open GL** (open graphics language)   A free 3D graphics programming language developed by silicon graphics.

**Operating system** (OS)   A suite of programs that allow communication between the BIOS and application software. It allows the user to communicate with the computer through the keyboard.

**OS2**   A GUI OS developed by IBM and Microsoft Corporation.

**P**   **Parallel port**   On a PC system at least one parallel port is provided. This is called LPT1. Additional ports LPT2 and LPT3 can also be added if needed. LPT1 is essentially an 8-bit Centronics port. Most modern motherboards supply an enhanced parallel port.

**Partition**   An area of the hard-disk space. The disk space can be divided up into smaller areas or treated as one large area. These areas are known as partitions. In an MS-DOS-based PC – also applies to Windows – there must be a primary DOS partition present. The primary partition is the boot partition containing the essential system files. In a PC two partitions are normally used the *primary* partition and an *extended* partition. The extended partition can be further divided into a maximum of 23 logical drives D to Z.

**PC** (personal computer)   A name coined by IBM in 1981, when it launched the first IBM PC. Its name indicates its original design purpose, as a computer for use by individuals.

**PCI bus** (peripheral component interconnect bus)   A bus designed to improve the bandwidth of the ISA bus, without the loading restrictions of the VESA local bus. It achieves this using a bridging circuit between the CPU buses and the PCI bus. The bus is clocked at 33 MHz and has a data bus width of 32 bits. This offers an inherent bandwidth increase of at least eight times that of the ISA bus. The PCI specification allows clock rates up to 66 MHz and a data bus width up to 64 bits, but this has not been implemented on the majority of PC systems.

**PCMCIA (PC cards)** (Personal Computer Memory Card International Association)   The PCMCIA specifies a bus standard that allows devices such as fax modems, network interfaces and disk-drive adapters, to be conveniently slotted into a laptop or notebook computer. It is now also supported on several desktop PCs.

**PIC**   See problem isolation chart.

**PIO mode 3 and 4**   These modes are a range of protocols for data transfer between the hard drive and controller, using programmed I/O rather than DMA. These modes are defined in the ATA-2 specification as 11.1 MB/s for mode 3 and 16.6 MB/s for mode 4.

**Pipeline**   Used to describe the instruction fetch mechanism of most modern CPUs. When the EU is busy, the BIU looks ahead and fetches more instructions from memory and stores them in an internal RAM array known as the instruction stream queue.

**Pixel** (from *picture element*)   The smallest dot that can be resolved on a monitor screen. Theoretically on a colour monitor, the smallest discernible pixel consists of a triad of phosphor dots – one each of red, green and blue. When struck with the electron beam, these dots can be made to fluoresce at different intensities producing a full range of possible colours.

Depending on the resolution set on the graphics card, a single pixel may contain several triads.

**Photo CD**   A digital imaging system developed by Eastman Kodak that allows normal photographic images to be digitized and stored on CD-ROM. Films and pictures are processed at a Kodak approved laboratory and stored on a multi-session CD-R. The stored images can be subsequently edited using an image editing program.

**Plug and play** (PnP)   This is a specification designed to make the process of installing a new hardware device as simple as possible, so that a users with a minimum amount of technical knowledge can install their own devices. The idea of PnP is to remove the need to manually select IRQs, port addresses and DMAs to prevent device conflicts – a chore that even a hardened PC buff often finds frustrating.

To make this possible several parts of the PC system must be PnP compatible, including the OS, BIOS, adapter card and driver software. Windows 95 is PnP compatible but Windows 3.1 is not. Current Pentium motherboards are also equipped with a PnP aware BIOS and chipset. The PnP system is not perfect but on the whole, modern PnP implementations work well.

**POST** (power on self-test)   The POST is a series of diagnostic routines stored in the BIOS ROM that are automatically executed when the PC is cold booted. The POST checks the main parts of the system hardware, including the motherboard and essential peripheral devices such as the disk drives and I/O ports and issues an error message or beep code if an error is detected.

**Power supply unit** (PSU)   A unit inside the case that isolates the AC mains and converts it to a set of low-voltage DC outputs at +5, −5, +12 and −12 V. The larger the maximum power output of the PSU, the greater the range of add-ons that can be accommodated without overload. Most PSUs have a capacity between 150 and 250 W.

**Primary DOS partition**   See partition.

**Primary cache**   See cache.

**Problem isolation chart** (PIC)   A step-by-step system of fault diagnosis where the service technician answers pre-set questions and pre-set routes to effect a logical fault deduction.

**Protected mode**   Advanced feature of the 286 and higher processors. In these devices the CPU can operate in one of two modes: *real mode* and *protected mode*. In *real mode* the CPU behaves like a conventional 8086 chip. However in protected mode the full power of the CPU can be realized with its virtual addressing capability and ability to run more than one program at a time.

Protected mode assigns each program its own unique portion of memory, protected from use by other programs. OSs and application programs must be specially written to handle these advanced features. Windows 3.1, Windows 95/98, Windows NT and OS2 are typical examples. See *virtual real mode*.

**R**   **RAID** (redundant array of inexpensive disks)   A way of improving fault tolerance in a disk storage system by using several drives in a mirroring or duplexing arrangement to make multiple copies of the saved data. When a drive fails it can usually be replaced on the fly using a so called 'hot swapping' system.

**RAM** (random access memory)   This is electronic memory designed to store data temporarily. RAM unlike ROM can be written to and read from, but any stored data is lost when the computer is switched off, that is, it is *volatile*.

**RAMDAC**   In a graphics card the RAMDAC converts the data stored in the frame buffer into the separate red, green and blue signals required by the monitor.

**Random access**   Memory access method in which memory locations can be accessed in any order. Each memory location is accessed in the same amount of time, irrespective of its position in the memory array. This is achieved using a row and column addressing technique. Both RAM and ROM use this addressing method.

**RDRAM** (read direct RAM)   A high-performance DRAM device using a built-in bus technology, allowing transfer rates of several hundred MHz. Typically sustained data transfer rates of 16 bytes in 10 ns can be achieved.

**Real mode**    The CPU mode of operation normally associated with the 8086/8088 CPU.

**Red book**    Defines the standard for audio CDs. See Section 1.

**Refresh**    The process of continually toping up the charge in the storage cells of a DRAM device to prevent data loss.

**Refresh rate**    The number of times that the video card refreshes an entire screen display in 1 s. Expressed as a frequency in Hz.

**Registry**    A crucial database used by Windows 95/98 for storing system and application configuration data.

**Removable storage device**    A high-capacity auxiliary storage device that can be removed from the main drive assembly and stored for safekeeping. The recordable media is inside a dust proof cartridge or cassette. Examples include Bernoulli drives and removable hard disks.

**Resolution**    The smallest amount of detail that can be resolved on a particular screen setting. Also a measurement of the quality of an image expressed in the number of horizontal and vertical dots making up the image. Each dot represents the smallest unit of information discernible. It is used to define the print quality of printers and the graphics resolution of display systems. In the latter case, each dot is called a pixel. Generally, the larger the resolution, the better the image quality.

**RIMM** (RamBus$^©$ in-line memory module)    A form of DRAM with a built-in bus system, for super fast read/write cycles. A special RAM slot to house RDRAM modules. Similar in look to conventional DIMM slots but totally incompatible, with 184 edge connectors arranged in two rows of 46, on both sides of the module. Some motherboards support both RIMM and DIMM slots.

**RISC** (reduced instruction set CPU)    A CPU with a simple instruction set. Complex instructions can be built up from several simple ones. The advantage of RISC over CISC is that instructions execute faster as they are inherently less complex. The Intel 80 series CPUs have traditionally been of the CISC variety but the later ones have used a combination of RISC and CISC technology.

**RLL** (run length limited)    It is a form of data coding used on hard disks, where groups of bits are assigned specific patterns of flux reversals, thus reducing the overall number of flux changes, therefore increasing the overall storage capacity. Modern IDE/EIDE and small computer system interface (SCSI) standards use various forms of RLL coding.

**ROM** (read only memory)    An ROM chip is an IC designed to permanently store digital data (programs). In an ordinary ROM chip, data is programmed into the device when it is manufactured. An *EPROM* however, can be programmed (burnt), using a special device called an EPROM programmer. Once programmed the contents remain intact even when power is removed from the chip. Under normal operating conditions it cannot be reprogrammed. It is therefore a read only device. However back in the LAB, it can be erased by shining ultra-violet (UV) light through a clear window on the top of the device, and then reprogrammed.

Another type of ROM known as Flash ROM is similar to an EPROM but it can be erased and reprogrammed much more easily. This is done by applying a higher than normal voltage to the device.

It does not need an elaborate programmer and UV light source, so it can be reprogrammed in situ by the user. Nowadays Flash ROMs are used as BIOS ROMs. This enables the BIOS to be updated whenever a new upgrade becomes available. The user simply has to download the new BIOS program via the Internet or from floppy disk and run a small Flash ROM programming routine, to install it.

**ROM BIOS**    A ROM device containing the BIOS.

**Root directory**    The main directory on a disk/disc. All other directories are sub-directories of the root directory.

**RS-232**    A serial interface standard designed to transfer data down medium length cables. The digital data from the computer is converted (level shifted) to provide a bidirectional current; logic 1 is represented by a negative voltage from 3 to 25 V and logic 0 is a positive voltage from −3 to 25 V.

**S**

**SATA**    In 1999 an association of seven PC manufacturers set out to define a new serial interface standard for ATA hard drives and ATAPI drives. The new standard is called serial ATA (SATA). This will be the dominant ATA hard-drive interface in a few years. Currently (2003), it is gaining increasing support by motherboard manufacturers.

**SCSI** (small computer system interface)    A parallel bus system that accepts up to seven SCSI peripheral devices in addition to the SCSI host adapter card. Each device including the host adapter is assigned a unique ID number from 0 to 7. The devices are connected in a daisy-chain fashion using a 50-way cable and connector system. Devices at each end of the bus must be terminated with a resistor network of 137 ohms to prevent signal reflections. See Unit 3, *Installing a CD-R* and Unit 4, *Installing a scanner*.

**Secondary cache**    See cache.

**Sector**    The small data portion that each track is divided into. A sector usually holds 512 bytes. See allocation unit.

**Seek time**    The time taken in milliseconds to move the heads of a disk drive to a required cylinder. Drive manufacturers often quote seek times as the *average seek time* (usually defined as the time it takes to move the head over one-third of the total number of cylinders), *track-to-track time* (the time it takes to move from one track to an adjacent track) and *full-stroke time* (the time it takes to move the head from the innermost to the outermost track).

**Serial interface**    An I/O channel that allows serial data transfer to take place. On the PC there are two serial RS232 compatible channels named COM1 and COM2.

**SIMM** (single in-line memory module)    A set of DRAM chips soldered to a small PCB with edge connectors on one edge; 30-pin SIMMs were common on 386 and 486 systems. Pentium-based systems use 72-pin SIMM or 168-pin DIMM modules. See *RAM* in Unit 2.

**SIPP** (single in-line pin package)    These are 30-pin DRAM modules almost identical to 30-pin SIMMs with pins instead of PCB edge connections. They soon went out of favour when SIMMs became the norm.

**Slot 1**    A CPU module mounting system, developed by Intel for the Pentium II processor range. The CPU and second-level cache are soldered to a PCB. The 242 connections are brought out to a double-sided PCB edge connector.

**Slot A**   A CPU module mounting system, developed by AMD for the Athlon series of processors. the CPU and second-level cache are soldered to a PCB. The 242 connections are brought out to a double-sided PCB edge connector.

**Slots**   A colloquial name for the motherboard edge connectors. See the expansion buses, AGP, AMR, ISA, PCI and the CPU slots, Slot 1 and Slot A.

**Socket 7**   This is the generic Pentium CPU socket specification for a range of Pentium and Pentium equivavlent CPUs based on the P54/P55 and 321-pin design. See Unit 2 and Appendix E.*

**Socket A**   A new socket specification by AMD featuring 453 pins first used by their Duron CPU (2000).

**Software**   Essentially the binary commands and information executed and processed by the CPU. In other words any programs and data that can be handled by a computer.

**Speech coil**   A coil suspended in the field of a strong magnet. When a DC current is applied to the coil it moves over an angular distance proportional to the current. Used to position the read/write heads of a hard disk. Name derived from the magnet/coil assembly used in a loudspeaker.

**Static RAM** (SRAM)   A RAM device that uses a flip-flop (bistable) to store a binary digit, instead of the storage capacitor used on a DRAM. The flip-flop is an electronic circuit composed of several transistors that can remain in one of two stable states without the need for a refresh signal. SRAM can operate at a higher-data rate than DRAM but it uses more transistors per storage bit. So it takes up more chip space and tends to be more expensive. SRAM is used mainly for *cache* memory where a high speed is essential.

**Supercomputer**   A supercomputer is an ultra-fast computer capable of performing billions of calculations per second with a massive primary and secondary storage area measured in billions of bytes.

**Superscalar**   A CPU architecture incorporating more than one data pipeline enabling more than one instruction to be executed per clock cycle. The 80486 and below can only execute one instruction at a time. The Pentium and clone CPUs use superscalar techniques, so pairs of instructions can be executed at once.

**Sustained transfer rate**   The read/write speed in bytes per second that the disk can maintain continuously.

**T**

**TCP/IP** (transfer control protocol/Internet protocol)   A widely used Internet protocol that is independent of the controlling hardware.

**Texel**   Similar to a pixel but used to define a single element of a texture map.

**Texture mapping**   In 3D graphics, texture mapping adds realistic looking textures such as sky, sea, grass, etc. to a basic scene.

**TFI head** (thin film inductive head)   A type of read/write head used on a hard drive. TFI read/write heads are made by depositing a tiny coil at the end of the head arm using photolithography – a process similar to that used to make ICs. See *MR head*.

**TFT** (thin film transistor)   A special ultra-thin, transparent transistor fabricated on glass. Used in LCDs.

**Track**   The thin concentric circles on a magnetic disk where data is recorded. On an optical disc, the track is one long continuous spiral – analogous to the track on a vinyl gramophone record.

* Information on Socket can be obtained from Appendix E on Elsevier website at http://books.elsevier.com/ companions/0750660740

**Track-to-track time**   The time it takes to move from one track to an adjacent track on a hard disk.

**TTL** (transistor/transistor logic)   A widely used chip technology based on bipolar transistors. Used for basic logic ICs. CMOS ICs based on MOSFET technology are now more widely used in PCs, due to their lower power consumption.

**U**

**UART** (universal asynchronous receiver transmitter)   A chip that contains shift registers and a coder/decoder. Used to convert parallel data at the computer to an asynchronous serial form and vice versa. In conjunction with a pair of TTL to RS232 and RS232 to TTL converter chips, it forms the basis of the RS232 port.

**USB** (universal serial bus)   USB 1.0 is a relatively fast (12 Mb/s) serial bus specification that supports Intel's PnP scheme, allowing peripherals like scanners and digital cameras to be added quickly and easily to the PC system. USB 2.0 supports speeds up to 480 Mb/s.

**V**

**Vector graphics**   Graphics in which images are created by calculating the position of certain points in an image using a mathematical algorithm. The position of these points is calculated as a 'vector', that is, a value which has magnitude and direction. For example, a straight line can be defined by a start position, $Xs, Ys$; angle; and end position $Xe, Ye$.

**Vertical refresh rate**   The number of fields (on an interlaced display) or frames (on a non-interlaced display) that are displayed in 1 s. This is measured in Hz (cycles per second).

**VESA**   Acronym of Video Electronics Standards Association.

**VGA**   Acronym of video graphics array.

**Video memory**   The RAM used in the frame buffer. It is also used in the *Z-buffer* and the *texture memory* fitted on some top-end graphics cards. The most common types of RAM used on a graphics card are SGRAM, WRAM and VRAM.

**Viewable area**   The area of the monitor screen that actually forms the display. Tube sizes are usually stated as the diagonal distance from the bottom left corner to the corresponding top right corner of the glass envelope. The actual screen area is of course several centimetres less than this, to allow for the thickness of the tube wall. The aspect ratio (width:height) of a monitor screen is normally 4:3, so the screen width is 0.8 times the diagonal distance and the height is 0.6 times the diagonal distance.

*Example*: On a 17-in. monitor with a diagonal viewing distance of 15.75 in., the width is $15.75 \times 0.8 = 12.6$ in., and the height is $15.75 \times 0.6 = 9.45$ in. The viewable area is therefore $12.6 \times 9.45 = 119$ in.$^2$.

**Virtual memory**   This is a technique that uses a portion of the hard-disk space as extra-RAM. For example, when running an image editor in Windows, if a large image is imported, that requires more RAM than currently available, the virtual memory manager kicks in and uses the allocated area of the hard disk (the virtual memory) as RAM. The only noticeable effect is a drastic slowing down of any action that accesses the virtual memory area. This is always accompanied by continuous activity of the hard disk. If this happens a lot, then more RAM is called for.

**Virtual real mode**   The 80386 CPU and above can run in protected mode while simulating the real mode operation of the 8086. For example, this facility allows several DOS programs to run independently of each other in Windows. Each program being assigned to its own area of protected memory. So if one program hangs, the others can run quite happily without the whole system being brought down.

**Volatile memory**   Memory devices whose stored data is lost when power is removed. RAM is an example of volatile memory. However it can be made non-volatile with battery backup as used with the CMOS setup RAM.

**VRAM** (video random access memory)   A specialized type of DRAM, VRAM is dual ported, meaning it can be read from and written to at the same time.

**W**   **Warm start**   A warm start resets the CPU, and therefore the whole system, without performing a full POST. This is the recommended way to restart the system as continuously switching power on and power off, stresses certain electronic components and inevitably shortens their life. Only use the *on–off* switch at the beginning and end of a computing session. See cold start.

**Write protect notch or tab**   Disks can be write protected by covering the square notch on the side of a 5.25-in. disk or by sliding up the write protector on 3.5-in. diskettes. With the protector in place, data can be read from this disk but not recorded, thus protecting the contents from accidental erasure.

**X**   **XT**   One of the first IBM PCs based on the Intel 8088 CPU, to incorporate a hard disk.

**Y**   **Y cable**   A Y-shaped splitter cable that provides an extra-connection point. A Y cable is commonly used to provide an extra-DC power connection for an additional hard drive or CD-ROM drive.

**Yellow book**   The ISO 10149 standard for computer CD-ROMs. Most application software in use today comes on CD-ROMs conforming to the yellow book standard.

**Z**   **Z-buffer**   Used on 3D graphics cards, the Z-buffer stores the depth and position of objects. In collaboration with the graphics engine, it can decide the parts of an object to show on screen, and the parts to be hidden behind other objects.

**ZIF socket** (zero insertion force)   This is an IC socket specially designed to apply minimum force to the pins of the IC. These are used on removable chips that are regularly inserted and removed from their sockets, or for chips with a large numbers of pins. The CPU socket on a modern motherboard is usually a ZIF device.

**ZIP™ drive**   The trade name of a popular type of floptical drive made by IOMEGA.

**Zone recording**   A method of placing more sectors on the outer tracks of a hard disk where the tracks are longer and less on the inner tracks where the tracks are shorter. Older drives use the same number of sectors for all tracks.

# Index